Oscar Hertwig, Henry J. Campbell, M. Campbell

The Cell

outlines of general anatomy and physiology

Oscar Hertwig, Henry J. Campbell, M. Campbell

The Cell

outlines of general anatomy and physiology

ISBN/EAN: 9783337311476

Printed in Europe, USA, Canada, Australia, Japan

Cover: Foto ©Andreas Hilbeck / pixelio.de

More available books at **www.hansebooks.com**

THE CELL

OUTLINES OF

GENERAL ANATOMY AND PHYSIOLOGY

BY

DR. OSCAR HERTWIG

*Professor Extraordinarius of Anatomy and Comparative Anatomy, Director of the
II. Anatomical Institute of the University of Berlin*

Translated by M. CAMPBELL, and Edited by

HENRY JOHNSTONE CAMPBELL, M.D

*Assistant Physician to the City of London Hospital for the Diseases of the Chest
and to the East London Hospital for Children
Senior Demonstrator of Biology and Demonstrator of Physiology in Guy's Hospital*

WITH 168      ILLUSTRATIONS

London
SWAN SONNENSCHEIN & CO
NEW YORK: MACMILLAN & CO
1895

BUTLER & TANNER,
THE SELWOOD PRINTING WORKS,
FROME, AND LONDON.

TO HIS FRIEND AND COLLEAGUE

W. WALDEYER

AUTHOR'S PREFACE

" Each living being must be considered a microcosm, a small universe, which is formed from a collection of organisms, which reproduce themselves, which are extremely small, and which are as numerous as the stars in heaven."

Darwin.

A GLANCE at the numerous text-books on histology shows us that many questions of great interest in scientific investigation are scarcely mentioned in them, whilst many branches of knowledge which are closely connected with histology are more or less excluded. The student is taught the microscopic appearances which are presented by the cell and the tissues, after these have been prepared according to the different methods which are most suitable to each, but he is taught very little of the vital properties of the cell, or of the marvellous forces which may be said to slumber in the small cell-organism, and which are revealed to us by the phenomena of protoplasmic movements, of irritability, of metabolism, and of reproduction. With regard to the different subjects which he studies, if he wish to be in touch with the progress of science, and to understand the nature and attributes of the cell-organism, he must read the works of specialists.

It is not difficult to discover the reason for this; it is chiefly due to the division of what was previously one subject into two, namely, into anatomy and physiology. This sub-division has been extended to the cell, and, it seems to me, with rather unfortunate results; for the separation which, in spite of the many disadvantages which are naturally attached to it, is in many respects a necessity in the investigation of the human body as a whole, is not practicable in the study of cells, and has in reality only brought about the result, that the physiology of the cell has been dogmatically treated as a part of descriptive anatomy, rather than as a science, and that in consequence much that the diligence of scientists has brought to light is barren of results. In this book I have avoided the beaten track, and in order to emphasise this fact, I have added to the principal title of the whole work, "The Cell and the Tissues," the secondary title "Outlines of General Anatomy and Physiology." Further, I am able to say, as I said of my *Text-book of Embryology: Man and Mammals*, that it has been produced in close connection with my academical labours. The contents of the first part, in which I have endeavoured to sketch a comprehensive picture of the structure and life of the cells, were the subject of two lectures which I delivered at the University of Berlin four years ago, under the titles of "The Cell and its Life," and "The Theory of Generation and Heredity."

Besides wishing to communicate to a larger circle of readers the views which I had often expressed verbally, I had the further desire of giving a comprehensive review of results obtained by private research, some of which were recorded in various Journals, whilst others appeared in the six papers on "The Morphology and Physiology of the Cell," which I wrote in conjunction with my brother.

Finally, a third reason which induced me to write this book was, that it should supplement my *Text-book of Embryology: Man and Mammals*. In it I have endeavoured to state the laws which underlie animal formation, according to which cells, formed from the fertilised egg-cell by repeated division, split up, as a result of unequal growth, the complicated layers and outgrowths into germinal folds, and finally into individual organs.

In addition to the distribution of cell-masses and to the arrangement of cells, that is to say, in addition to the morphological differentiation, a second series of processes, which may be grouped together under the term histological differentiation, takes place during development. By means of histological differentiation, the morphologically separated cell material is capable of performing the different functions into which the vital processes of the developed collective organism may be divided.

In my *Text-book of Embryology*, it was impossible to deal exhaustively with the second or more physiological side of the process of development. *The Anatomy and Physiology of the Cell*, forms a necessary complement to it, as I mentioned above. This will be especially noticed by the student in the first part of the book, which deals with the cell alone. For not only is there, in the seventh chapter, a detailed description of the anatomy and physiology of reproduction, which is ultimately a cell phenomenon, but at the end of the book, in the ninth chapter, there is a section entitled "The Cell as the Elemental Germ of an Organism," in which both the older and more recent theories of heredity are dealt with.

The second part of the complete work, which is to deal with the tissues, will be of about the same length, and will form to a greater extent a supplement to the *Text-book of Embryology*. For in addition to a description of the tissues, especial emphasis will be laid upon their origin of histogenesis and upon the physiological causes which underlie the formation; the other side of the process of development, that is to say, histological differentiation, will also be discussed.

In the account, which I have endeavoured to make as intelligible as possible, scientific views have primarily guided me. What I have striven to do to the best of my ability is, to fix the scientific stand-point occupied at present by the doctrines of cell and tissue formation. Further, I have tried to delineate the historical course of the development of the more important theories. With regard to disputed points I have frequently compared various opinions. If, as is natural, I have placed my own views in the foreground, and, moreover, if I have occasionally differed from the views and explanations of prominent and highly-esteemed scientists whose opinions I value extremely, it is only due to them to say that I do not on that account consider the conceptions preferred by me to be unconditionally correct, still less do I wish to belittle the views from which I differ. Antagonistic opinions are necessary to the life and development of science; and, as I have observed in studying the history of the subject, science progresses most rapidly and successfully in proportion to the diversity of the opinions held by different authorities. As is only human, almost all observations and the conclusions deduced from them are one-sided, and hence continually need correction. How necessary then must this be in the subject of the present inquiry, that is to say, in the cell, which is a marvellously complicated organism, a small universe, into the construction of which we can only laboriously penetrate by means of microscopical, chemico-physical and experimental methods of inquiry.

OSCAR HERTWIG.

Berlin, October, 1892.

EDITOR'S PREFACE

THE translation of Professor Hertwig's book has been no easy task. The extreme complexity of much of the matter treated, in addition to the large number of subjects referred to, has often rendered it extremely difficult to express the author's meaning in readable English. Of one thing there can be no doubt, and that is, that the subject matter is of very great importance; moreover, it cannot but prove most useful to the student who does not read German fluently, to possess in English so comprehensive an account of the Anatomy and Physiology of the Cell, as the one contained in Professor Hertwig's book.

In many cases it has been extremely difficult to find equivalents for terms used in the German. Amongst these the word "Anlage" may be specially mentioned. Various terms have been used by different translators to express the meaning of this word, but none of them seems to be applicable to all cases. Professor Mark has introduced the word "fundament," and Mr. Mitchell has suggested the term "blast," but neither of these appears to express the meaning of the German word sufficiently accurately to justify the use of either of them exclusively. Hence, we thought it best in some cases to employ the somewhat cumbrous expression, "elemental germ," although it is undoubtedly open to objection; however, it frequently seemed to us to convey the author's idea most correctly. On other occasions we have thought better to make use of a paraphrase.

Several additions have been made to the Bibliography of papers that the English student might wish to consult. The frequent quotations from English authors have in most cases been verified by reference to the originals; but in some cases, despite careful search, we have been unable to find the passages in question.

H. JOHNSTONE CAMPBELL.

54, Welbeck Street, London, W.

CONTENTS

CHAPTER I.

	PAGE
Introduction	1
The History of the Cell Theory	2
The History of the Protoplasmic Theory	6
Literature	9

CHAPTER II.

THE CHEMICO-PHYSICAL AND MORPHOLOGICAL PROPERTIES OF THE CELL . 11

I. The Chemico-physical and Morphological Properties of the Protoplasm 11
    (a) Justification of the Use of the Term Protoplasm . . . 12
    (b) General Characteristics of Protoplasm 13
    (c) Chemical Composition of Protoplasm 15
    (d) The more minute Structure of Protoplasm 18
    (e) Uniformity of Protoplasm. Diversity of the Cell . . . 26
    (f) Various examples of the Structure of the Cell-body . . 27
        1. Cells consisting almost entirely of Protoplasm . . 27
        2. Cells which contain several different substances in their Protoplasm 31
II. The Chemico-physical and Morphological Properties of the Nucleus 30
    (a) The form, size and number of Nuclei 37
    (b) Nuclear Substance 40
    (c) The Structure of the Nucleus. Examples of its various Properties 45
III. Are there Elementary Organisms existing without Nuclei? . . 54
IV. The Central or Pole Corpuscles of the Cell 55
V. Upon the Molecular Structure of Organised Bodies . . 58
Literature 61

CHAPTER III.

THE VITAL PROPERTIES OF THE CELL 65

The Phenomena of Movement 65
  I. Protoplasmic Movements 66
    (a) The Movements of naked Protoplasm 66
    (b) The Movements of Protoplasm inside the Cell-Membrane . 71
    (c) Theories concerning Protoplasmic Movements . . . 73

		PAGE
II.	Movements of Flagella and Cilia	77
	(a) Cells with Flagella	79
	(b) Cells with numerous Cilia	83
III.	The Contractile Vacuoles, or Vesicles, of Unicellular Organisms	85
IV.	Changes in the Cell during passive movement	88
	Literature	89

CHAPTER IV.

THE VITAL PROPERTIES OF THE CELL 91

Phenomena of Stimulation 91
   I. Thermal Stimuli 94
  II. Light Stimuli 99
 III. Electrical Stimuli 106
Phenomena produced by Galvanotropism 108
 IV. Mechanical Stimuli 110
  V. Chemical Stimuli 111
    (a) Chemical Stimuli which affect the whole body . . 112
    (b) Chemical Stimuli which come into contact with the Cell-body at one spot only 115
      1. Gases 115
      2. Liquids 117
Literature 123

CHAPTER V.

THE VITAL PROPERTIES OF THE CELL 126

Metabolism and Formative Activity 126
  I. Absorption and Excretion 128
    1. The Absorption and Excretion of Gaseous Material . 128
    2. The Absorption and Excretion of Fluid Substances . 133
    3. The Absorption of Solid Bodies 141
 II. The Assimilative and Formative Activity of the Cell . . 145
    1. The Chemistry of Assimilation 146
    2. The Morphology of Metabolism 154
      (a) Internal Plasmic Products 154
      (b) External Plasmic Products 166
Literature 174

CHAPTER VI.

THE VITAL PHENOMENA OF THE CELL 177

Reproduction of the Cell by division 177
  I. History of Cell-formation 177
 II. Nuclear Division 179

1. Nuclear Segmentation. Mitosis (Flemming); Karyokinesis
   (Schleicher) 179
   (a) Cell division as it occurs in *Salamandra maculata* . . 179
      First Stage. Preparation of the Nucleus for Division. 182
      Second Stage of Division 185
      Third Stage of Division 187
      Fourth Stage of Division 188
   (b) Division of the Egg-cells of *Ascaris megalocephala* and
      Toxopneustes lividus 189
   (c) Division of Plant Cells 196
   (d) Historical remarks and unsolved problems concerning
      Nuclear Segmentation 199
2. Direct Nuclear Division. Fragmentation. Amitosis . . 207
3. Endogenous Nuclear Multiplication, or the Formation of
   Multiple Nuclei 211
III. Various methods of Cell Multiplication 213
   1. General Laws 213
   2. Review of the Various Modes of Cell Division . . . 223
      1*a*. Equal Segmentation 224
      1*b*. Unequal Segmentation 225
      1*c*. Cell-Budding 228
         2. Partial or Meroblastic Segmentation . . . 230
         3. So-called Free Cell Formation . . . 232
         4. Division with Reduction 235
IV. Influence of the Environment upon Cell Division. Degeneration . 239
Literature 246

CHAPTER VII.

THE VITAL PROPERTIES OF THE CELL 252

The Phenomena and Methods of Fertilisation 252
   I. The Morphology of the Process of Fertilisation . . . 256
      1. The Fertilisation of the Animal Egg 256
         (a) Echinoderm Eggs 257
         (b) Eggs of *Ascaris megalocephala* 259
      2. The Fertilisation of Phanerogamia 263
      3. The Fertilisation of Infusoria 265
      4. The various forms of Sexual Cells; equivalence of partici-
         pating Substances during the Act of Fertilisation; Con-
         ception of Male and Female Sexual Cells . . . 272
      5. Primitive and Fundamental modes of Sexual Generation
         and the first appearance of Sexual Differences . . 278
   II. The Physiology of the Process of Fertilisation . . . 290
      1. The Need of Reproduction of Cells 291
         (a) Parthenogenesis 295
         (b) Apogamy 300
      2. Sexual Affinity 300

		PAGE
(a) Sexual Affinity in general		301
(b) More minute discussion of Sexual Affinity, and its different gradations		305
α. Self-fertilisation		306
β. Bastard Formation, or Hybridisation		310
γ. The Influence of Environment upon Sexual Affinity		313
δ. Recapitulation and Attempted Explanations		316
Literature		320

CHAPTER VIII.

METABOLIC CHANGES OCCURRING BETWEEN PROTOPLASM, NUCLEUS AND CELL PRODUCTS 323

I. Observations on the Position of the Nucleus, as an indication of its participation in Formative and Nutritive Processes . . . 324
II. Experiments proving Reciprocal Action of Nucleus and Protoplasm 330
Literature 332

CHAPTER IX.

THE CELL AS THE ELEMENTARY GERM OF AN ORGANISM. THEORIES OF HEREDITY 334

I. History of the older Theories of Development 335
II. More Recent Theories of Reproduction and Development . . 339
III. The Nucleus as the Transmitter of Hereditary Elemental Germs . 344
    1. The Equivalence of the Male and Female Hereditary Masses 345
    2. The equal Distribution of the Multiplying Hereditary Mass 346
    3. The Prevention of the Summation of the Hereditary Mass . 350
    4. Isotropy of Protoplasm 354
IV. Development of the Elemental Germs 357
Literature 361
Index 363

THE CELL

CHAPTER I

INTRODUCTION

BOTH plants and animals, although they differ so widely in their external appearance, are fundamentally similar in their anatomical structure; for both are built up of similar elementary units, which, as a rule, are only to be seen with the microscope. These units, in consequence of a hypothesis which was once believed in, but is now discarded, are called cells; and the view that plants and animals are built up in a similar manner of these extremely minute particles is called the *cell-theory*. The cell-theory is rightly considered to be one of the most important and fundamental theories of the whole science of modern biology. In the study of the cell, the botanist, the zoologist, the physiologist, and the pathologist go hand in hand, if they wish to search into the vital phenomena which take place during health and disease. For it is in the cells, to which the anatomist reduces both plant and animal organisms, that the vital functions are executed; they, as Virchow has expressed it, are the vital elementary units.

Regarded from this point of view, all the vital processes of a complex organism appear to be nothing but the highly-developed result of the individual vital processes of its innumerable variously functioning cells. The study of the processes of digestion, of the changes in muscle and nerve cells, leads finally to the examination of the functions of gland, muscle, ganglion, and brain. And just as physiology has been found to be based upon the cell-theory, so has the study of disease been transformed into a *cellular pathology*.

Hence, in many respects, the cell-theory is the centre around which the biological research of the present time revolves.

Further, it forms the basis of the study of minute anatomy, now more commonly called histology, which consists in the examination of the composition and minute structure of the organism.

The conception or idea connected with the word "cell," used scientifically, has been considerably altered during the last fifty years. The history of the various changes in this conception, or the history of the cell-theory, is of great interest, and nothing could be more suitable than to give a short account of this history in order to introduce the beginner to the series of conceptions connected with the word "cell"; this, indeed, may prove useful in other directions. For whilst, on the one hand, we see how the conception of the cell, which is at present accepted, has developed gradually out of older and less complete conceptions, we realise, on the other hand, that we cannot regard it as final or perfect; but, on the contrary, we have every ground to hope that better and more delicate methods of investigation, due partly to improved optical instruments, may greatly add to our present knowledge, and may perhaps enrich it with a quite new series of conceptions.

The History of the Cell-Theory. The theory, that organisms are composed of cells, was first suggested by the study of plant-structure. At the end of the seventeenth century the Italian, Marcellus Malpighi (I. 15), and the Englishman, Grew (I. 9), gained the first insight into the more delicate structure of plants; by means of low magnifying powers they discovered, in the first place, small room-like spaces, provided with firm walls, and filled with fluid, the cells; and in the second, various kinds of long tubes, which, in most parts of plants, are embedded in the ground tissue, and which, from their appearance, are now called spiral ducts or vessels.

Much greater importance, however, was attached to these facts after the investigations, which were carried on in a more philosophical spirit by Bahn towards the end of the eighteenth century, were published.

Caspar Friedrich Wolff (I. 34, 13), Oken (I. 21), and others, raised the question of the development of plants, and endeavoured to show that the ducts and vessels originated in cells. Above all, Treviranus (I. 32) rendered important service by proving in his treatise, entitled *Vom inwendigen Bau der Gewächse*, published in 1808, that vessels develop from cells; he discovered that young cells arrange themselves in rows, and become transformed, by the breaking down of their partition walls, into elongated tubes; this discovery was confirmed and established as a scientific fact by the subsequent researches of Mohl in 1830.

The study of the lowest plants has also proved of the greatest importance in establishing the cell-theory. Small algæ were observed, which during their whole lifetime remain either single cells, or consist of simple rows of cells, easily to be separated from one another. Finally, the study of the metabolism of plants led investigators to believe that, in the economy of the plant, it is the cell which absorbs the nutrient substances, elaborates them, and gives them up in an altered form (Turpin, Raspail).

Thus, at the beginning of our century, the cell was recognised by many investigators as the morphological and physiological elementary unit of the plant. This view is especially clearly expressed in the following sentences, quoted from the *Text-book of Botany* (I. 16), published by Meyen in 1830: " Plant-cells appear either singly, so that each one forms a single individual, as in the case of some algæ and fungi, or they are united together in greater or smaller masses, to constitute a more highly-organized plant. Even in this case each cell forms an independent, isolated whole; it nourishes itself, it builds itself up, and elaborates the raw nutrient materials, which it takes up, into very different substances and structures." In consequence, Meyen describes the single cells as " little plants inside larger ones."

These views, however, only obtained general acceptance after the year 1838, when M. Schleiden (I. 28), who is so frequently cited as the founder of the cell-theory, published in Müller's *Archives* his famous paper "Beiträge zur Phytogenesis." In this paper Schleiden endeavoured to explain the mystery of cell-formation. He thought he had found the key to the difficulty, in the discovery of the English botanist, R. Brown (I. 5), who, in the year 1833, whilst making investigations upon orchids, discovered nuclei. Schleiden made further discoveries in this direction; he showed that nuclei are present in many plants, and as they are invariably found in young cells, the idea occurred to him, that the nucleus must have a near connection with the mysterious beginning of the cell, and in consequence must be of great importance in its life-history.

The way in which Schleiden made use of this idea, which was based upon erroneous observations, to build up a theory of phytogenesis, must now be regarded as a mistake (I. 27); on the other hand, it must not be forgotten that his perception of the general importance of the nucleus was correct up to a certain point, and that this *one* idea has in itself exerted an influence far beyond the narrow limits of the science of botany, for it is owing to this that the cell-theory was first applied to animal tissues. For it is just in animal cells that the nuclei stand out most distinctly from amongst all the other cell-contents, thus showing most evidently the similarity between the histological elements of plants and animals. Thus this little treatise of Schleiden's, in 1838, marks an important historical turning-point, and since this time the most important work, in the building up of the cell-theory, has been done upon animal tissues.

Attempts to represent the animal body as consisting of a large number of extremely minute elements had been made before Schleiden's time, as is shown by the hypotheses of Oken (I. 21), Heusinger, Raspail, and many other writers. However, it was impossible to develop these theories further, since they were based upon so many incorrect observations and false deductions, that the good in them was outweighed by their errors.

It was not until after some improvements had been made in optical instruments, during the years from 1830–1840, that work justifying the application of the cell-theory to animal tissues was accomplished.

Purkinje (I. 22) and Valentin, Joh. Müller (I. 20) and Henle (I. 11), compared certain animal tissues with plant tissues, and recognized that the tissue of the chorda dorsalis, of cartilage, of epithelium and of glands, is composed of cells, and in so far is similar in its construction to that of plants. Schwann (I. 31), however, was the first to attempt to frame a really comprehensive cell-theory, which should refer to all kinds of animal tissues. This was suggested to him by Schleiden's "Phytogenesis," and was carried out by him in an ingenious manner.

During the year 1838 Schwann, in the course of a conversation with Schleiden, was informed of the new theory of cell-formation, and of the importance which was attached to the nucleus in plant-cells. It immediately struck him, as he himself relates, that there are a great many points of resemblance between animal and vegetable cells. He therefore, with most praiseworthy energy, set on foot a comprehensive series of experiments, the results of which he published in 1839, under the title, *Mikroscopische untersuchungen über die Uebereinstimmung in der Structur und dem Wachsthum der Thiere und Pflanzen*. This book of Schwann's is of the greatest importance, and may be considered to mark an epoch, for by its means the knowledge of the microscopical anatomy of animals was, in spite of the greater difficulty of observation, immediately placed upon the same plane as that of plants.

Two circumstances contributed to the rapid and brilliant result of Schwann's observations. In the first place Schwann made the greatest use of the presence of the nucleus in demonstrating the animal cell, whilst emphasizing the statement that it is the most characteristic and least variable of its constituents. As before mentioned, this idea was suggested to him by Schleiden. The second, no less important circumstance, is the accurate method which Schwann employed in carrying out and recording his observations. As the botanists by studying undeveloped parts of plants traced the development of the vessels, for instance, from primitive cells, so he, by devoting especial attention to the history of the development of the tissues, discovered that the embryo, at its earliest stage, consists of a number of quite similar cells; he then traced the metamorphoses or transformations, which the cells undergo, until they develop into the fully-formed tissues of the adult animal. He showed that whilst a portion of the cells retain their original spherical shape, others become cylindrical in form, whilst yet others develop into long threads or star-shaped bodies, which send out numerous radiating processes from various parts of their surface. He showed how in bones, cartilage, teeth, and other tissues, cells become surrounded by firm walls of varying thickness; and, finally, he explained the appearance of a number of the most atypical tissues by the consideration that groups of cells become, so to speak, fused together; this again is analogous to the development of the vessels in plants.

Thus Schwann originated a theory which, although imperfect in many respects, yet is applicable both to plants and animals, and which, further, is easily understood, and in the main correct. According to this theory, every part of the animal body is either built up of elements, corresponding to the plant cells, massed together, or is derived from such elements which have undergone certain metamorphoses. This theory has formed a satisfactory foundation upon which many further investigations have been based.

However, as has been already mentioned, *the conception which Schleiden and Schwann formed of the plant and animal element was incorrect in many respects.* They both defined the cell as *a small vesicle, with a firm membrane enclosing fluid contents, that is to say,* as a small chamber, or *cellula*, *in the true sense of the word*. They considered the membrane to be the most important and essential part of the vesicle, for they thought that in consequence of its chemico-physical properties it regulated the metabolism of the cell. According to Schwann, the cell is *an organic crystal*, which is formed by a kind of *crystallisation process from an organic mother-substance (cytoblastema)*.

The series of conceptions, which we now associate with the word "cell," are, thanks to the great progress made during the last fifty years, essentially different from the above. Schleiden and Schwann's cell-theory has undergone a radical reform, having been superseded by *the Protoplasmic theory*, which is especially associated with the name of Max Schultze.

The History of the Protoplasmic theory is also of supreme interest. Even Schleiden observed in the plant cell, in addition to the cell sap, a delicate transparent substance containing small granules; this substance he called plant slime. In the year 1846 Mohl (I. 18) called it Protoplasm, a name which has since become so significant, and which before had been used by Purkinje (I. 24) for the substance of which the youngest animal embryos are formed. Further, he presented a new picture of the living appearances of plant protoplasm; he discovered that it completely filled up the interior of young plant cells, and that in larger and older cells it absorbed fluid, which collected into droplets or vacuoles. Finally, Mohl established the fact that protoplasm, as had been already stated by Schleiden about the plant slime, shows strikingly peculiar movements; these were first discovered in the year 1772 by Bonaventura Corti, and later in 1807 by C. L. Treviranus, and were described as "the circulatory movements of the cell-sap."

By degrees further discoveries were made, which added to the importance attached to these protoplasmic contents of the cell. In the lowest algæ, as was observed by Cohn (I. 7) and others, the protoplasm draws itself away from the cell membrane at the time of reproduction, and forms a naked oval body, the *swarm-spore*, which lies freely in the cell cavity; this swarm-spore soon breaks down the membrane at one spot, after which it creeps out through the opening, and swims about in the water by means of its cilia, like an independent organism; but it has no cell membrane.

Similar facts were discovered through the study of the animal cell, which could not be reconciled with the old conception of the cell. A few years after the enunciation of Schwann's theory, various investigators, Kölliker (I. 14), Bischoff (I. 4), observed *many animal cells, in which no distinct membrane* could be discovered, and in consequence a lengthy dispute arose as to whether these bodies were really without membranes, and hence not cells, or whether they were true cells. Further, movements similar to those seen in plant protoplasm were discovered in the granular ground substance of certain animal cells, such as the lymph corpuscles (Siebold, Kölliker, Remak, Lieberkühn, etc.). In consequence Remak (I. 25, 26) applied the term protoplasm, which Mohl had already made use of for plant cells, to the ground substance of animal cells.

Important insight into the nature of protoplasm was afforded by the study of the lowest organisms, Rhizopoda (Amœbæ), Myxomycetes, etc. Dujardin had called the slimy, granular, contractile substance of which they are composed Sarcode. Subsequently, Max Schultze (I. 29) and de Bary (I. 2) proved, after most careful investigation, that *the protoplasm of plants and animals and the sarcode of the lowest organisms are identical.*

In consequence of these discoveries, investigators, such as Nägeli, Alexander Braun, Leydig, Kölliker, Cohn, de Bary, etc., considered the cell membrane to be of but minor importance in comparison to its contents; however, the credit is due to Max Schultze, above all others, of having made use of these later discoveries in subjecting the cell theory of Schleiden and Schwann to a searching critical examination, and of founding a protoplasmic theory. He attacked the former articles of belief, which it was necessary to renounce, in four excellent though short papers, the first of which was published in the year 1860. He based his theory that the cell-membrane is not an essential part of the elementary organisms of plants and animals on the following three facts: first, that a certain substance, the protoplasm of plants and animals, and the sarcode of the simplest forms, which may be recognised by its peculiar phenomena of movement, is found in all organisms; secondly, that although as a rule the protoplasm of plants is surrounded by a special firm membrane, yet under certain conditions it is able to become divested of this membrane, and to swim about in water as in the case of naked swarm-spores; and finally, that animal cells and the lowest unicellular organisms very frequently possess no cell-membrane, but appear as naked protoplasm and naked sarcode. It is true that he retains the term "cell," which was introduced into anatomical language by Schleiden and Schwann; but he defines it (I. 30) as: *a small mass of protoplasm endowed with the attributes of life.*

Historical accuracy requires that it should be mentioned that in this definition Max Schultze reverted to the older opinions held by Purkinje (I. 22–24) and Arnold (I. 1), who endeavoured to build up a theory of granules and masses of protoplasm, but without much result, for the cell theory of Schwann was both more carefully worked out, and more adapted to the state of knowledge of the time.

The term, a small mass of protoplasm, was not intended by Max Schultze and other investigators even then to mean so simple a matter as appears at first. The physiologist, Brücke (I. 6), especially came to the correct conclusion, gathered with justice from the complexity of the functions of life, which are inherent in protoplasm, that the protoplasm itself must be of a complex construction, that is must possess "an extremely intricate structure," into which, as yet, no satisfactory insight has been gained owing to the imperfections of our means of observation. Hence Brücke very pertinently designated the "ultimate particle" of animals and plants, that is the mass of protoplasm, an *elementary organism.*

Hence it is evident that the term "cell" is incorrect. That it, nevertheless, has been retained, may be partly ascribed to a kind of loyalty to the vigorous combatants, who, as Brücke expresses it, conquered the whole field of histology under the banner of the cell-theory, and partly to the circumstance, that the discoveries which brought about the new reform were only made by degrees, and were only generally accepted at a time when, in consequence of its having been used for several decades of years, the word cell had taken firm root in the literature of the subject.

Since the time of Brücke and Max Schultze, our knowledge of the true nature of the cell has increased considerably. Great insight has been gained into the structure and the vital properties of the protoplasm, and in especial, our knowledge of the nucleus, and of the part it plays in cell-multiplication, and in sexual reproduction, has recently made great advances. The earlier definition, "the cell is a little mass of protoplasm," must now be replaced by the following: "*the cell is a little mass of protoplasm, which contains in its interior a specially formed portion, the nucleus.*"

The history of these more recent discoveries will be entered into later, being only incidentally mentioned here and there in the following account of our present knowledge of the nature of the elementary organism.

The enormous amount of knowledge which has been acquired through a century of investigation will be best systematically arranged in the following manner:—

In the *first* section the chemico-physical and morphological properties of the cell will be described.

The second section will treat of the vital properties of the cell. These are, (1) its contractility, (2) its irritability, (3) the phenomena of metabolism, (4) its power of reproduction.

Further, in order to complete and amplify our account of the nature of the cell, two sections more speculative in character will be added, one treating of the relationship between the protoplasm, the nucleus, and the cell products, and the other of the cell considered as the germ of an organism.

Literature I.

1. Fr. Arnold. *Lehrbuch der Physiologie des Menschen.* 2 Theil. Zürich. 1842. *Handbuch der Anatomie des Menschen.* 1845.
2. de Bary. *Myxomyceten. Zeitschrift f. wissenschaftl. Zool.* 1859.
3. Lionel S. Beale. *On the Structure of the Simple Tissues of the Human Body.* 1861.
4. Bischoff. *Entwicklungs-geschichte des Kanincheneies.* 1842.
5. R. Brown. *Observations on the Organs and Mode of Fecundation in Orchideæ and Asclepiadeæ. Transactions of the Linnean Soc., London.* 1833.
6. Brücke. *Die Elementarorganismen. Wiener Sitzungsber. Jahrg.* 1861. XLIV. 2. Abth.
   Cleland. *On Cell Theories. Quar. Jour. Microsc. Sc.* XIII., p. 255.
7. Cohn. *Nachträge z. Naturgeschichte des Protococcus pluvialilis. Nova acta.* Vol. XXII., pp. 607–764.
8. Bonaventura Corti. *Osservazioni microsc. sulla Tremella e sulla circolazione del fluido in una pianta acquaiola.* 1774.
   Dallinger and Drysdale. *Researches on the Life History of the Monads. Month. Mic. Journ.* Vols. X.–XIII.
9. Grew. *The Anatomy of Plants.*
10. Haeckel. *Die Radiolarien.* 1862. *Die Moneren.*
11. Henle. *Symbolæ ad anatomiam villorum intestinalium.* 1837.
12. Oscar Hertwig. *Die Geschichte der Zellentheorie. Deutsche Rundschau.*
13. Huxley. *On the Cell Theory. Monthly Journal.* 1853.
14. Kölliker. *Die Lehre von der thierischen Zelle. Schleiden u. Nägeli Wissenschaftl. Botanik.* Heft 2, 1845.
    Kölliker. *Manual of Human Histology,* trans. Sydenham Society. 1853.

15. MALPIGHI. *Anatome plantarum.*
16. MEYEN. *Phytotomie. Berlin.* 1830.
17. H. v. MOHL. *Ueber die Vermehrung der Pflanzenzellen durch Theilung. Dissert. Tübingen.* 1835. *Flora.* 1837.
18. H. v. MOHL. *Ueber die Saftbewegung im Innern der Zellen. Botanische Zeitung.* 1846.
19. H. v. MOHL. *Grundzüge der Anatomie und Physiologie der vegetabilischen Zelle. Wagners Handwörterbuch der Physiologie.* 1851.
20. J. MÜLLER. *Vergleichende Anatomie der Myxinoiden.*
21. OKEN. *Lehrbuch der Naturphilosophie.* 1809.
22. PURKINJE. *Bericht über die Versammlung deutscher Naturforscher und Aertzte in Prag im September,* 1837. *Prag,* 1838, pp. 174, 175.
23. PURKINJE. *Uebersicht der Arbeiten und Veränderungen der schlesischen Gesellschaft für vaterländische Cultur im Jahre,* 1839. *Breslau,* 1840.
24. PURKINJE. *Jahrbücher für wissenschaftliche Kritik.* 1840. Nr 5, pp. 33–38.
25. REMAK. *Ueber extracelluläre Entstehung thierischer Zellen und über Vermehrung derselben durch Theilung. Müllers Archiv.* 1852.
26. REMAK. *On the Embryological Basis of the Cell Theory* (translated). Q. J. M. S. II., p. 277.
27. SACHS. *Geschichte der Botanik.* 1875.
28. MATTHIAS SCHLEIDEN. *Beiträge zur Phytogenesis. Müllers Archiv.* 1838. *Principles of Scientific Botany,* translated by Lankester. 1849.
29. MAX SCHULZE. *Das Protoplasma der Rhizopoden und der Pflanzenzelle.*
30. MAX SCHULZE. *Ueber Muskelkörperchen und was man eine Zelle zu nennen habe. Archiv für Anatomie und Physiologie.* 1861.
31. TH. SCHWANN. *Mikroscopische Untersuchungen über die Uebereinstimmung in der Structur und dem Wachsthum der Thiere und Pflanzen.* 1839.
    SCHWANN und SCHLEIDEN. *Microscopical Researches, trans. Sydenham Soc.* 1837.
32. C. L. TREVIRANUS. *Vom inwendigen Bau der Gewächse,* 1805.
33. R. VIRCHOW. *Cellular Pathology as based upon Physiological and Pathological Histology,* trans. by Chance. 1860.
34. CASP. FRIEDR. WOLFF. *Theorie von der Generation.* 1764.

CHAPTER II

THE CHEMICO-PHYSICAL AND MORPHOLOGICAL PROPERTIES OF THE CELL

The cell is an organism, and by no means a simple one, being built up of many different parts. To ascertain with accuracy the true nature of all these constituents, which, for the greater part, elude our observation at present, will remain a problem for biological research for a long time. Our position, with regard to the cell, is similar to that of investigators towards the whole animal or vegetable body a hundred years ago, before the discovery of the cell theory. In order to penetrate more deeply into the secrets of the cell, optical instruments, and, above all, methods of chemical examination, must be brought to a much higher degree of perfection than they have attained at present. It seems best to me to lay stress on these points to start with, in order that the student may have them always before his mind's eye in reading the following account.

In each cell there is invariably to be seen one specially well-defined portion, the nucleus, which throughout the whole of the animal and vegetable kingdom is very uniform in appearance; evidently the nucleus and the remaining portion of the cell have different functions to perform in the elementary organism. Hence the examination of the chemico-physical and morphological properties of the cell becomes naturally divided into two sections, the examination of the protoplasm and of the nucleus.

To these, three short sections are added. The first deals with the question, Are there cells which possess no nuclei? The second treats of the pole or central corpuscles, which are at times found as special cell-structures in addition to the nucleus; and in the third a short account is given of Nägeli's theory of the molecular structure of organic bodies.

I. The Chemico-physical and Morphological Properties of the Protoplasm. Some animal and plant-cells appear to differ so much from one another as to their form and contents, that, at first sight, they seem to have nothing in common, and that hence it is impossible to compare them. For instance, if a cell at the growing-point of a plant be taken and compared with one filled with starch granules from the tuber of a potato, or if the contents of an embryo cell from a germinal disc be compared with those of a fat cell, or of one from the egg of an Amphibian filled with yolk granules, the inexperienced observer sees nothing but contrasts. Nevertheless, all these exceedingly different cells are seen on closer examination to be similar in one respect, *i.e.* in the possession of a very important, peculiar mixture of substances, which is sometimes present in large quantities, and sometimes only in traces, but which is never wholly absent in any elementary organism. In this mixture of substances the wonderful vital phenomena, which are dealt with later, may very frequently be observed (contractility, irritability, etc.); and, moreover, since in young cells, in lower organisms, and in the cells of growing-points and germinal areas, it is in the cell-substance alone (the nucleus of course being excepted) that these properties have been observed, this substance has been recognised as the chief supporter of the vital functions. It is the protoplasm or "forming matter" of the English histologist, Beale (I. 3).

a. **Justification of the Use of the Term Protoplasm.** In order to know what *protoplasm* is, it is advisable to examine it in those cells in which it is present in large quantities, and in which it is as free as possible from admixture with other bodies; and amongst such the most suitable are those organisms from the study of which the founders of the protoplasmic theory formed their conception of the nature of this substance. Such organisms are, young plant-cells, Amœbæ, and the lymph corpuscles of vertebrates. After the student has learnt to recognise the characteristic properties of protoplasm in such bodies, he will be able to discover it in others, in which it is only present in small quantities and is more or less concealed by other substances.

It has been proposed (II. 10) to give up altogether the use of the term protoplasm, since it has been associated with such mistaken views; for the word has now come to be used in so indefinite and vague a manner, that it may be questioned whether it is not at present more misleading than useful.

However, this proposition cannot be considered to be advisable or even justifiable in the present condition of affairs, for, although it must be admitted that the word is frequently used incorrectly;

and that further, it is impossible in a short phrase to give an adequate definition of its meaning; and finally, that frequently it is difficult to determine what part of the cell really consists of protoplasm, and what does not; yet, in spite of all this, the necessity of the conception remains. Similar objections could be raised against a number of other words which we use for certain definite compounds present in organic bodies. For instance, to designate a certain portion of the nucleus we use the term nuclein or chromatin, which is considered fairly adequate by many people. And yet the microscopist is bound to admit that it is impossible to state exactly which part of a resting nucleus consists of linin, and which of nuclein, or to determine in any special case whether too much or too little has been stained.

Now the term protoplasm is quite as necessary in speaking about the constituent parts of a cell. Only it must be stipulated that the word protoplasm must not be understood to designate a substance of definite chemical composition.

The word protoplasm is a morphological term (the same is true in a greater or less degree of the word nuclein, and of many others); it is an expression for a complex substance, which exhibits a variety of physical, chemical, and biological properties. Such expressions are absolutely necessary in the present state of our knowledge. Any one who is acquainted with the history of the cell knows what a number of observations and how much logical thought were necessary before this conception was arrived at, and further is quite aware that with the creation of this expression the whole theory of cells and tissues gained in depth and significance. How much wordy warfare was necessary before it was established that the cell contents, and not the cell membrane, constitute the essential portion of the cell, and further that amongst these cell contents a peculiar substance is invariably present, which takes part in the vital processes in quite a different way from the cell sap, the starch granules, and the fat globules.

Thus we see that the use of the word protoplasm is not only justifiable from an historical point of view, but also from a scientific one, and we will now proceed to endeavour to explain what is meant by the term.

b. **General Characteristics of Protoplasm.** The protoplasm of unicellular organisms, and of plant and animal cells (Figs. 1 and 2), appears as a viscid substance, which is almost always colourless, which will not mix with water, and which, in consequence of a certain resemblance to slimy substances, was called by Schleiden the slime of the cell. Its refractive power is greater than that of water, so that the most delicate threads of protoplasm, although colourless, may be distinguished in this medium. Minute granules, the microsomes, which look only like dots, are always present in greater or less numbers in all protoplasm, and may be seen with a low power of the microscope to be embedded in a homogeneous ground substance. According to whether there are few or many of these microsomes in the protoplasm, it is more transparent (hyaline) or darker and more granular in appearance.

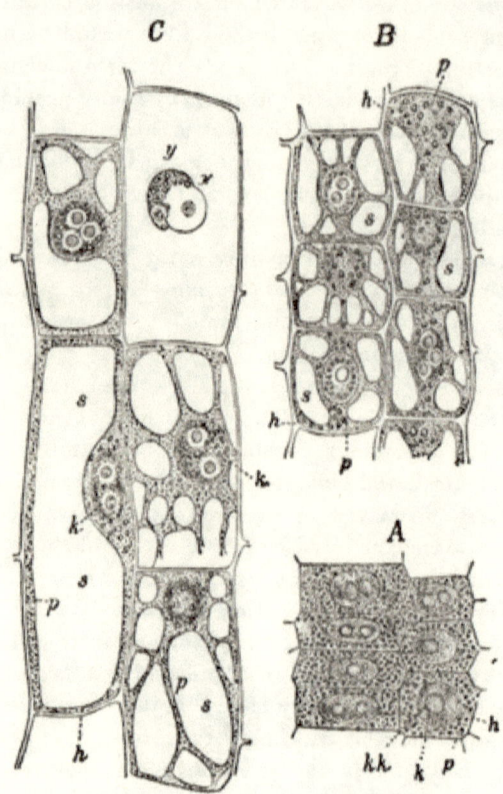

FIG. 1.—Parenchyma cells, from the cortical layer of the root of *Fritillaria imperialis*; longitudinal sections (× 650); after Sachs (II. 38), Fig. 75: *A* very young cells, as yet without cell-sap, from close to the apex of the root; *B* cells of the same description, about 2mm above the apex of the root; *C* cells of the same description, about 7–8 mm. above the apex; the two lower cells on the right-hand side are seen in a front view, the large cell on the left side is seen in optical section, the upper right-hand cell is opened by the section; the nucleus (*xy*) has a peculiar appearance, being distended with water which it has absorbed; *k* nucleus; *kk* nucleolus; *h* membrane.

The distribution of these granules in the body of the cell is rarely regular. Generally a more or less thin outer zone remains free from granules. Now as this layer appears to be somewhat firmer in consistence than the more watery granula mass, it has been thought advisable to distinguish two kinds of protoplasm, the *ectoplasm* or *hyaloplasm*, and the *endoplasm* or *granularplasm* (Fig. 2, *ek*, *en*).

Many investigators, such as Pfeffer, de Vries, etc., are inclined to consider that *this peripheral layer is a specially differentiated organ of the cell and is endowed with special functions*. The following experiment which I have made seems to bear out this view.

Some ripe eggs of *Rana temporaria*, which had entered the oviduct and were surrounded with a gelatinous coating, were carefully pierced with the exceedingly fine point of a glass needle. The puncture thus made was not visible externally after the operation, nor was any yolk seen to exude through the holes. However, some time after fertilisation of the eggs had taken place, a fair quantity of yolk began to make its way out of all the punctured eggs, and to form a more or less large ridge (extraovat, Roux) between the membrane of the egg and the yolk. This welling out of the yolk substance was induced by the act of fertilisation, for the entrance of the spermatozoon stimulates the surface layer to contract energetically, as may be easily demonstrated under suitable conditions. Hence the puncture must have caused a wound in the peripheral layer, which had not time to heal before fertilisation took place, and through which the yolk was only pressed out after the contraction caused by the fertilisation had taken place. Now since between the piercing of the eggs and their fertilisation a fairly long interval, which however I did not accurately measure, had elapsed, this experiment seems to show that the peripheral layer possesses a structure differing somewhat from that of the rest of the cell contents, and also that it has properties peculiar to itself.

c. **Chemical Composition of Protoplasm.** Our knowledge of the *chemical nature of protoplasm* is most unsatisfactory. It has sometimes been described as an albuminous body, or as "living albumen." Such expressions may give rise to utterly incorrect conceptions of the nature of protoplasm. On this account I will recapitulate what I said in section *a*: Protoplasm is not a chemical, but a morphological conception; it is not a single chemical substance, however complex in composition, but is composed of a large number of different chemical substances, which we have to picture to ourselves as most minute particles united together to form a wonderfully complex structure.

Chemical substances exhibit similar properties under different circumstances (as, for instance, hæmoglobin, whether present as a constituent of the blood corpuscles, or dissolved in water, or in the form of crystals). Protoplasm, on the other hand, cannot be placed under different conditions without ceasing to be protoplasm, for its essential properties, in which its life manifests itself, depend upon a fixed organisation. For as the principal attributes of a marble statue consist in the form which the sculptor's hand has given to the marble, and as a statue ceases to be a statue if broken up into small pieces of marble (Nägeli II. 28), so a body of protoplasm is no longer protoplasm after the organisation, which constitutes its life, has been destroyed; we only examine the considerably altered ruins of the protoplasm when we treat the dead cells with chemical reagents.

It is possible that after a time our knowledge of chemistry may have advanced sufficiently to enable us to produce albuminous bodies artificially by synthesis. On the other hand, the attempt to make a protoplasmic body would be like Wagner's endeavour to crystallise out a homunculus in a flask. For, as far as we know at present, *protoplasmic bodies are only reproduced from existing protoplasm, and in no other way; hence the present organisation of protoplasm is the result of an exceedingly long process of development.*

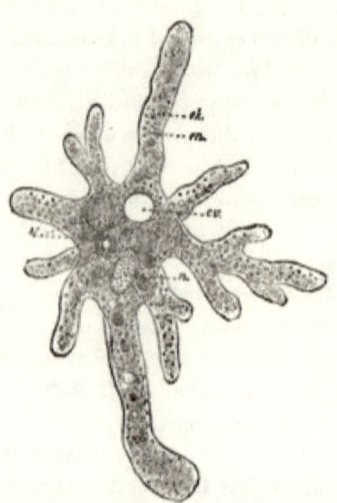

Fig. 2.—*Amœba Proteus* (after Leidy; from Rich. Hertwig): *n* nucleus; *cv* contractile vacuole; *n* food vacuoles; *en* endoplasm; *ek* ectoplasm.

It is very difficult to determine the chemical nature of the substances which are peculiar to living protoplasm. For setting aside the fact that the bodies are so unstable that the least interference with them essentially alters their constitution, the difficulty in analysing them is considerably increased by the presence in each cell of various waste products of metabolism, which it is not easy to separate from the rest of the cell contents. Amongst these complex substances the proteids, as the true sustainers of the vital processes, are of especial importance; these proteids are the most complex of all known organic substances, but up till now very little has been determined as to their chemical structure. This complex structure depends, in the first place, upon the very remarkable chemical properties of carbon (Haeckel II. 15). In proteids carbon occurs combined with four other elements, hydrogen, oxygen, nitrogen, and sulphur, in proportions which, it has been endeavoured to express by the following formula: $C^{72}H^{106}N^{18}SO^{22}$ (*composition of a molecule of egg-albumen*).

Amongst the various kinds of proteid bodies (albumins, globulins, fibrins, plastins, nucleins, etc.) plastin alone seems to be peculiar to protoplasm (Reinke II. 32; Schwarz II. 37; Zacharias II. 44); plastin is insoluble in water, in 10 per cent. salt solution, and in 10 per cent. solution of sulphate of magnesia; it is precipitated by weak acetic acid, whilst concentrated acetic acid causes it to swell up; it is precipitated in concentrated salt solution; it resists both pepsin and trypsin digestion. It is hardly, or not at all, stained by basic aniline dyes, but is stained by acid ones (eosin and acid fuchsine).

In addition, globulins and albumins are present in smaller quantities; these are also found in solution in the cell-sap of plants.

Protoplasm is very rich in water, which, as Sachs (II. 33) states, is built up into the structure of its molecule, in the same sense as, for example, the water of crystallisation is a necessary constituent of many crystals, which lose their characteristic form if the water of crystallisation is withdrawn. Reinke (II. 32) found 71·6 per cent. of water and 28·4 per cent. of solid substances in fresh sporangia of the *Æthalium septicum* (66 per cent. of this water could be squeezed out).

Further, a number of various salts are present in protoplasm; these remain as ash when the protoplasm is burnt; in the case of the *Æthalium septicum* the ash contains the following elements: chlorine, sulphur, phosphorus, potassium, sodium, magnesium, calcium, and iron.

Living protoplasm is distinctly alkaline in reaction; red litmus paper is turned blue by it, as is also a red colouring matter, which is obtained from a species of cabbage, and which has been used by Schwarz. This is also the case with plants, although the cell-sap, as a rule, has an acid reaction. According to the investigations of Schwarz (II. 37) on plants, this alkaline reaction is due to the presence of an alkali, which is united with the proteid bodies in living protoplasm. Reinke (II. 32) states that the *Æthalium septicum* gives off ammonia after it has been dried.

Moreover, the most different metabolic products are always to be demonstrated in protoplasm; these are produced either by progressive or retrogressive metamorphosis. There is a great similarity shown between the substances occurring in plant and in animal cells. For example, the following substances are found in both,—pepsin, diastase, myosin, sarcin, glycogen, sugar, inosit, dextrin, cholesterin and lecithin, fat, lactic acid, formic acid, acetic acid, butyric acid, etc.

As an example of the quantitative composition of a cell including its nucleus, Kossel (II. 35) quotes in his text-book, the analysis of pus-corpuscles which was made by Hoppe-Seyler. According to this statement, 100 parts by weight of organic substance contain:

Various albuminous substances	13·762
Nuclein	34·257
Insoluble substances	20·566
Lecithin and fat	14·383
Cholesterin	7·400
Cerebrin	5·199
Extractives	4·433

Phosphorus, sodium, iron, magnesium, calcium, phosphoric acid and chlorine were found in the ash.

As regards the physical properties of protoplasm, streaming protoplasmic threads are sometimes noticed in which double refraction is seen, the movements being for the most part in a direction such that their optical axes coincide (Engelmann).

d. **The more minute Structure of Protoplasm.** Protoplasm was defined above as a combination of substances, the most minute particles of which we must picture to ourselves as united together to form a complex structure. Investigators have endeavoured to discover more about this marvellous structure, partly by speculation, and partly by microscopical observation.

As to the first, Nägeli has made some important suggestions, a more detailed account of which is given later in the section entitled "The Molecular Structure of Organised Bodies."

As to the second, numerous investigators, amongst whom Frommann, Flemming, Bütschli and Altmann are conspicuous, have recently been working at the subject. Living protoplasm, as well as that which has been killed by special reagents, has been examined; in the latter, its most minute structure has been rendered visible, by means of various staining reagents; thus we have already a considerable amount of literature on the subject of the structure of protoplasm.

Starting with the assumption that protoplasm consists of a mixture of a small quantity of solid substances with a large quantity of fluid, to which circumstance it owes its peculiar viscid property as a whole, the question might be raised as to whether it be possible, by using the strongest lenses, to distinguish optically the solid particles from the fluid which contains them, and to recognise their arrangement into special structures. *A priori*, it does not seem to be necessary to distinguish them from one another, since the solid particles are so very small, and since they differ so little from the fluid in their refractive power. Thus, according to Nägeli's micellar theory, which will be described in detail later on, they are supposed to be arranged as a framework, *which, however, in consequence of the minute size of the hypothetical micellæ, escapes our observation*. In a word, it is possible that protoplasm may have a very complicated structure, although it appears to us to be a homogeneous body. Hence the expression homogeneous protoplasm does not necessarily imply that protoplasm does not possess a definite structure or organisation.

Recent observations, for which powerful oil immersion lenses have been successfully used, point more and more to the conclusion that protoplasm possesses a structure which may be optically demonstrated; however, individual microscopists differ so essentially in their views upon the nature of this structure, that it is impossible to come to any definite decision upon the subject.

At the present time, at least four conflicting theories hold the field; these may be described as the framework theory, the foam or honeycomb theory, the filament theory, and the granula theory.

The *framework theory* has been advocated by Frommann (II. 14), Heitzmann (II. 17), Klein (II. 21), Leydig (II. 26), Schmitz (II. 36), and by others. According to this theory, protoplasm consists of a very fine network of fibrillæ or threads, in the interstices of which the fluid is held. Thus, roughly speaking, it is like a sponge, or, shortly expressed, its structure is spongiose. The microsomes, which are seen in the endoplasm (granular plasma), are nothing but the points where the fibrillæ intersect.

A glance over the literature on this subject shows the reader that very different appearances are sometimes described under the title, "The spongiose structure of protoplasm." Sometimes the description refers to coarser frameworks, which, being due to the deposition in the protoplasm of various kinds of substances, should not be considered as pertaining to protoplasm, nor should they be included in its description. This holds true, for example, of the description of the goblet cells of List (II. 48) (see p. 36, fig. 17). This subject is more fully discussed later on.

Sometimes net-like structures are described and depicted, which, as they are evidently caused by coagulation (due to some precipitation process), must be considered as artificial products. For instance, artificial framework structures may be easily produced, if a solution of albumen or gelatine be caused to coagulate by the addition of chromic acid, picric acid, or alcohol. Thus Heitzmann (II. 17) demonstrates, in a somewhat diagrammatic manner, the presence of networks in the most various cells of the animal body, which does not correspond to actual fact. Bütschli also remarks in his abstract of the literature on the subject (II. 7b, p. 113): "Above all, it is frequently very difficult to determine whether the net-like appearances described by earlier observers are really delicate protoplasmic structures, or whether they are caused by coarser vacuolisation. Since the same appearance is produced in either case, it is only possible to form a fairly correct opinion by considering their relative sizes." Bütschli found that in all cases the spaces in the meshes of the protoplasm measured barely 1 μ.

Thus, although no doubt many statements may be legitimately questioned, yet it is undeniable that many investigators (Frommann, Schmitz, Leydig) have really based their descriptions upon the more delicate structures of the cell.

In the explanation of these so-called net-work appearances, Bütschli takes up a position which is different from that of the other observers who have been mentioned, and which has caused him to advance a *foam or honeycomb theory of protoplasm* (II. 7a, 7b).

He succeeded in producing a very delicate emulsion by mixing inspissated olive oil with K_2CO_3, common salt, or cane-sugar.

This emulsion consists of a groundwork of oil, containing an exceedingly large number of spaces, which are completely closed in and filled with watery liquid ; if the emulsion is too fine to be seen except under the microscope, the diameter of the spaces is generally less than ·001 mm. In appearance they are very like the cells of a honeycomb, being in the form of very varying polyhedra; they are separated from one another by the most delicate lamellæ of oil, which refract the light somewhat more strongly than the watery liquid does. As a result of physical laws, only three lamellæ can touch at one edge. Hence it appears in optical section, that only three lines meet in any one point. If before the formation of the emulsion fine particles of lamp-black are distributed throughout the oil, these collect at the point of intersection. Finally,

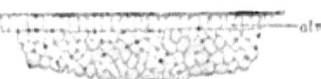

Fig. 3.—Optical section of the edge of a drop of an emulsion made with olive oil and salt; the alveolar layer (alv.) is very distinct, and relatively deep. (× 1250: after Bütschli, Pl. III., Fig. 4.)

the superficial layer is composed of a delicate froth, the framework of which is arranged in a peculiar fashion, the partition walls of oil, which touch the surface, being perpendicular to it, and thus appearing parallel to one another in optical section. Bütschli describes this as the *alveolar layer* (Fig. 3 *alv.*).

Bütschli considers that the protoplasm of all plant and animal cells (Figs. 4, 5) possesses a structure which is similar to this.

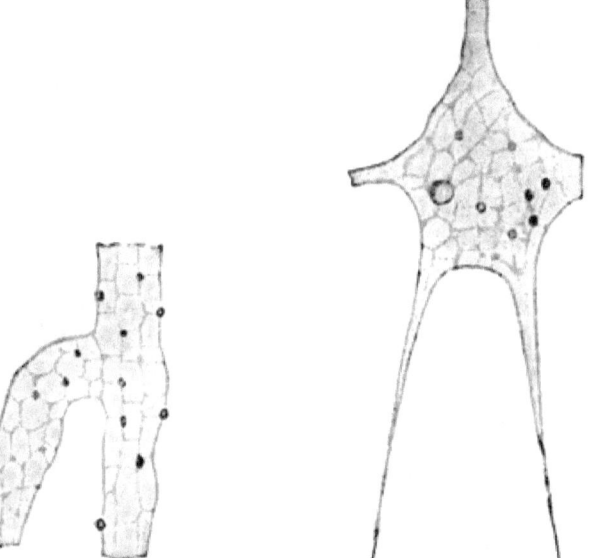

Fig. 4.   Fig. 5.

Fig. 4.—Two living strands of plasma from a hair-cell of a *Mallow*. (× about 3,000: after Bütschli, Pl. II., Fig. 14.)

Fig. 5.—Web-like extension, very distinct in structure, from the pseudopodic net of a *Miliola* from life. (× about 3,000: after Bütschli, Pl. II., Fig. 5.)

His opinion is based upon his experiments on living objects, which he treated with various reagents. In his opinion there is a framework of plasma corresponding to the lamellæ of oil, which, in the artificial emulsion, separate the droplets of fluid from one another. Similarly here also granules (microsomes) are collected together at the points of intersection. Further the protoplasmic body is frequently differentiated externally to form an alveolar layer. The appearance, described by other observers as a thread or net-like structure with spaces which communicate and contain fluid, Bütschli considers to be due to the presence of a froth or honey-comb structure, in which the cavities are closed in on all sides; he himself, however, remarks that, in consequence of the minuteness of the structures in question, it is impossible to decide finally, simply by the appearance under the microscope, whether a net-like or honeycomb structure really exists (II. 7b, p. 140), since " in either case the appearance under the microscope is the same."

Now it seems hardly justifiable, that this similarity to an artificially prepared froth, although it has caused Bütschli finally to make up his mind, should be allowed to settle the question.

Two objections to this theory of Bütschli's must be mentioned. The first is that it does not apply to nuclear substance, which without doubt is similar in its organisation to protoplasm. For during the process of nuclear division threadlike arrangements in the form of spindle-threads and nuclein-threads are so distinctly to be seen, that their existence certainly cannot be questioned by any one.

The second objection is more theoretical in nature. The oil lamellæ are composed of a fluid which does not mix with water. Now if the comparison between the structure of this emulsion and that of protoplasm is to depend upon something more than a mere superficial similarity, the plasma lamellæ, corresponding to the oil lamellæ, must be composed of a solution of albumen or of liquid albumen. Now this cannot be the case, for a solution of albumen is capable of mixing with water, and hence would of necessity mix with the contents of the spaces; hence the albuminous froth would have to be prepared with air. In order to get over this difficulty, Bütschli assumes that the chemical basis of the framework substance is a fluid, composed of molecules of albumen combined with those of a fatty acid (II. 7b, p. 199); this supposition, and especially the theory that the framework substance is a fluid, is not likely to meet with much support. For on many accounts it seems to be true that the structural elements of protoplasm, whether they form the threads of a net, or the lamellæ of a honey-comb, or granules, or what not, must be solid in their nature. Protoplasm does not consist of two non-miscible fluids, such as water and oil, but of a combination of solid organic particles with a large quantity of water. Hence quite different physical conditions are necessarily present. (Compare section on molecular structure, p. 58.)

The third of the above-mentioned views, or the *filament theory*, is connected with the name of Flemming (II. 10).

Whilst examining a large number of living cells (cartilage, liver, connective tissue, and ganglion cells, etc.), Flemming observed in the protoplasm (Fig. 6) the presence of extremely delicate threads which have somewhat greater refractive power than the intervening ground substance. These threads vary in length, being longer in some cells than in others; sometimes larger numbers are present than at others. It seemed impossible to determine with certainty whether they are separated from one another all along their length, or whether they join together to form a net; if they do form a net, then its meshes must be very uneven in size. Hence Flemming considers that two different substances occur in protoplasm, a *thread substance* and an *interstitial substance*, or a *filamentous* and an *interfilamentous substance* (mitome and paramitome); upon the chemical nature of these substances and upon their general condition Flemming does not enlarge. How much importance should be attached to this structure, about which at present nothing further can be stated, it remains for the future to reveal.

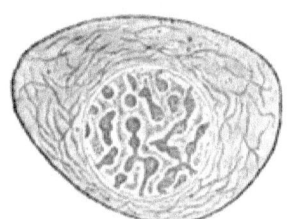

FIG. 6.—Living cartilage cell of a *Salamander* larva, much magnified, with clearly marked filamentous substance; after Flemming (from Hatschek, Fig. 2).

In this section, "On the Structure of Protoplasm," the ray-like arrangement of the protoplasm which is observed at certain stages of the division of the nucleus, or the striated appearance which is exhibited by the protoplasm of secretory cells, might be more fully described. Since, however, such structures only occur under special conditions, it has been considered more advisable to defer their consideration to a later period.

Fourthly, and finally, come the attempts of Altmann (II. 1) to demonstrate a still more minute structure of protoplasm (*granula theory*). By means of a special method of treatment, this investigator has succeeded in rendering minute particles visible in the body of the cell; these he calls granula. He preserves the organ in a mixture of 5 per cent. solution of potassium bichromate with 2 per cent. solution of perosmic acid; he then prepares thin sections of the organ and stains them with acid fuchsine, finally treating them with alcoholic solution of picric acid, by means of which the differentiation is rendered more distinct. The result of these staining reactions is to render visible a large number of very minute dark-red granules. Sometimes they are seen to be isolated, sometimes more densely packed; sometimes they are near together, sometimes further apart; or they may be united in rows to form threads.

In consequence of these observations, Altmann has propounded a very important and far-reaching hypothesis. He considers these granules to be still more minute elementary organisms, of which the cell itself is composed; he calls them *bioblasts*, attributes to them the structure of organised crystals, and looks upon them as equivalent to the micro-organisms which, as individuals, arrange themselves in masses to form a zooglea, or in rows to form threads. "As in a zooglea the single individuals are connected together by means of a gelatinous substance secreted by themselves, and at the same time are separated from one another by it, so in the cell the same might occur with the granula; in this case also we must not consider that there is merely water and salt solution surrounding the granula, but similarly that a more gelatinous substance (intergranula substance) is present; this is sometimes liquid, and sometimes fairly viscid in consistency. The great mobility, peculiar to most protoplasm, renders the former probable. If this intergranula substance becomes collected without granula at any point in the cell, a true hyaloplasm may be formed, which, being free from living elements, does not really deserve the name of protoplasm."

Thus Altmann defines protoplasm as "a colony of bioblasts, the individual elements of which are grouped together either in a zooglea condition or in the form of threads, and which are connected by an indifferent substance." "Hence the bioblast is the much-sought-after, morphological unit of all organic substances, with which all biological investigation must finally deal." However, the bioblast is not able to live alone, but dies with the cell in which, according to Altmann, it multiplies by fission (*omne granulum e granulo*).

Many objections may be raised to this hypothesis of Altmann's, in so far as it refers to the interpretation of recorded observations. Firstly, the most minute micro-organisms of a zooglea are connected by means of a great number of forms, which are intermediate as to size, with the larger fission and yeast fungi; and since these are not to be distinguished from cells in their construction, they also must, according to Altmann, be colonies of bioblasts. Further, Bütschli has shown that the larger micro-organisms are most probably divided into nucleus and protoplasm, and hence are similar in structure to other cells. The flagella, also, which have been demonstrated in many micro-organisms, must be considered to be cell organs. Secondly, we have not been sufficiently enlightened upon the nature and function of the granula in the cell, excepting that for some reason or other we are to conclude that they are its true vital elements. According to Altmann's hypothesis, the relative importance which has been attached to cell-substances is completely reversed. The substance which he calls intergranula substance, and which in its physiological importance he considers to correspond to the gelatinous substance of the zooglea, is to all intents and purposes the protoplasm of the generally accepted cell theory, that is to say, the substance which is considered to form the most important generator of the vital processes; on the other hand, the granula belong to the category of protoplasmic contents, and as such have had a much less important *rôle* ascribed to them. Thus Altmann designates the melanin granules of a pigment cell as the bioblasts, and the connecting protoplasm as the intergranula substance. Similarly he completely reverses the physiological importance of the substances in the nucleus, as will be shown later on, in that he considers that his granula are contained in the nuclear sap, whilst his intergranula substance corresponds to the nuclear network, containing the chromatin.

Under the term granula, Altmann has, according to our opinion, classed together substances of very different morphological importance, some of which should be considered as products of the protoplasm. However, he has rendered important service by facilitating the investigation of protoplasm by means of new methods, although his bioblastic theory, which is based upon these experiments, is not likely to attract many supporters. (See the conclusion of the ninth chapter.)

e. **Uniformity of Protoplasm. Diversity of the Cell.**
A great uniformity of appearance is manifested by protoplasm in all organisms. With our present means of investigation we are unable to discover any fundamental difference between the protoplasm present in animal cells and that in plant cells, or unicellular organisms. This uniformity is of necessity only apparent, being due to the inadequacy of our methods of investigation. For since the vital processes occur in each organism in a manner peculiar to itself, and since the protoplasm, if the nucleus be excepted, is the chief site of the individual vital processes, these differences must be due to differences in the fundamental substance, that is to say, in the protoplasm. We must therefore accept, as a theory, that the protoplasm of different organisms varies in its material, composition and structure. Apparently, however, these important differences are due to variations in molecular arrangement.

In spite of the uniform appearance of the protoplasm, the individual cell, of which after all the protoplasm forms only a more or less important part, when taken as a whole, may vary very much in appearance; this is due partly to variations in external form, but chiefly to the fact, that sometimes one, and sometimes another substance is stored up in the protoplasm, in such a manner as to be distinguishable from it. Sometimes this occurs to so great a degree that the whole cell appears to be composed almost entirely of substances which under other circumstances are not present in protoplasm at all. If we imagine that these substances have been eliminated, a number of larger and smaller gaps would be naturally produced in the cell, between which the protoplasmic groundwork of the cell would be seen as partition walls and frameworks, which are sometimes extremely delicate. This arrangement of the protoplasm, as has been already mentioned (p. 19), must not be confused with the network structure, which, according to the opinion of many investigators, is inherent to protoplasm itself, and which was more fully described in the chapter on the structure of protoplasm.

The names deutoplasm (van Beneden) and paraplasm (Kupffer, II. 24) have been proposed for these adventitious substances. Since, however, the idea of an albuminous substance is always connected with the word plasm—and these substances may consist of fat, carbohydrates, sap, and of many other bodies—the use of the above terms does not seem desirable, and it is better either to class them generally as *intraplasmic products* and *adventitious cell contents* or, according to their significance, as *reserve material* and *secretions*, or indeed to specify them, as yolk granules, fat globules, starch granules, pigment granules, etc.

The difference between the protoplasm and these substances, which may be classed together as cell contents, is the same as that between the materials of which the organs of our body are composed and those substances which in the first place are taken up as food by our bodies, and which later on are circulated in a liquid form as a nutrient fluid through all the organs; the former, which are less dependent upon the condition of nourishment of the body for the time being, and hence are less subject to variations, are called in physiological language *tissue substances*, the latter *circulating substances*. The same distinction may be applied to the substances which compose the cell. *Protoplasm is the tissue material, whilst the adventitious bodies are circulating substances.*

f. **Various examples of the structure of the cell body.** In connection with the chemico-physical and morphological properties of the cell, a few especially pertinent examples may be of use in order to explain the general statements. For this purpose we will compare various lower unicellular organisms, both plant and animal, choosing first, cases in which the body consists almost entirely of protoplasm, and secondly, those in which the cells also contain considerable quantities of various adventitious substances, and hence are very much altered in appearance.

Unicellular organisms, which live in water or on damp earth, such as Amœbæ, Mycetozoa, and Reticularia, form very useful subjects for examination in studying the cell; in addition, lymph corpuscles, the white blood corpuscles of vertebrates, and young plant cells are most suitable objects for investigation.

1. **Cells consisting almost entirely of Protoplasm.** An *Amœba* (Fig. 7) is a small mass of protoplasm, from the surface of which, as a rule, a few short irregular processes (*pseudopodia*) or foot-like organs are extended. The body is quite naked, that is to say, it is not separated from the surrounding medium by any special thin coating or membrane; the only differentiation being that the superficial layer of the protoplasm (*ectoplasm*), *ek*, is free from granules, and hence is transparent, like glass; this ectoplasm is most marked in the pseudopodia; below the ectoplasm lies the darker and more liquid endoplasm (*en*), in which the vesicular nucleus (*n*) is embedded.

Very similar in appearance to the Amœba, but much smaller in size, are the *white blood corpuscles and the lymph corpuscles of the vertebrates* (Fig. 8). If they are examined just after they have been taken from the body of the living animal, they are seen to be more or less globular masses of protoplasm, each one consisting of a scarcely visible hyaline layer, enclosing a granular internal portion in which the nucleus is situated. However, whilst the specimen is fresh, this nucleus can hardly be distinguished, and sometimes even is quite invisible. After a time, the little body begins to push out from its surface, processes similar to the pseudopodia of the Amœba.

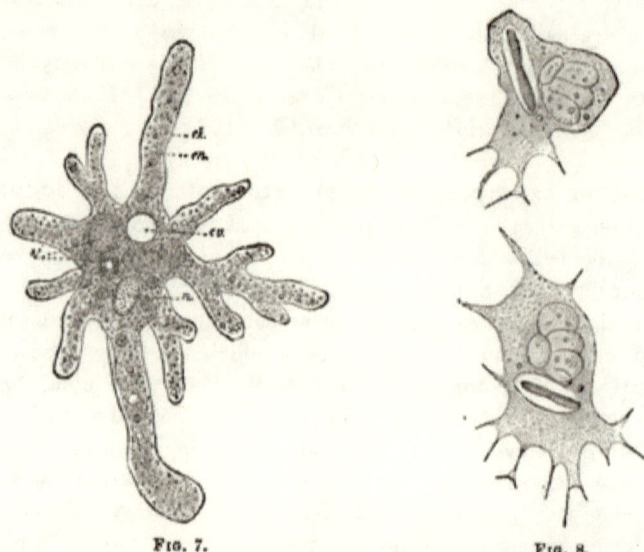

FIG. 7. FIG. 8.

FIG. 7.—*Amœba proteus* (after Leidy: from R. Hertwig, Fig. 16): n nucleus; cv contractile vacuole; n food vacuoles; en endoplasm; ek ectoplasm.

FIG. 8.—A leucocyte of the *Frog*, containing a *Bacterium* which is undergoing the process of digestion; the *Bacterium* has been stained with vesuvine. The two figures represent two successive changes of shape in the same cell. (After Metschnikoff, Fig. 54.)

Myxomycetes and Reticularia, which also consist of naked protoplasm, are very different in appearance. The Myxomycete, which is best known to us, is the *Æthalium septicum*, which forms the so-called *flowers of tan* and grows over large portions of the surface of tan-pits, during its vegetative condition, like a thin coherent skin of protoplasm (plasmodium).

Chondrioderma is another slime fungus which is nearly allied to the above. A small piece of its edge is represented in Fig. 9.

Towards its edge the plasmodium becomes broken up into a number of threads of protoplasm, which are sometimes exceedingly thin, and sometimes somewhat thicker, and which unite together to form a fine network. In the thicker threads it is possible to distinguish both a thin layer of homogeneous ectoplasm, and also the endoplasm which it encloses; these cannot, however, be made out in the thinner ones. Throughout the whole mass of protoplasm, which is sometimes very extensive, a large number of minute nuclei are seen to be distributed.

Amongst the Reticularia, of which many different kinds occur in fresh and salt water, *Gromia oviformis* (Fig. 10) is especially well known, in consequence of the experiments which have been made upon it by Max Schultze (I. 29). Part of the granular protoplasm, which contains a few small nuclei, lies within the oval shell, in which there is a wide opening at one pole, whilst the remainder protrudes through this opening, covering the surface of the shell with a thin layer. If the organism has not been disturbed, very delicate threads of protoplasm (pseudopodia) stretch out from this layer into the water in every direction; sometimes these pseudopodia are exceedingly long, many become forked, others break up into numerous minute threads, whilst yet others send off side branches, which unite with neighbouring pseudopodia.

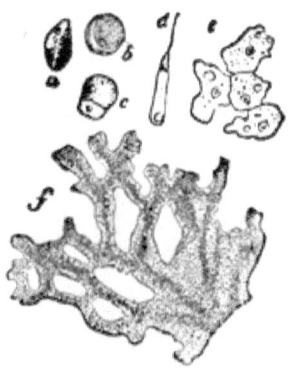

FIG. 9.—*Chondrioderma difforme* (from Strasburger): *f* part of a fairly old plasmodium; *a* dry spore; *b* the same, swollen up in water; *c* spore, the contents of which are exuding; *d* zoospore; *e* amœboid forms, produced by the transformation of zoo-spores which are commencing to unite together to form a plasmodium. (In *d* and *e* the nuclei and contractile vacuoles may be seen.)

Dujardin gave the name of *sarcode* to the peculiar substance of which the bodies of the lower organisms, described above, are composed, because, like the muscle-substance of the higher animals, it is capable of exhibiting movements. Influenced by Schleiden and Schwann's cell theory, investigators attempted to prove that sarcode was composed of a number of minute cells, so that the sarcode organisms might be included in the cell hypothesis. However, the solution to the difficulty was found to be in quite another direction. Investigators like Cohn (I. 7) and Unger were the first to compare sarcode with the protoplasmic contents of a plant-cell, in consequence of the similarity of the vital phenomena. Finally, Max Schultze (I. 29), de Bary (I. 2), and Haeckel (I. 10) established beyond a doubt the identity of sarcode with the protoplasm of plant and animal cells; and this discovery was most helpful to Max Schultze in working out his cell theory, and in establishing his theory of protoplasm (p. 6).

In Amœba, lymph cells, Mycetozoa, and Reticularia, we have learnt to recognise naked cells; those of plants on the contrary are almost invariably enclosed by a well-defined layer, which is sometimes very thick and firm; this is also very frequently the case with animal cells (membrane, intercellular substance), and thus in such cases a little chamber, or cell, in the true sense of the word is formed. Young cells from the neighbourhood of the growing point of a plant, and cartilage cells from a Salamander larva, are very good examples of this.

The cells at the growing point of a plant (Fig. 12 *A*), where they multiply very rapidly, are very small, and are very similar to animal cells. They are only separated from one another by very thin

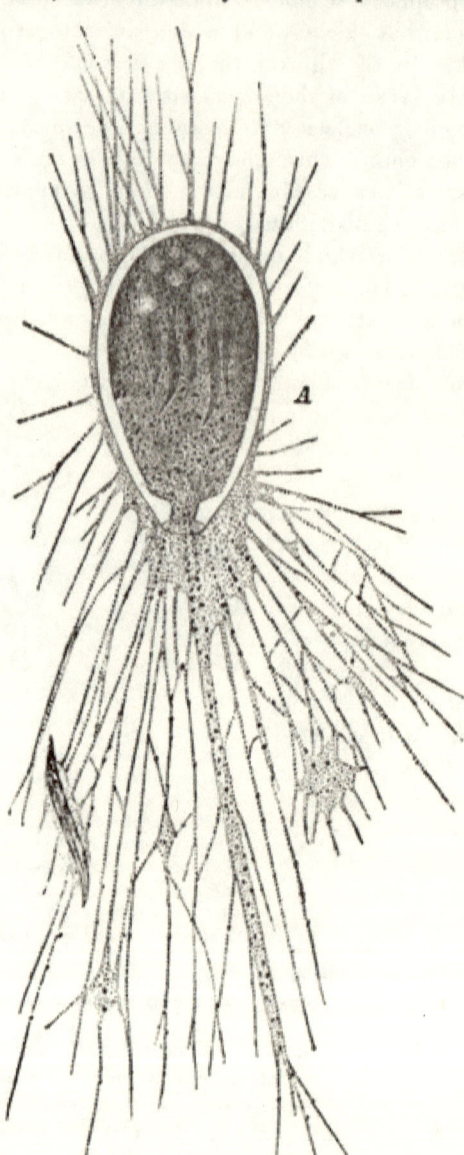

FIG. 10.—*Gromia oviformis*. (After M. Schultze.)

cellulose walls. The small cell spaces are completely filled up with the cell-substance, which, with the exception of the nucleus and chlorophyll, consists solely of finely granular protoplasm.

Flemming recommends cartilage cells from young Salamander larvæ as affording the best and most reliable material for the study of the structure of living protoplasm (Fig. 11). The cell-substance, which during life, as in the young plant-cells, completely fills the spaces in the cartilaginous ground-substance, is traversed by wavy threads of fairly high refractive power; these are less than 1 μ in diameter, and are generally most numerous, and at the same time most wavy, in the neighbourhood of the nucleus; sometimes the periphery of the cell is nearly, if not entirely, free from threads, but sometimes they are present in great numbers here also.

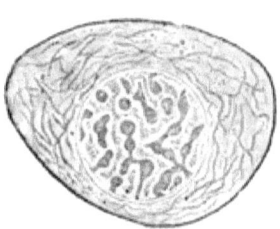

FIG. 11.—Living cartilage cell of a *Salamander larva*, much magnified, with distinctly marked threads. (After Flemming: from Hatschek, Fig. 2.)

2. **Cells which contain several different substances in their protoplasm.** In plants, and in unicellular organisms, the protoplasm frequently contains drops of fluid, in which salt, sugar, and albuminates are dissolved (circulating substances). The further we go (Fig. 12 *A*) from the growing-point of a plant, where the minute elementary particles of pure protoplasm as described above are grouped, the larger do the individual cells (*c*) appear, until they are frequently seen to be more than a hundred times as large as they were originally, whilst, in addition, their cellulose wall has become considerably thicker. However, this growth depends only to a very small extent upon any marked increase of the protoplasmic substance. The cavity of such a large plant cell is never seen to be completely filled with granular protoplasmic substance. The increase in the size of the cell is due much more to the way in which the small amount of protoplasmic substance, which was originally present at the growing point, takes up fluid, which in the form of cell-sap separates out into small spaces in the interior, called *vacuoles*. By this means a frothy appearance is produced (Fig. 12 *B*, *s*).

More or less thick protoplasmic strands stretch out from the mass of protoplasm in which the nucleus is embedded. These strands serve to separate the individual sap vacuoles from one another, and in addition they unite together on the surface to form a continuous layer (primordial utricle), which adheres closely to the inner surface of the enlarged and thickened cellulose membrane.

Two different conditions which are found in the fully grown plant cell are the result of this arrangement. Through the further increase of the cell-sap, the vacuoles are enlarged, and the partition wall attenuated. Finally the latter partially breaks down, so that the separate spaces are connected by openings, and thus form one continuous vacuole. Consequently part of the protoplasmic substance becomes transformed into a fairly thin layer lying close to the cellulose membrane, and the rest into more or less numerous strands and threads traversing the large continuous vacuole which is filled with fluid (Fig. 12, right side, and

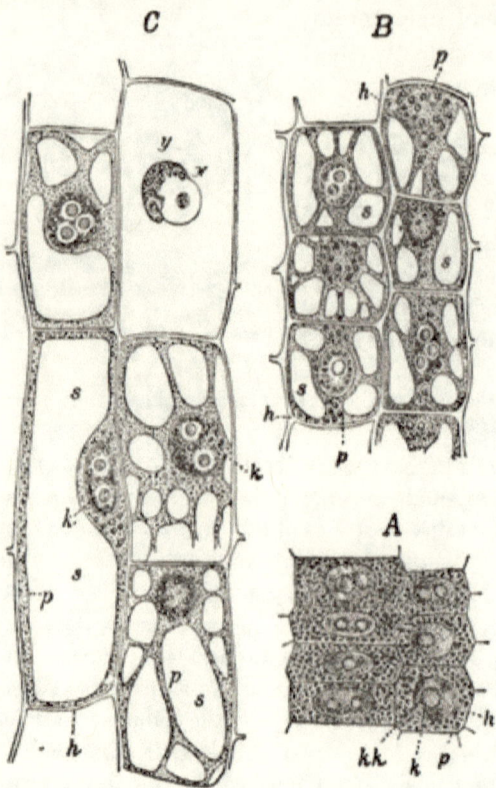

Fig. 12.—Parenchyma cells from the cortical layer of the root of *Fritillaria imperialis* (longitudinal sections, × 550; after Sachs II. 33, Fig. 75): *A* very young cells, as yet without cell-sap, from close to the apex of the root; *B* cells of the same description, about 2 mm. above the apex of the root; the cell-sap (*s*) forms in the protoplasm (*p*) separate drops between which are partition walls of protoplasm; *C* cells of the same description, about 7-8 mm. above the apex; the two lower cells on the right hand side are seen in a front view; the large cell on the left hand side is seen in optical section; the upper right hand cell is opened by the section; the nucleus (*xy*) has a peculiar appearance, in consequence of its being distended, owing to the absorption of water; *k* nucleus; *kk* nucleolus; *h* membrane.

Fig. 13). Finally, in other cases, even these strands of protoplasm in the interior of the cell may disappear. Then the protoplasmic substance is represented solely by a thin skin, which lines the interior of the little chamber, to use an expression of Sachs (II. 33), as the paper covers the walls of a room, and which contains one single large sap vacuole (Fig. 12 *C*, left lower cell, and Fig. 59). In very large cells this coating is sometimes so thin that, except for the nucleus, the presence of protoplasm can hardly be demonstrated at all in the cell, even when a high power of the microscope is used, so that special methods of investigation are necessary in order to render it visible.

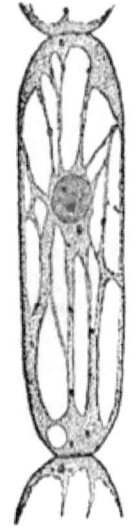

FIG. 13.—A cell from a hair on a staminal filament of *Tradescantia virginica* (× 240: after Strasburger, *Practical Botany*, Fig. 15).

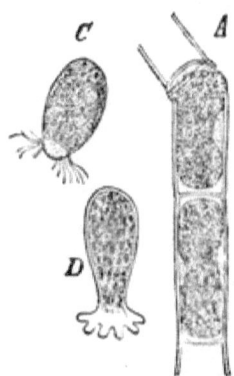

FIG. 14.—*Œdogonium*, during process of forming zoospores (after Sachs; from R. Hertwig's *Zoologie*, Fig. 110): *A* a portion of the thread of the alga, with the cell contents just escaping; *C* zoospore, which has reached the exterior; *D* stationary spore undergoing germination.

It was by the study of such cells, that the earlier investigators, such as Treviranus, Schleiden, and Schwann, arrived at their conception of the cell. Hence it is not surprising that they considered that the cell membrane and the nucleus constituted the essential portions of the cell, and quite overlooked the importance of the protoplasm. That this latter is the true living body in the plant-cell too, and that it is able to exist independently of the

D

membrane, has been proved beyond a doubt by the following observation, which has played such an important part in the history of the cell theory (I. 7). In many algæ (*Œdogonium*, Fig. 14), at the time of reproduction, the protoplasmic substance becomes detached from the cellulose cell-wall, and, whilst parting with some of its fluid contents, contracts up into a smaller volume, so that it no longer quite fills up the cavity; it thus forms a naked swarmspore, which is either globular or oval in shape (*A*). After a time this swarmspore breaks down the original cell-wall, and, escaping through the opening it has made, reaches the exterior. It then develops cilia (*C*) upon its surface, by means of which it moves about pretty quickly in the water, until after a time it comes to rest (*D*), when it differentiates a delicate new membrane upon its surface. Thus Nature herself has afforded us the best evidence that the protoplasmic body is the true living elementary organism.

A similarly great formation of vacuoles and separation of sap, as is found in plant-cells, is also seen in the naked protoplasm of the lower unicellular organisms, especially in certain Reticularia and Radiolarians; thus the *Actinosphærium*, which is depicted in Fig. 15, presents quite a frothy appearance, resembling the fine froth which is produced when albumen or soap-suds are beaten up. An immense number of larger and smaller vacuoles, filled with fluid, are distributed throughout the whole body. These are only separated from one another by delicate partition walls of protoplasm, which are sometimes too thin to be measured. The protoplasm consists of a homogeneous ground substance, in which granules are embedded.

The result of this formation of vacuoles is that the protoplasmic substance becomes broken up, so that surfaces of it become exposed to the nutrient solutions in the vacuoles, in consequence of which diffusion can take place between them. Evidently the whole arrangement adds considerably to the facility with which materials are taken up and given out. This internal increase of surface may be compared with the external increase of surface, which is shown in the formation of many-branched pseudopodia (Fig. 10), and indeed it answers the same purpose.

In animal-cells, on the contrary, the formation of vacuoles and the secretion of sap only take place extremely rarely, for instance, in notochordal cells; on the other hand, adventitious substances, such as glycogen, mucin, fat globules, albuminous substances, etc., are more frequently found; these either distend the cell or render it somewhat solid. When there has been a considerable development of such substances, the protoplasm may again assume a frothy appearance, as in *Actinosphærium* (Fig. 15), or it may become transformed into a network structure, as in a *Tradescantia* cell (Fig. 13), the only difference being that the interstices are filled with substances denser than sap.

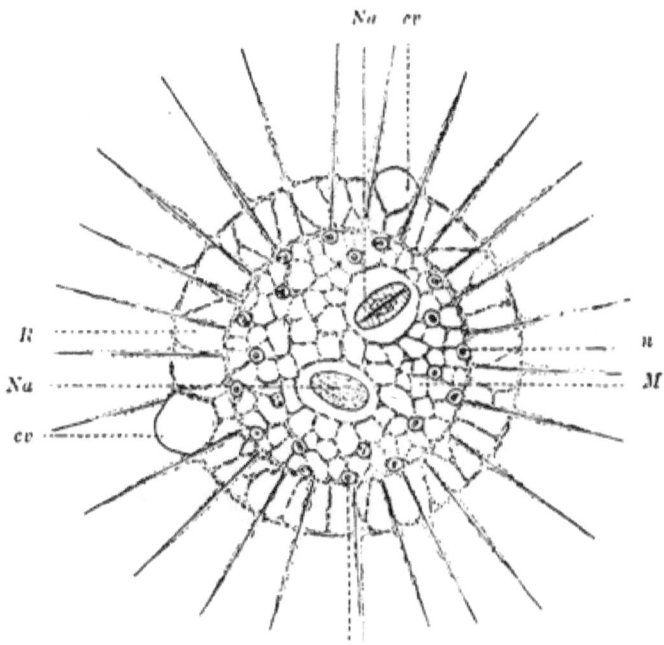

FIG. 15.—*Actinosphærium Eichhorni* (after R. Hertwig, *Zoologie*, Fig. 117): *M* medullary substance, with nuclei (*n*); *R* peripheral substance, with contractile vacuoles (*cv*); *Na* nutrient substances.

The most perfect examples are often seen in animal egg-cells. The exceedingly large size, which is attained by many of these, is not so much caused by an increase of protoplasm, as by the storing up of reserve materials, which vary very much as to their chemical composition, being sometimes formed and sometimes unformed substances, and which are intended for future use in the economy of the cell. Very often the egg-cell appears to be almost entirely composed of such substances. The protoplasm only fills up the small spaces between them, like the mortar between the stones of a piece of masonry (Fig. 16); if a section be made of an egg, the protoplasm is seen to be present in the form of a delicate network, in the larger and smaller meshes of which these reserve substances are deposited. The only place where it is collected together into a thick, cohesive layer is on the surface of the egg, and in the neighbourhood of the nucleus.

Another good example of a protoplasmic framework structure, caused by the deposition of various substances, is afforded us by the mucous cells of vertebrates (Fig. 17) and invertebrates. The section varies according as to whether it is taken from the epithelial surface, or from the base of the goblet. In the former case it is wider, and is seen to consist chiefly of homogeneous shining secretion, the mucilaginous substance, which is evacuated

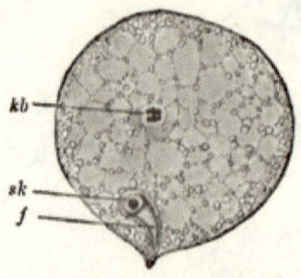

Fig. 16.—An egg of *Ascaris megalocephala*, which has just been fertilised (after Van Beneden; from O. Hertwig, Fig. 22): *sk* spermatozoon, with its nucleus which has just entered; *f* glistening fatty material of spermatozoon; *kb* female pronucleus.

Fig. 17.— Goblet-cell from the bladder epithelium of *Squatina vulgaris*, hardened in Müller's fluid. (After List, Plate I., Fig. 8.)

from time to time by the cell, through a small opening at its free end, and transformed into mucin. The protoplasm traverses the mass of secretion in the form of fine threads, which join together to make a wide meshed network, only forming a compact body at the lower extremity of the cell, in which also the nucleus is situated.

II. **The Chemico-physical and Morphological Properties of the Nucleus.** The nucleus is quite as important as the protoplasm in the economy of the cell. It was first discovered, in 1833, by Robert Brown (I. 5), in plant-cells; soon afterwards Schleiden (I. 28) and Schwann (I. 31) made it the foundation stone of their theory of cell formation; after that the study of the nucleus remained for some time in the background, as the interesting vital phenomena of the protoplasm became more fully known. During the last thirty years, however, one discovery after another has been made about the nucleus, the result of which is that this neglected body has been shown to be of as much importance to the elementary organism as the protoplasmic substance.

It is of interest that the history of the nucleus is analogous in some respects to that of the cell. The nucleus was also considered at first to consist of a vesicle; indeed, it was even held to be a smaller cell inside the larger one. But just as it came to be recognised that the protoplasm is the vital substance of the cell, so by degrees it came to be seen that the form of the nucleus is of minor importance, and that its vitality depends far more upon the presence in it of certain substances, the arrangement of which may vary very considerably according as to whether the nucleus is in an active or a passive condition.

Richard Hertwig (II. 18) was the first to enunciate this clearly in a short paper entitled, "Beiträge zu einer einheitlichen Auffassung der verschiedenen Kernformen," in the following words: "It is necessary to state at the commencement of my observations, as the most important point to be considered in classifying the various nuclear forms, that they all possess a certain uniformity in composition. Whether the nuclei of animals, plants, or Protista be under examination, it is invariably seen that they are composed of a larger or smaller quantity of a material which, like the earlier writers, I shall call nuclear substance (nuclein). We must commence with the properties of this substance in the same way as he who wishes to describe the important characteristics of the cell must begin with the cell substance, i.e. protoplasm."

Hence the nucleus is now defined, not, according to Schleiden and Schwann's idea, as a vesicle in the cell, but as *a portion of a special substance which is distinct from the protoplasm, and to a certain extent separate from it, and which may vary considerably, as to form, both in the resting and in the actively dividing condition.*

We will now consider the form, the size, and the number of nuclei in a cell, and then the substances contained in the nucleus, and their various modes of arrangement (the structure of the nucleus).

a. **The form, size and number of Nuclei.** As a rule the nucleus in plant and animal-cells appears as a round or oval body (Figs. 1, 2, 6, 16), situated in the middle of the cell. Since it is frequently richer in water than protoplasm is, it may be distinguished from the latter even in the living cell, appearing as a bright spot with indistinct outlines, or as a vesicle or vacuole. But this is not always the case. In many objects, such as lymph corpuscles, corneal cells, and the epithelial cells of gills of Salamander larvæ, no nuclei can be distinguished during life, although they immediately become visible when coagulation, induced either by the death of the cell, or by the addition of distilled water or weak acids, occurs.

In many kinds of cells, and in the lower organisms, the nucleus may assume very various shapes. Sometimes it is in the shape of a horse-shoe (many Infusoria), sometimes of a more or less twisted

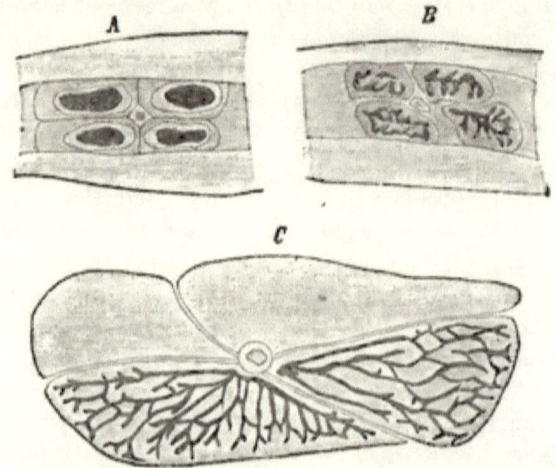

FIG. 18.—(After Paul Mayer, from Korschelt, Fig. 12.) *A* A piece of the seventh appendage of a young *Phronima*, 5 mm. in length (× 90). *B* A piece of the sixth appendage of a half-grown *Phronimella* (× 90). *C* A group of cells from a gland in the sixth appendage of a *Phronimella*; the nucleus is only shown in two cells (×90).

strand (Vorticella), and sometimes it is very much branched, stretching into the protoplasm in every direction (Fig. 18 *B, C*). This latter form chiefly appears in the large gland-cells of many insects (in the Malpighian tubes, in the spinning and salivary glands, etc.), and similarly in the gland-cells of the crustacean *Phronima*.

The size to which the nucleus attains is generally proportional to the size of the mass of protoplasm surrounding it; the larger this is, the larger is the nucleus. Thus, in the great ganglionic cells of the spinal cord, extremely large vesicular nuclei are seen. Similarly, enormously large nuclei occur in immature egg-cells, which themselves are of a great size. Sometimes the nuclei of immature eggs of Fishes, Amphibians, and Reptiles are perceptible to the naked eye as small spots; under these circumstances they can be easily extracted with needles and isolated. Yet there are exceptions to this rule; for even these same eggs which, when immature, have such immense nuclei, when they are mature and fertilised contain such minute nuclei, that they can only be demonstrated with the greatest difficulty.

The lowest organisms, when of a considerable size, frequently possess one single large nucleus. It is sometimes enormously large in the central capsules of many Radiolarians.

As regards the number present, as a general rule there is only one nucleus in each cell in plants and animals. To this rule, however, there are some exceptions; there are frequently two nuclei in liver cells, whilst a hundred or more have been observed in the giant cells of bone marrow. Osteoclasts and the cells of many tumours, the cells of several Fungi, and of many of the lower plants, such as Cladophora (Fig. 19) and Siphoneæ (*Botrydium, Vaucheria, Caulerpa*, etc.), are remarkable for this *plurality of nuclei*, as has been described by Schmitz.

Similarly, a large number of the lowest organisms, such as Myxomycetes, many Mono- and Poly-thalamia, Radiolarians, and Infusoria (*Opalina ranarum*), possess many nuclei in each cell. Frequently in these cases the nuclei are so minute, and are distributed in such numbers throughout the protoplasm,

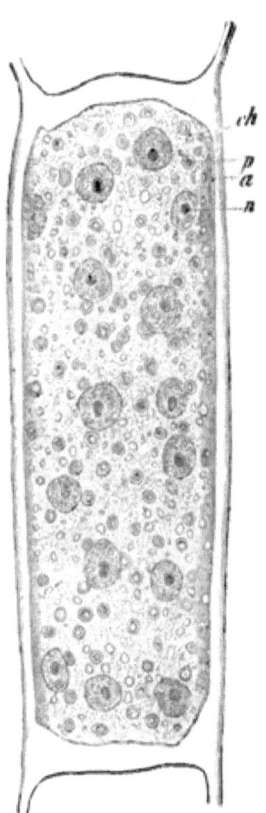

FIG. 19.—*Cladophora glomerata*. A cell from a thread in a chromic acid carmine preparation (after Strasburger, *Pract. Botany*, Fig. 75): *n* nucleus; *ch* chromatophores; *p* amyloid bodies (pyrenoids); *a* starch granules (× 540).

that they have only been demonstrated quite recently by means of the most improved methods of staining (Myxomycetes).

b. **Nuclear Substance.** As regards its composition, the nucleus is a fairly fixed body. Two chemically distinct proteid substances, which can be distinguished from one another with the microscope, are always present; very often there are more. The two constant ones are nuclein or chromatin, and paranuclein, or pyrenin; in addition, linin, nuclear sap, and amphipyrenin are generally to be found.

Of these, NUCLEIN, or chromatin, is the most characteristic proteid of the nucleus, and it generally preponderates as regards quantity. When fresh it resembles non-granular protoplasm (hyaloplasm), but it can be easily distinguished from this substance by its behaviour towards certain staining solutions. After it has been caused to coagulate by means of reagents, it takes up the colouring matter from suitably prepared staining solutions (solutions of carmine, hæmatoxylin, aniline dyes), as has been discovered by Gerlach. This occurs to a more considerable extent during the stages preceding division, and during division itself, than when the nucleus is in a resting condition. Whether this is due to chemical or to physical causes has not yet been worked out. The art of staining is now so fully understood that it is quite easy to make the nuclein of the nucleus stand out clearly from the rest of the nucleus and the protoplasm, which are either quite colourless or are only slightly stained. In this manner even small particles of nuclein, only about as large as Bacteria, may be rendered visible in comparatively speaking large masses of protoplasm, as, for example, the minute heads of spermatozoa, or the chromosomes of the direction spindles in the centres of large egg-cells.

The following fact, which is emphasised by Fol (II. 13), may at some future period prove to be of far-reaching importance: "that the staining of the nucleus with neutral staining solutions always produces the same shade of colour as the dye in question assumes when a small quantity of a substance of basic reaction is added to it. For example, red alum carmine becomes lilac when the solution is rendered slightly alkaline, Böhmer's violet hæmatoxylin becomes blue, red ribesia (black currant juice) bluish-green, whilst the red dye made from red cabbage turns green. Now, it has been observed that nuclei of tissue-cells, stained with neutral solutions of these substances, exhibit a corresponding colouration; that is to say, they become lilac in alum carmine, blue in hæmatoxylin, light blue in ribesia, green in the colouring matter of red cabbage. *That part of the nucleus which can be stained (the nuclein) behaves,* as a rule, towards the staining substance united to it, like a weakly alkaline body " (Fol).

Further, nuclein exhibits characteristic chemical reactions, which must not be forgotten in preparing nuclear structures for preservation (Schwarz II. 37, Zacharias II. 43, 45). It swells up in distilled water, in very dilute alkaline solutions, and in 2 or more per cent. solution of common salt, of sulphate of magnesia, or of monopotassium phosphate and of lime-water. If solutions of from 10 per cent. to 20 per cent. of the above-named salts are used, the nuclein, whilst swelling gradually, becomes quite dissolved. Similarly, it dissolves completely in a mixture of ferrocyanide of potassium and acetic acid, or in concentrated hydrochloric acid, or if it is subjected to pancreatic digestion. It becomes precipitated in a fairly unaltered form if treated with acetic acid from 1 to 50 per cent. in strength, when it can be very clearly distinguished from the protoplasm by its greater refractive power, and by a glistening appearance which is peculiar to it.

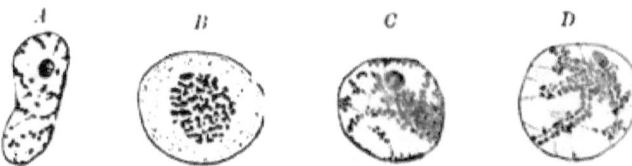

FIG. 20.—*A* resting nucleus of a spermato-genetic cell of *Ascaris megalocephala biralens*. *B* Nucleus of a sperm-mother-cell from the commencement of the growth-zone of *Ascaris megalocephala biralens*. *C* Resting nucleus of a sperm-mother-cell of the growth zone of *Ascaris megalocephala biralens*. *D* Bladder-like nucleus of a sperm-mother-cell of *Ascaris megalocephala biralens*, from the commencement of the dividing zone, shortly before division.

In the nuclear vesicle (Fig. 20), the nuclein sometimes appears as isolated granules (*A*), or as delicate network (*B*, *C*), or as threads (*D*).

Miescher (II. 49) has attempted to obtain pure nuclein from pus corpuscles and from spermatozoa, in the heads of which it is present. An important ingredient in its composition is phosphoric acid, of which at least 3 per cent. is always present. Several facts seem to indicate that the nuclein of the nucleus "consists of a combination of an albuminous body with a complex organic compound containing phosphoric acid (Kossel II. 35). This latter has been called nucleic acid, and Miescher has calculated its formula to be $C_{29}H_{49}N_9P_3O_{22}$.

"If subjected for a long time to the action of weak acids or alkalies, or even if kept in a damp condition, nuclein becomes decomposed, albumen and nitrogenous bases being formed, whilst in addition phosphoric acid separates out. The two latter decomposition products are also formed from nucleic acid. The bases are: adenin, hypoxanthin, guanin, and xanthin."

PARANUCLEIN, or pyrenin, is a proteid substance, which is always present in the nucleus; however, the part it plays in the vital functions of the latter has not yet been worked out, much less being known about it than about nuclein. It occurs in the nucleus in the form of small granules, which are described as true nucleoli or nuclear corpuscles (Fig. 20).

These paranuclein bodies resist the action of all the media (distilled water, very dilute alkaline solutions, solutions of salt, sulphate of magnesia, potassium phosphate, lime-water) in which nuclein substances swell up. Whilst the latter disappear from view in the nuclear cavity, which has become homogeneous in appearance, the former often stand out with greater clearness. They are invariably more easily seen after death than during life.

This explains the fact that these nuclear corpuscles were well known long ago to the older histologists, Schleiden and Schwann, who always examined their tissues in water.

Osmic acid is a very useful reagent for rendering these corpuscles visible, for it very much increases their refractive power, whilst rendering the nuclein structures paler.

Paranuclein and nuclein behave quite differently towards acetic acid (1 to 50 per cent.). Whilst the latter coagulates, and increases in refractive power, the nuclear corpuscles swell up more or less, and may become quite transparent; however, they do not become dissolved, for if the acetic acid is washed away, they shrink up, and become visible again.

In addition, it must be pointed out that paranuclein, in contradistinction to nuclein, is insoluble in 20 per cent. solution of common salt, in a saturated solution of sulphate of magnesia, in 1 per cent. and 5 per cent. solutions of potassium phosphate, of ferrocyanide of potassium plus acetic acid, and of copper sulphate; finally, it is very resistent to the action of the pancreatic juice.

Further distinct differences are shown in their behaviour towards staining solutions. As Zacharias has observed, and as I can corroborate as a general rule from my own experience, nuclein bodies become especially clearly and intensely coloured in acid staining solutions (aceto-carmine, methyl green, and acetic acid), whilst paranuclein bodies remain almost unaffected; on the other hand, the latter become better stained in ammoniacal staining solutions, such as ammonia, carmine, etc. Many substances, such as eosin, acid fuchsine, etc., have a greater affinity for paranuclein. Hence it is possible, by using two staining solutions at the same time, to stain the nuclein bodies a different colour from the paranuclein ones, thus bringing about a so-called contrast staining (fuchsine and solid green, hæmatoxylin and eosin, Biondi's stain); however, since the nature of staining processes is as yet very imperfectly understood by us, it is not possible at present to lay down general rules concerning the staining properties of these two nuclear substances.

I consider that *nuclein and paranuclein are the essential constituents of the nucleus*, and that its physiological action depends in the first instance upon their presence. They seem to me to be correlated in some way or other. Flemming (II. 10) has suggested, that the nucleoli may consist of nuclein in a special condition of development and density, thus representing a preliminary chemical phase of it. The material that we have at present for examination is not sufficient to enable us to decide these questions.

The three other substances which may be distinguished in the nucleus, *linin*, *nuclear sap*, and *amphipyrenin*, appear to me to be of much less importance; it is possible also that they are not always present.

The name LININ has been applied by Schwarz (II. 37) to the material of which the threads, which frequently form a network or framework in the nuclear cavity, consist; these threads are not affected by the ordinary staining reagents used for the nucleus, and can by this means, as well as by their different chemical reactions, be easily distinguished from the nuclein, which is deposited upon them in the form of small particles and granules (Fig. 20 *A*, *C*). In many respects it resembles the plastin of protoplasm, and indeed Zacharias has called it by that name.

NUCLEAR SAP may be present in larger or smaller quantities; it fills up the interstices left in the structures composed of nuclein, linin, and paranuclein. It may be compared to the cell-sap which is contained in the vacuoles of the protoplasm, and no doubt functions in a similar manner, by nourishing the nuclear substances, just as the cell-sap nourishes the protoplasm. By the action of several reagents, such as absolute alcohol, chromic acid, etc., finely granular precipitates are caused to make their appearance in the nuclear sap; these, being artificial products, must not be confused with the normal structures. Hence cell-sap must contain various substances in solution, amongst which albuminates are probably present; Zacharias has grouped these together under the common name of paralinin, a term which may well be dispensed with.

The name AMPHIPYRENIN has been applied by Zacharias to the substance of the membrane which separates the nuclear space from the protoplasm, just as this latter is separated from the exterior by the cell membrane. In many cases it is as difficult to demonstrate the presence of this nuclear membrane, as to decide the vexed question whether a large number of cells are enclosed by membranes or no. It is most easily seen in the large germinal vesicles of many eggs, such as those of Amphibians, where it is at the same time somewhat dense in consistency. It is on this account that it is so easy to extract the nucleus quite intact from immature eggs with a needle. The nuclear membrane can be ruptured, as a result of which its contents flow out, and may be spread out in the liquid in which the examination is taking place. But it seems to me to be equally certain that, in other cases, a true nuclear membrane is absent, so that the nuclear substance and protoplasm come into direct contact. Thus Flemming (II. 10), in the blood cells of Amphibians, and I myself, in the sperm-mother-cells of Nematodes at a certain stage of their development (Fig. 20 *B*), have failed to discover a nuclear membrane.

ALTMANN has endeavoured, by means of a special staining process with cyanin, to demonstrate a granula structure in the nucleus as well as in the protoplasm. By means of this process he has succeeded in intensely staining the sap which fills up the interstices in the nuclear network, and in thus showing up granula, whilst the nuclear network remains uncoloured, and is designated intergranula substance. In this manner Altmann has obtained a, so to speak, negative impression of the nuclear structure, as it becomes revealed by staining the nuclear network with the usual nuclear staining reagents. Since he considers that the granula form the most important part of the nucleus, his opinion of the relative importance of the nuclear substances differs from the one which is generally accepted, and according to which the nuclear sap is of less importance than the nuclein and paranuclein.

c. **The Structure of the Nucleus. Examples of its various Properties.** The above-mentioned substances, of which nuclein and paranuclein at any rate are never absent, occur in very different forms in the nuclei of various plant and animal cells; this is especially true of nuclein, which may be present as fine granules, as large masses, as fibrils, as a framework, or in the form of a honeycomb structure. Further, one such structure may develop into another during the various vital phases of the cell's life-history.

Hence in formulating a definition of the nucleus, its varying form must be quite disregarded; the difficulty consists in defining the active substances contained in it, similarly as, in defining the cell, the difficulty lies in describing protoplasm.

The nucleus consists of a mass of substances, which are peculiar to it, and which, to a certain extent, differ from protoplasm, and may be distinguished from it. On this account, in all definitions of the nucleus, more importance should be attached to the properties of its structural components than is usually the case.

The following selection of typical examples will serve to show what a multiplicity of forms may be assumed by the internal structure of the resting nucleus.

It is beyond dispute that *the simplest structure*—disregarding the molecular conditions discussed later—is seen in the *nuclei of mature sperm-cells*. When the sperm-cells, as is the rule, assume a thread-like form, being the one most suitable for boring their way into the egg-cells, the nuclei constitute the anterior ends or heads of the threads. In the *Salamandra maculata* the head is like a sword, terminating in a sharp point (Fig. 21 *k*); it consists of dense nuclein which, even when most highly magnified, is still homogeneous in appearance. A short cylindrical body, the so-called middle portion (*m*), which also appears homogeneous, is joined on to the head; this portion reacts like paranuclein. Hence, apparently, it must be considered to form part of the nuclear portion of the sperm-thread; this, however, can only be finally proved when its further development has been observed.

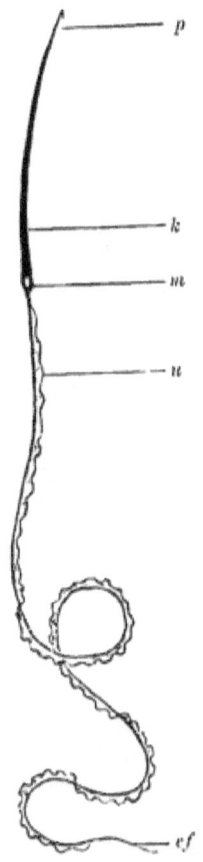

Fig. 21. — Spermatozoon of *Salamandra maculata*: *k* head; *m* middle portion; *cf* terminal portion; *sp* apex; *u* undulating membrane.

Further, in sperm elements, where the form of the cell has been retained, the nucleus appears as a compact globular mass of nuclein; this is the case in the sperm elements of *Ascaris megalocephala* (Fig. 22), which, when immature, are shaped like fairly large, round cells, and when mature assume the form of a thimble.

Fig. 22. — Sperm-cell of *Ascaris megalocephala* (after Van Beneden; from O. Hertwig's Embryology, Fig. 21): *k* nucleus; *b* base of cone, by which it attaches itself to the egg; *f* shining substance resembling fat.

Having examined this simple condition of the nucleus, as it occurs in sperm-cells, and where it is composed almost entirely of active nuclear substances, being nearly free from the admixture of other substances, we may now proceed to examine other nuclear forms. In these we see that *the chief cause for the variety in form, which has been observed in plant and animal cells, is the fact, that the active nuclear substances evince a great inclination to take up liquid, with the substances dissolved in it, and to store it up, generally to such an extent, that the whole nucleus acquires the appearance of a bladder enclosed in protoplasm.*

Thus in the nucleus, a process takes place similar to that which occurs in protoplasm, where the cell-sap becomes collected in vacuoles or large sap-cavities. This circumstance bears the same significance in either case. These vacuoles are concerned in the metabolism both of the cell and of the nucleus, for they contain in solution nutrient materials, which can be easily taken up by the active substances, in consequence of the great superficial development of the vacuoles.

This *process of sap absorption* may be directly observed when, after fertilisation has taken place, the nucleus of the spermatozoon, in performance of its function, enters the egg-cell. In many cases it begins to swell up gradually, until it becomes ten to twenty times as large as it was originally; this is not due to any increase of its active substances, which remain absolutely unaltered in quantity, but entirely to the absorption of fluid substances which were held in solution in the yolk. In such a nucleus, which has become transformed into a vesicular body, the nuclein is spread out in fine threads to form a net; in addition, one or two globules of paranuclein (nucleoli) are now to be seen. A similar process occurs each time a nucleus divides, when the daughter nuclei are being reconstructed.

According as to whether the nucleus has absorbed a greater or less quantity of nuclear sap, its solid constituents, which on account of their chemical properties have been distinguished above as linin and nuclein, arrange themselves in the form of a *more or less fine framework structure*. Figs. 23–26 show us examples of the various modifications which may occur.

Fig. 23 represents the *nucleus of a cilio-flagellate organism*. It consists, like the chief nucleus of the Infusoria, of a small-meshed framework of nuclein. Bütschli (II. 5) considered that it is in the form of a small delicate honeycomb; in his opinion the nucleus is composed of extended faviform chambers, with three or more sides, separated from one another by very delicate partition walls of nuclein, and enclosing the nuclear sap, which is only slightly affected by staining reagents. Similarly their upper surfaces are separated from the protoplasm by means of a delicate layer of nuclein, there being no distinct true nuclear membrane. The points

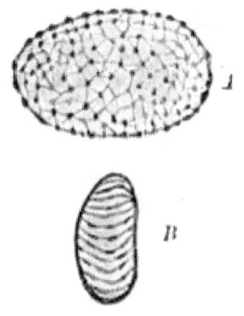 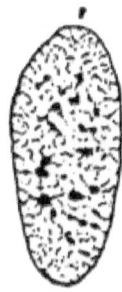

FIG. 23. — Nucleus of *Ceratium tripos*, in which the faviform structure is very plainly shown (after Bütschli, Pl. 26, Fig. 14): *A* ventral view; *B* lateral view. Both illustrations represent optical sections only.

FIG. 24. — Nucleus of a connective tissue cell from the peritoneum of a *Salamander larva*, with central corpuscles lying near it. (After Flemming, Fig. 4.)

where the partition walls meet are thickened like columns. The appearance varies according to the point of view from which the nucleus is seen, in consequence of the extended form of the faviform chambers, which lie parallel to one another; a glance at Fig. 23 *A*, *B*, explains this. One or two nucleoli are to be seen in the cavity.

Fig. 24 represents the *nuclear framework of a connective tissue cell* of a *Salamander larva*. It consists of a fairly close network composed of extremely delicate threads. A few denser swellings occur here and there, usually where several threads cross; these swellings retain the stain with especial tenacity. They consist of collections of nuclein, and may look very like true nucleoli, which consist of paranuclein, and on this account Flemming has called them *net-knots*, in order to distinguish them from nucleoli.

The framework of the nuclei of the various animal tissue cells may be fine or coarse. In the latter case it consists of only a few strands, so that "it hardly deserves the name of a net or framework." As a rule, the nuclei of young, embryonic and growing tissues possess, as Flemming has observed, networks coarser than those of similar tissues in the adult.

For the most part the nuclear framework is composed of two different substances, linin and nuclein; of these the latter alone is capable of absorbing and retaining the ordinary staining reagents. The two substances are generally so arranged that the nuclein, in the form of coarser and finer granules, is evenly distributed upon and throughout the colourless linin. When the meshes of the framework are very fine (Fig. 24) it may be very difficult, or indeed impossible, to distinguish the two substances from one another. In a coarser network, such as is represented in Fig. 25, it is much easier to do so; here a *resting nucleus from the protoplasmic lining of the wall of the embryo-sac of Fritillaria imperialis* is portrayed. According to Strasburger's description, the delicate framework threads as a rule do not become stained; hence they must consist of linin. Coloured nuclein granules of varying size are seen to be deposited upon them. In addition a number of variously sized nucleoli are to be seen.

FIG. 25.—*Fritillaria imperialis. A resting nucleus* (after Strasburger, Fig. 191 *A*).

If any one should wish to convince himself of the fact that a special framework of linin is present in the nucleus, he cannot do better than examine the nuclei of the sperm-mother-cells, of the round worm of the horse (Fig. 26). During the early stages of division, all the nuclein is gathered into eight bent hook-shaped rods, which collect together into two bundles; they are, as it were, suspended in the nuclear cavity, for colourless threads of linin connect them both to the nuclear membrane and to one another. It is impossible for these threads to be coagula in the nuclear sap, produced by the use of reagents, since they are invariably regularly arranged. Similarly their chemical reaction and their behaviour during the process of division show that they are composed of a substance which differs somewhat from nuclein and paranuclein.

Moreover, the nuclein is not always spread out upon a framework. For example, the large *vesicular nuclei of Chironomus larvæ* (Fig. 27) enclose, as Balbiani (II. 2) has discovered, a single thick nuclear thread; this is variously twisted, and in stained preparation is seen to be composed of regular alternately stained and unstained layers. This has also been observed by Strasburger in some plants. The two ends of the thread terminate in nucleoli.

Further, in other cases the greater part of the nuclein is collected into a large round body, which looks like a nucleolus, but which is really very different from the above-described true nucleoli, which contain paranuclein (p. 42). In order to avoid confusion it is best to call such bodies nuclein corpuscles. As an example of this class the *nucleus of Spirogyra* may be mentioned; the nuclei of many of the lower organisms are very similar to it in structure. It consists

Fig. 26.   Fig. 27.

Fig. 26.—Nucleus, about to divide, of a cell from *Ascaris megalocephala bivalens*, with the eight nuclear segments arranged in two bundles, and with two pole corpuscles. (Hertwig II. 19 b, Tab. II., Fig. 18.)

Fig. 27.—Structure of the nucleus of a cell from the salivary gland of *Chironomus*. (After Balbiani, *Zoolog. Anzeiger*, 1881, Fig. 2.)

of a vesicle which is separated from the protoplasm by a delicate membrane, and which contains a fine nuclear framework. Since this is incapable of retaining the dyes of staining solutions, it is evident that it consists chiefly of linin, upon which only a few nuclein granules are deposited. One large nuclein body is present in the framework; occasionally, however, it is divided into two smaller ones. That this body really consists of nuclein is proved partly by its behaviour towards staining solutions, but chiefly by the fact that during nuclear division its substance breaks up into granules, thus forming the nuclear segments.

Similar nuclein bodies, which in literature generally go under the name of nucleoli, play a very important part in the *structure of the germinal vesicles of animal egg-cells.* These germinal vesicles differ considerably in their structure from the nuclei met with in ordinary tissues, as may be seen from Figs. 28, 29, 30.

Fig. 28 represents the immature egg of a sea urchin; if it is examined when alive, an exceedingly coarse network of rather thick isolated threads can be distinguished. These, as is shown by their micro-chemical properties, consist chiefly of linin. The stained material is nearly all collected into a single large round body, the "*germinal spot*," which lies in a net-knot of the framework, where the greatest number of linin threads intersect.

In the enormously large germinal vesicles, for which the large eggs of Fishes, Amphibians, and Reptiles, which are so rich in yolk, are remarkable, the number of germinal spots increases consider-

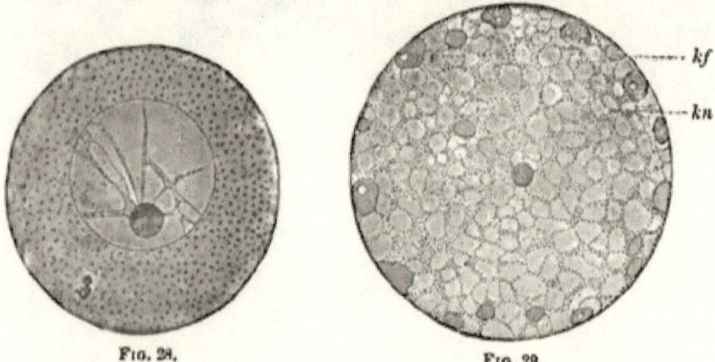

Fig. 28.   Fig. 29.

Fig. 28.—Immature egg from the ovary of an Echinoderm. In the large germinal vesicle there is a network of threads, the nuclear net, in which the germinal spot can be seen. (O. Hertwig, Embryology, Fig. 1.)

Fig. 29.—Germinal vesicle of a small immature egg from the Frog. In a dense nuclear net (*kn*) a very large number of germinal spots, mostly peripheral (*kf*), are to be seen. (O. Hertwig, Embryology, Fig. 2.)

ably during the growth of the cell, until finally they may number some hundreds; whether this multiplication takes place by division or in some other fashion is not yet known. The position of the germinal spots varies at different times; generally, however, they are situated on the surface of the vesicle, being distributed at even distances over the membrane, as is shown in Fig. 29, where the nucleus of a rather small immature egg of a frog is depicted.

The *shape of the germinal spots* also varies; they may be round — this is especially the case when they are isolated — or oval; sometimes they are somewhat extended, at others they are constricted in the middle; occasionally they are irregular in outline, and when they are very numerous, they show considerable differences in their size. Very frequently a few small vacuoles filled with fluid are to be seen. The examination of living egg-cells shows that these vacuoles are not artificially produced. Additional vacuoles may be formed after the death of the egg, whilst those already present may increase in size, as has been pointed out by Flemming (II. 10, p. 151).

These germinal spots differ in their chemical properties from true nucleoli, which consist of paranuclein and do not become stained with the usual nuclear staining reagents. On the other hand, it has not yet been discovered whether their substance is quite identical with the nuclein of the framework. Up to the present this point has not yet been satisfactorily worked out, in spite of the numerous experiments which have been made upon the nucleus. One thing alone can be accepted as certain—that the more or less rounded bodies present in various plant and animal nuclei, which in scientific literature are classed together, for the most part incorrectly, under the name of nucleoli, show material differences amongst themselves. This has been proved beyond a doubt by the investigations made by Flemming (II. 10), Carnoy (II. 8), myself (II. 19a), Zacharias (II. 45), and others. Either such very different bodies should not be called by the same name, or if, merely on account of their similarity in form, the common name of nucleolus or nuclear body is retained for all round nuclear contents, at any rate in each case an accurate description of the chemical nature of the nucleolus in question should be given. Above all, as has been already remarked, in all examinations of the nucleus, more attention should be paid to the chemical properties of its individual constituents than to their form and arrangement, which are always of comparatively little importance. For the function of a framework in the nucleus composed of linin threads differs considerably from that of one consisting of nuclein, or of a combination of the two substances, and similarly the function of the nucleolus varies according to the material of which it is composed.

I will conclude this discussion of nucleoli with the remark that *germinal spots* exist which are most evidently built up of *two different substances*. This circumstance was first observed by Leydig in a lamellibranchiate Mollusc, and his statement has since been verified by Flemming (II. 10) from observations on the same animal, and by myself (II. 19) from those on other objects. I here quote the description as it is given by Flemming.

In *Cyclas cornea* and in the Naiadeæ a principal nucleolus, in addition to a few smaller secondary nucleoli, is present in the germinal vesicle. "The former consists of two differently constituted portions; these may be seen in Fig. 30 as a smaller, strongly stained more refractive part, and a larger, paler, less chromatic one, which swells up more in acids. In *Anodon* these two portions are closely coherent; in *Unio* they very frequently only just touch each other, or, indeed, may lie apart. The smaller secondary nucleoli, which lie in the meshes of the framework, show the same power of refracting light, of swelling up, and of becoming stained, as the larger portion of the principal nucleolus. If water is added, this larger portion disappears, as well as the small nucleoli, amongst the strands of the framework; the small, strongly chromatic portion of the principal nucleolus alone remains; this becomes more sharply defined, shrinking up somewhat, and developing a clearly marked outline. The addition of strong acetic acid (5 per cent. or more) causes the larger paler portion of the principal nucleolus to swell up rapidly and to disappear, whilst the smaller shining portion, though also swelling up somewhat, remains visible." "When nuclear staining reagents are used, both portions of the nucleolus, and also the secondary nucleoli, become coloured to a considerable extent; the most strongly refractive part of the former, however, is especially intensely stained." "Such a differentiation of the principal nucleolus into two parts occurs in the egg-cells of many animals. In *Dreissena polymorpha* the strongly refractive chromatic portion covers the paler one like a hollow cap."

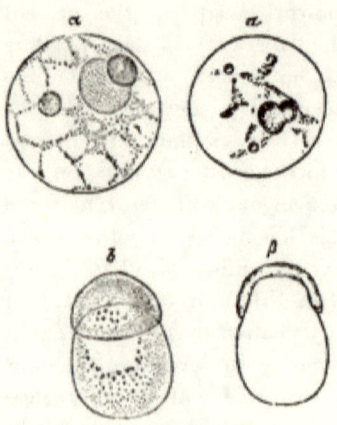

FIG. 30.—(After Flemming, Fig. F, p. 104.) *a* Nucleus of an egg from the ovary of Unio; it has just emerged from the cell into the ovarian fluid. Nucleolus with two protuberances. A small portion of the nuclear framework is visible; *a* a similar nucleus after 5 per cent. acetic acid has been added. The framework strands stand out more clearly; the larger paler portion of the principal nucleolus, as well as the minor nucleoli, have similarly become swollen up and faded; the smaller portion of the principal nucleolus is also swollen up, but to a less degree. *b* Nucleolus of an egg of *Tichogonia polymorpha*; the principal glistening portion rests like a cap upon the larger one. β Diagrammatic representation of an optical section of above.

I have observed (II. 19) that the germinal spot is composed of two substances in *Helix*, *Tellina*, and *Asteracanthion*, as well as in *Anodon*. *Asteracanthion* (Fig. 31) is of special interest, as the separation into two substances (*p n, n n*) only becomes distinctly visible when the germinal vesicle commences to break up and to form the polar spindle out of its contents.

Finally, in the description of the structure of the resting nucleus, attention must be drawn to one other important point. *According to the age or stage of development of a cell, the resting nucleus may present very considerable variations in all its separate parts: as to the appearance of its framework, and as to the number, size, and peculiarities of its nucleoli.* Thus, as Flemming (II. 10) remarks, "In young eggs from the ovaries of Lamellibranchs, this twofold composition of the large nucleolus is not to be seen; it only develops in the mature egg." Above all, the germinal vesicles of the eggs undergo during their development important metamorphoses, which at present have been but little investigated, whilst their significance is still less understood. The same is true of the nuclei of sperm mother cells. I have endeavoured to follow accurately these changes of form in such cells obtained from the testis of *Ascaris megalocephala* (II. 19 *b*), which are very suitable for the purpose.

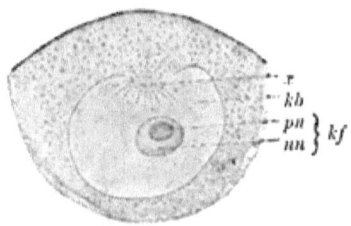

Fig. 31.—Section from an egg of *Asterias glacialis* showing the degeneration of the germinal vesicle. This begins to shrink up, whilst a mass of radiated protoplasm (*r*) forces its way into the interior, breaking down the membrane. The germinal spot (*kf*) is still visible, but is divided into two substances, nuclein (*nn*) and paranuclein (*pn*). (O. Hertwig, Embryology, Fig. 12.)

As is shown in Fig. 32, form *A* gradually becomes transformed into form *B*, and this during the process of development of the spermatozoon into form *C*; the youngest sperm mother cells (*B*) have naked nuclei containing dense nuclear frameworks, and superficially-placed nucleoli; this form develops in older cells (*C*) into a vesicular nucleus with a distinctly marked membrane. In the vesicle a few linin threads are extended through the nuclear sap, the nuclein heaped up into one or two irregular masses, amongst which the more or less globular nucleolus is situated. In cells which are not yet mature, the nuclein is collected chiefly at one spot of the nuclear membrane in the form of a thick layer, whilst granules of varying size lie upon the surface of the linin threads, a few of which are extended throughout the nuclear space. A considerable time before division occurs, the nuclein

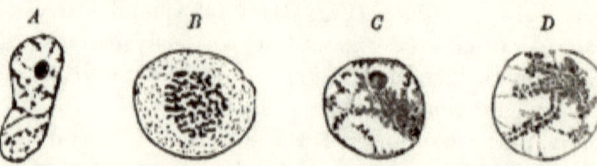

FIG. 32.—*A* Resting nucleus of a primitive sperm cell of *Ascaris megalocephala bivalens*; *B* nucleus of a sperm mother cell from the commencement of the growth zone; *C* resting nucleus of a sperm mother cell from the growth zone; *D* vesicular nucleus of a sperm mother cell from the commencement of the dividing zone just before division.

becomes arranged in definite threads (*D*). A nucleolus is always present in the meshes of the framework.

III. **Are there Elementary Organisms existing without Nuclei?** The important question, as to whether the nucleus is an indispensable portion of every cell, follows naturally on the description of the chemical and morphological properties of the nucleus. Are there elementary organisms without nuclei? Formerly investigators were not at a loss to answer this question. For since, in consequence of the inadequacy of former methods of examination, no nuclei had been discovered in many of the lower organisms, the existence of two different kinds of elementary cells was assumed: more simple ones, consisting only of a mass of protoplasm, and more complex ones, which had developed in their interior a special organ, the nucleus. The former were called cytodes by Haeckel (I. 10; II. 15), to the simplest, solitary forms of which he gave the name of Monera; the latter he called cellulæ, or cytes. But since then the aspect of the question has become considerably changed. Thanks to the improvements in optical instruments, and in staining methods, the existence of organisms without nuclei is now much questioned.

In many of the lower plants, such as Algæ and Fungi, and in Protozoa, Vampyrella, Polythalamia, and Myxomycetes, all quoted formerly as examples of non-nucleated cells, nuclei may now be demonstrated without much trouble. Further, since the nucleus has been discovered in the mature ovum (Hertwig II. 19 *a*), we may safely say that, in the whole animal kingdom, there is not a single instance where the existence of a cell without a nucleus has been proved. I shall probably be confronted with the red corpuscles of Mammals. It is true that they contain no nuclei, but then neither do they contain any true protoplasm, and hence the theory, more fully described later, that the blood discs of Mammals are not true cells, but only the products of the metamorphosis, or of the development of former cells, may be defended for many reasons.

The only remaining instance of cells in which, on account of their extreme minuteness, no differentiation into protoplasm and nuclear substance can be demonstrated, is furnished by Bacteria and other allied forms. However, even here Bütschli (II. 6) has endeavoured to prove the existence of a nuclear-like body. Thus in Oscillaria and in others (Fig. 33 *A*, *B*), he has pointed out bodies which are not digested by gastric juice, and which contain a few granules, which stain intensely (probably nuclein granules); these make up the greater part of the cell substance, the protoplasm being present only as a delicate envelope. Bütschli's views are for the most part shared by Zacharias (II. 47).

Even if it is objected that the above statement is at present unproven, it cannot be denied that the supposition that Bacteria consist entirely, or principally, of nuclear substance, seems at any rate as probable, if not more so, as the one that they are minute masses of pure protoplasm. The extraordinary affinity of these organisms for staining reagents is very much in favour of the first view.

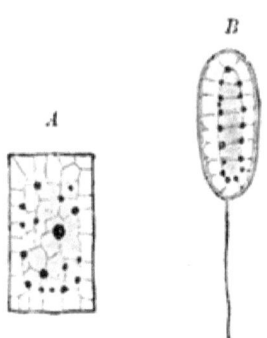

FIG. 33.—*A Oscillaria*: Optical section of a cell from a thread, killed with alcohol and stained with hæmatoxylin (after Bütschli, Fig. 12 *a*). *B Bacterium lineola* (Cohn), in optical section, killed with alcohol and stained with hæmatoxylin (after Bütschli, Fig. 3 *a*).

IV. **The Central or Pole Corpuscles of the Cell.** Long ago an exceedingly minute object, which, on account of its function, is of the greatest importance, was observed in addition to the nucleus in the protoplasm of some cells; this is the *central or pole corpuscle* (*centrosome*). This was first noticed during cell division (which is described later on in Chapter IV.), and here it plays a most important part, as it forms a central point for the peculiar radiated appearances, and above all functions as the centre of the cell, around which the various cell contents are, to a certain extent, arranged.

As to *size*, it is only just visible, and is frequently much smaller than the most minute micro-organism. As to its *composition*, it appears to consist of the same substance as the so-called neck or middle portion of the seminal thread, to which, further, during the process of fertilisation, genetic functions have been ascribed (*vide* Chap. VII., 1). When the ordinary methods for staining the nucleus are employed it does not absorb any of the dye; if, however, special reagents, especially acid aniline dyes, such as acid fuchsine, safranin, and orange, are used, it becomes vividly coloured. This is the only way to distinguish the central corpuscle from the other granules in the cell (microsomes) unless it is enclosed by a special radiation sphere or envelope. If we disregard the processes of cell division and of fertilisation, which are treated of in later sections, the central corpuscles have been, up till now, most frequently observed in lymph cells (Flemming II. 11, 12 *b*, and Heidenhain II. 16), in the pigment cells of the Pike (Solger II. 38), and in the flattened epithelial, endothelial, and connective tissue cells of *Salamander larvæ* (Flemming II. 12 *b*).

As a rule there is only one central corpuscle present in each lymph cell (Fig. 34); this can be seen without having been stained, since the protoplasm in its immediate neighbourhood assumes a distinctly ray-like appearance forming the radiation, or attraction sphere, which later on will occupy so much of our attention. The central corpuscle is sometimes situated in an indentation of the nucleus, or, if the latter has broken down into several pieces, a condition which is frequently seen in lymph cells, it lies between them and some portion or other of the protoplasmic body.

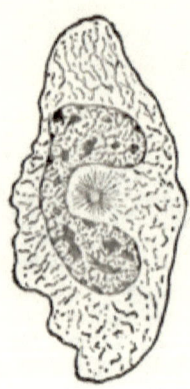

Fig. 34. — Leucocyte from the peritoneum of a *Salamander larva*. For the sake of clearness in the figure, the central corpuscle, surrounded by its radiation sphere, has been distinguished by a bright ring, which is not really present in nature. (After Flemming, Fig. 5.)

In pigment cells (Fig. 35), Solger (II. 38) was able to make out the radiation sphere as a bright spot between the pigment granules, and in consequence he concluded that the central corpuscle was present.

In the epithelium of the lung, and in the endothelium and connective tissue cells of the peritoneum of *Salamander larvæ* (Fig. 36 *A*, *B*), Flemming found, almost without exception, that instead of a single central corpuscle, two were present, lying close together, either in the immediate neighbourhood of the resting nucleus, or in an indentation of it, directly in contact with the nuclear membrane. As a rule no radiation sphere was to be seen in these cases; sometimes the two central corpuscles, instead of touching each other closely, were somewhat separated from one another, and under these circumstances the first commencement of a spindle formation between them was visible.

FIG. 35.—Pigment cell of the *Pike*, with two nuclei, and one pole corpuscle, surrounded by a radiation sphere. (After Solger, Fig. 2.)

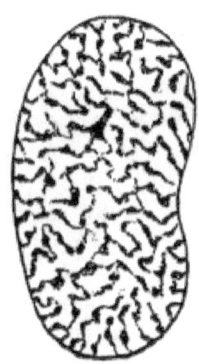
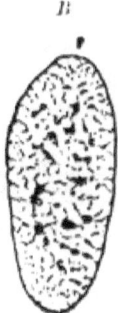

FIG. 36.—*A* Nucleus of an endothelial cell from the peritoneum of a *Salamander* larva, with the pole corpuscle lying near (after Flemming, Fig. 2). *B* Nucleus of a connective tissue cell from the peritoneum of a *Salamander* larva, with the pole corpuscle lying near (after Flemming, Fig. 1).

Van Beneden (II. 52) first advanced the theory that *the central corpuscle, like the nucleus, is a constant organ of each cell*, and that it must be present in the cell in some portion of the protoplasm near the nucleus. The property possessed by the central corpuscle of being able to multiply itself by *spontaneous division* (*vide* Chap. VI.) seems to be in support of the first part of this view, as is also the *rôle* it plays in the process of fertilisation (*vide* Chap. VII. I); but the second portion of this theory, although it is very generally accepted, that the central corpuscle belongs to the protoplasm, appears to me, on the contrary, less certainly true.

I have for some time held the opinion, which, for reasons that I will state later (*vide* Chap. VI.), I still hold to be worthy of consideration, that the central corpuscles are generally constituent parts of the resting nucleus, since after division has taken place they enter its interior, and whilst it is preparing for division come out again into the protoplasm. Only in rare cases do the central corpuscle or corpuscles remain in the protoplasm itself, whilst the nucleus is resting, and then to a certain extent they represent a subordinate nucleus in addition to the principal one. This theory would explain the fact that, even with the more recent methods and most improved optical instruments, *the central corpuscles as a rule cannot be demonstrated near the resting nucleus in the protoplasm of the cell.*

V. Upon the Molecular Structure of Organised Bodies.

In order to explain the chemico-physical properties of organised bodies, Nägeli (V. 17, 18; II. 27, 28) has advanced a micellar theory, which, although undoubtedly to a great extent hypothetical, is very useful in rendering many complicated conditions more easy of comprehension, and above all more easily pictured to the imagination. A short abstract of this micellar theory, which deserves attention, if only on account of the strictly logical manner in which it has been worked out, will not be out of place here.

One of the most remarkable properties of an organised body is its capacity of swelling up, that is to say, of absorbing into its interior a large, though not unlimited, quantity of water, with the substances dissolved in it. This may take place to such an extent that in an organised body only a small percentage of solid substances may be present.

The body increases in size in proportion to the amount of water absorbed, shrinking up again when the water is expelled. Hence the liquid is not stored up in a pre-existent cavity, which before was filled with air, as in a porous body, but becomes evenly distributed amongst the organised particles, which, as the body swells up, must become farther and farther pushed apart, being separated from one another by larger and larger envelopes of water. In spite of the absorption of so much water, none of the organised substance becomes dissolved. In this respect the phenomenon differs from that which takes place with a crystal of salt or sugar, which on the one hand does not possess the power of swelling up, and on the other becomes dissolved in the water, its molecules separating from one another, and distributing themselves evenly throughout the water.

Its power of swelling up and its non-solubility in water are the most important properties of an organised body, without which it is inconceivable that the vital processes could proceed.

Many organised bodies may be dissolved if treated according to special methods, as for example starch and gelatine-producing substances, which become dissolved when they are boiled in water. But even these starch and gelatine solutions differ very much in their chemical properties from solutions of salt or sugar. The latter diffuse easily through membranes, whilst the former either do not do so at all, or only to a very small extent, whilst their solutions are slimy or viscous. Graham distinguishes between the two groups of substances, which exhibit such different properties in solution, by calling them *crystalloids* and *colloids*.

Now Nägeli has attempted to explain all these phenomena as being due to differences in the molecular structure of the various bodies. As atoms combine together to form molecules, thus producing so great a variety of chemical substances, so he considers that the molecules unite together in groups to form still more complex units, the *micellæ*, and that in this manner the complex properties of organised bodies arise. *In comparison with that of the molecule, the size of the micella is considerable, although too small to be seen with the microscope; it may be built up, not only of hundreds, but even of many thousands of molecules.*

Nägeli ascribes a crystalline structure to these micellæ, in consequence of their power of double refraction, which further is exhibited by many organised bodies, such as cellulose, starch, muscular substance, and even protoplasm itself in polarised light. In addition, great differences may be present in their outward appearance as well as in their size.

The micellæ have an affinity for water as well as for each other; hence their power of swelling up. In a dry organised body the micellæ lie close together, being only separated by delicate envelopes of water; as more water becomes absorbed, these envelopes increase considerably in size, since at first the micellæ have a stronger affinity for water than for each other. Thus they become pushed apart from each other by the penetrating water as with a wedge; "however, organised bodies cannot become really dissolved, for the molecular attraction of the micellæ for the water diminishes with distance at a proportionally greater rate than that of the micellæ for each other, and hence when the envelopes have reached a certain size a condition of equilibrium, the limit of the power of the body to swell up, is reached."

When, however, by means of special methods of treatment, the attraction of the micellæ for each other is quite overcome, a *micellar solution* is obtained. This solution is cloudy and opalescent, which is an indication that the light is unevenly refracted. Nägeli compares this with the slimy opalescent masses produced when Schizomycetes are crowded together in large numbers.

Nägeli explains the differences, which Graham has described as existing between crystalloids and colloids, by the statement that in the former isolated molecules are distributed amongst the particles of water, whilst in the latter crystalline groups of molecules or isolated micellæ are so distributed. Hence numbers of the one group form molecular solutions, and those of the other micellar solutions (such as egg-albumen, glue, gum, etc.). The micellæ themselves have considerable power of preventing the substance from breaking down into molecules. Such a breaking down is generally accompanied by chemical transformation. Thus starch, after it has been converted into sugar, is capable of forming a molecular solution, as is also the case with proteids and gelatine-yielding substances after they have been converted into peptones.

In organised bodies the micellæ unite together to form regularly arranged colonies, in which the individual micellæ may consist of similar or different chemical substances, and may vary as to size and form; further, they may unite in smaller or larger groups of micellæ within the colony itself. *The micellæ within these micellar colonies appear as a rule to hang together in chains, which further unite together to form a frame or network structure with more or less wide meshes. In the gaps or micellar interstices the water is enclosed.* "Only in this manner is it possible to have a firm structure, composed of a large quantity of water and a small quantity of solid matter, such as is seen in a jelly."

The water, which is contained in organised bodies, may be found in three conditions, distinguished by Nägeli under the names *water of constitution or of crystallisation, water of adhesion,* and *capillary water.* By the first are understood the molecules of water, which, as in a crystal, are united firmly to the molecules of the substance in a fixed proportion, thus entering into the structure of the micella.

The water of adhesion consists of molecules of water, which are held closely to the surface of the micella by molecular attraction. "The concentric layers of water, which compose the spherical envelope surrounding the micella, vary considerably as to their density and their immobility; they are naturally most dense and firmly attached when they are in direct contact with the surface of the micella" (Pfeffer).

The capillary water finally is outside the sphere of attraction of the individual micellæ and fills up the gaps in the micellar network.

"These three kinds of water show considerable variation as to the degree of motility shown by their molecules. The molecules of capillary water are as free in their movements, as those of free water; in the water of adhesion the progressive movements of the molecules are more or less diminished, whilst the molecules of the water of constitution are fixed and non-motile." Hence only the waters of capillarity and of adhesion can pass through a membrane by osmosis.

Just as water particles may be firmly held upon the surface of the micellæ by molecular attraction, other substances (calcium and silicon salts, colouring matter, nitrogenous compounds, etc.), having been taken up in solution into the organised body, may be deposited upon them. The growth of organised matter by intussusception is explained by Nägeli, by the supposition that particles of material in solution make their way into the organised body, such as, for example, molecules of sugar into a cellulose membrane, where they may either become deposited upon the micellæ which are already present, thus adding to their size, or to a certain extent they may crystallise out to form new micellæ situated between the ones already present. As an example of this, the phenomenon of sugar molecules becoming converted into cellulose molecules may be quoted.

This micellar hypothesis of Nägeli is frequently referred to in later chapters, as it often is of great use in forming a mental picture of the complex arrangement of matter in the elementary organism.

Literature II.

1. Altmann. *Die Elementarorganismen u. ihre Beziehungen zu den Zellen.* Leipzig. 1890.
2. Jul. Arnold. *Ueber feinere Structur der Zellen unter normalen und pathologischen Bedingungen. Virchows Archiv.* Bd. 77, 1879, p. 181.

3. BALDIANI. *Sur la structure du noyau des cellules salivaires chez les larves de Chironomus.* Zoologischer Anzeiger, 1881, p. 637.
4. VAN BENEDEN et NEYT. *Nouvelles recherches sur la fécondation et la division mitosique chez l'ascaride mégalocéphale.* Leipzig. 1887.
5. BÜTSCHLI. *Einige Bemerkungen über gewisse Organisationsverhältnisse der sogenannten Cilioflagellaten und der Noctiluca.* Morph. Jahrbuch. Bd. X. 1885.
6. BÜTSCHLI. *Ueber den Bau der Bakterien und verwandter Organismen.* Leipzig. 1890.
7A. BÜTSCHLI. *Ueber die Structur des Protoplasmas.* Verhandlungen des Naturhist.-Med.-Vereins zu Heidelberg. N. F., Bd. IV., Heft 3. 1889. Heft 4. 1890. (See Quar. Jour. Mic. Soc., 1890.)
7B. BÜTSCHLI. *Untersuchungen über mikroskopische Schäume u. das Protoplasma.* 1892.
8. CARNOY. *Several papers in La Cellule. Recueil de Cytologie et d'histologie générale.*
   La cyto-liérèse chez les Arthropodes, T. I.
   La vésicule germinative et les glob. polaires chez divers nématodes.
   See also *Conférence donnée à la société belge de microscopie, T. III.*
   See also A. B. LEE. *On Carnoy's cell researches.* Quar. Jour. Mic. Soc. Vol. XXVI., pp. 481–497.
9. ENGELMANN. *Ueber den faserigen Bau d. contractilen Substanzen.* Pflügers Archiv. Bd. 26. *Physiology of Protoplasmic Movement,* trans. Quar. Jour. Mic. Soc. Vol. XXIV., p. 370.
10. FLEMMING. *Zellsubstanz, Kern und Zelltheilung.* Leipzig. 1882.
11. FLEMMING. *Ueber Theilung u. Kernformen bei Leukocyten und über deren Attractionssphären.* Archiv. f. mikroskop. Anat. Bd. 37, p. 249.
12A. FLEMMING. *Neue Beiträge zur Kenntniss der Zelle. II. Theil.* Archiv. f. mikroskop. Anat. Bd. 37, p. 685.
12B. FLEMMING. *Attractionssphären und Centralkörper in Gewebszell u und Wanderzellen.* Anatomischer Anzeiger. Bd. VI.
    See also JOHN E. S. MOORE. *On the Relationships and Rôle of the Archoplasm during mitosis in the Larval Salamander.* Quar. Jour. Mic. Soc. Vol. XXXIV., p. 181.
13. FOL. *Lehrbuch der vergleichen mikroskop. Anatomie.* Leipzig. 1884.
14A. FROMMANN. *Zur Lehre von der Structur der Zellen.* Jenaische Zeitschrift f. Med. und Naturw. Bd. 9. 1875.
14B. FROMMANN. *Zelle. Realencyklopädie der gesammten Heilkunde.* 2 Aufl. 1890.
15. HAECKEL. *Générale Morphologie.*
16. MARTIN HEIDENHAIN. *Ueber Kern und Protoplasma.* Festschrift für Kölliker. 1892.
    See also W. D. HALLIBURTON. *Gulstonian Lectures on the Chemical Physiology of the Animal Cell.* Brit. Med. Jour. Vol. I. 1893.
17. C. HEITZMANN. *Untersuch. über Protoplasma.* Wiener Sitzungsver. math. naturw. Classe. Bd. LXVII. 1873.
18. RICHARD HERTWIG. *Beiträge zu einer einheitlichen Auffassung der verschiedenen Kernformen.* Morphol. Jahrbuch. Bd. 2. 1876.

19a. Oscar Hertwig. *Beiträge zur Kenntniss der Bildung, Befruchtung und Theilung des Thierischen Eies. Morphol. Jahrbuch. Bd. I., II., IV.*
19b. Oscar Hertwig. *Vergleich der Ei- u. Samenbildung bei Nematoden. Archiv. f. mikroskop. Anatomie. Bd. 36. 1890.*
20. Hofmeister. *Die Lehre von der Pflanzenzelle. Leipzig. 1867.*
21. E. Klein. *Observations on the Structure of Cells and Nuclei. Quar. Jour. Mic. Soc. Vol. XVIII., 1878, p. 315.*
22. Kölliker. *Handbuch der Gewebelehre. 1889.*
23. Kossel. *Zur Chemie des Zellkerns. Zeitschrift für physiolog. Chemie von Hoppe Seyler. 1882. Bd. 7.*
  Untersuchungen über die Nucleine und ihre Spaltungsprodukte. Strassburg. 1881.
  Kanthack and Hardy. *Proceedings of the Royal Society. Vol. LII.*
24. C. Kupffer. *Ueber Differenzirung der Protoplasma an den Zellen thierischer Gewebe. Schriften des naturwissenschaftl. Vereins für Schleswig-Holstein. Bd. I., p. 229. Heft 3. 1875.*
25. Leydig. *Untersuchungen zur Anatomie u. Histologie der Thiere. Bonn. 1883.*
26. Leydig. *Zelle und Gewebe. Bonn. 1885.*
27. Nägeli u. Schwendener. *The Microscope. Theory and Practice, trans. London.*
28. C. Nägeli. *Mechanisch-physiologische Theorie der Abstammungslehre. München und Leipzig. 1884.*
29. Pfitzner. *Beiträge zur Lehre vom Bau des Zellkerns u. seinen Theilungserscheinungen. Archiv. f. mikrosk. Anatomie. Bd. 22. 1883.*
  J. Priestley. *Recent Researches on the Nuclei of Animal and Vegetable Cells. Quar. Jour. Mic. Soc. Vol. XVI., pp. 131-152.*
30. v. Rath. *Ueber eine eigenartige polycentrische Anordnung des Chromatins. Zoolog. Anzeiger. 1890.*
31. Rauber. *Neue Grundlegungen zur Kenntniss der Zelle. Morph. Jahrb. VIII. 1882.*
32. Reinke u. H. Rodewald. *Studien über das Protoplasma. Untersuchungen aus dem botanischen Institut der Universität. Göttingen. Heft 2. 1881.*
33. Sachs. *Textbook of Botany, Morphological and Physiological, trans. by S. H. Vines. 1882.*
34. Schäfer and E. Ray Lankester. *Discussion on the Present Aspect of the Cell Question. Nature. Vol. XXXVI. 1887.*
  See also Schäfer in *Quain's Anatomy, Vol. I., pt. 2. 1891.*
35. Schieferdecker u. Kossel. *Gewebelehre mit besondere Berücksichtigung des menschl. Körpers.*
36. Schmitz. *Untersuchungen über die Structur des Protoplasmas und der Zellkerne der Pflanzenzellen. Sitz. Ber. der Niedenh. Gesellsch. f. Natur u. Heilk. Bonn. 1880.*
37. Frank Schwarz. *Die morphologische und chemische Zusammensetzung des Protoplasmas. Beiträge zur Biologie der Pflanzen. Bd. IV. Breslau. 1887.*
38. Solger. *Zur Kenntniss der Pigmentzellen. Anatomischen Anzeiger. Jahrg. VI., p. 182.*

39. Strasburger. *Zellbildung und Zelltheilung.* 2 Aufl. Jena. 1876.
40. Strasburger. *Studien über das Protoplasma. Jenaische Zeitschrift.* 1876. Bd. X.
41. Strasburger. *Practical Botany, trans. by Hillhouse.* London.
42. Wiesner. *Elementarstructur und Wachsthum der lebenden Substanz.*
43. Zacharias. *Ueber den Zellkern. Botanische Zeitung,* 1882, p. 639.
44. Zacharias. *Ueber Eiweiss, Nuclein und Plastin, Botanische Zeitung.* 1883.
45. Zacharias. *Ueber den Nucleolus. Botanische Zeitung.* 1885.
46. Zacharias. *Beiträge zur Kenntniss des Zellkerns u. der Sexualzellen. Botan. Zeitung,* 1887. Bd. 45.
47. Zacharias. *Ueber die Zellen der Cyanophyceen. Botan. Zeitung.* 1890. See also Halliburton *loc. cit.*
48. List. *Untersuch. über das Cloakenepithel der Plagiostomen. Sitzungsber. der kaiserl. Acad. der Wissensch. zu Wien,* Bd. XCII. III., Abth. 1885.
49. Miescher. *Verhandl. der naturforschenden Gesellschaft in Basel.* 1871.
50. Anebach. *Organologische Studien.* Heft I. 1874.

CHAPTER III.

THE VITAL PROPERTIES OF THE CELL.

I. **The Phenomena of Movement.** All the mysteries of life, which are exhibited by plants and animals, are present, as it were in a rudimentary form, in the simple cell. Each individual cell, like the whole complex organism, has an independent life of its own. If we wish to study more deeply the true nature of protoplasm, we must above all things investigate its most important properties, its so-called vital properties. However, life, even the life of the simplest elementary organism, is a most complex phenomenon, which it is most difficult to define; it manifests itself, to use a wide generalisation, in this, that the cell in consequence of its own organisation, and under the influence of its environment, experiences continual changes and develops powers, by means of which its organic substance is being continually broken down and built up again. During the former process, energy is set free. The whole vital process, as Claude Bernard (IV. 1A) expresses it, depends upon the continual co-relation of this organic destruction and restoration.

It is most convenient to classify these most complex phenomena under four heads. Thus each living organism exhibits four different fundamental functions or properties, by means of which its life is made manifest: it can alter its form, and exhibit movements; it reacts to certain external stimuli in various ways, that is to say, it is irritable; it has the power of nourishing itself, it can by absorbing and transforming food material, and by giving up waste products, form substances, which it utilises for growth, for building up tissues, and for special vital functions; finally, it can reproduce itself.

Hence we will discuss the vital properties of the cell in four chapters, which we will take in the following order:

1. Phenomena of movement.
2. Phenomena of irritability.
3. Metabolism and formative activity.
4. Reproduction.

In addition there will be a special chapter on the process of fertilisation.

The cell may exhibit *several kinds of movement*, as is seen if an extensive comparative study is made. We will here distinguish between : (1) true protoplasmic movements ; (2) ciliary or flagellar movements ; (3) the movements of the pulsating vacuole ; (4) the passive movements and changes of shape exhibited by cells.

In addition to these four, there are a few special phenomena of motion, of which it will be best to treat in later chapters, for example, the formation of the receptive protuberance which appears in the egg-cell in consequence of fertilisation; the radiation figures which are seen in the neighbourhood of the spermatozoon after it has penetrated into the ovum, and those which occur during the process of cell division, when the cell body splits up into two or more parts.

Protoplasmic Movements. Although it is probable that movements take place in all protoplasm, yet in most cases, with our present means of observation, they cannot be perceived on account of their great slowness ; hence in only a few objects in the plant and animal kingdoms can this phenomenon be studied and demonstrated. The movement manifests itself partly in changes in the external form of the cell, and partly in the arrangement of the structure enclosed in the protoplasm, the nucleus, the granules, and the vacuoles.

These movements differ somewhat according as to whether they are manifested in naked protoplasm, or in that which is enclosed by a firm membrane.

a. **The Movements of naked Protoplasm.** Small unicellular organisms, white blood corpuscles, lymph corpuscles, connective tissue cells, etc., exhibit movements which, in consequence of their similarity to those seen in the Amœba, are termed amœboid.

If a *lymph corpuscle of a Frog* (Fig. 37) is observed under suitable circumstances, it is seen to undergo continual changes of form. Small processes of protoplasm, the foot-like processes, or pseudopodia, are protruded from its surface ; at first as a rule they consist of hyaloplasm alone, but after a time granular protoplasm streams into them. By this means the pseudopodia are increased in size ; they become broader, and may in their turn extend new, more minute processes from their surface. Or the protoplasm may flow back again, thus causing them to decrease in size, until finally they are completely withdrawn, whilst new processes are being protruded from another portion of the body. By means of these alternate protrusions and retractions of their pseudopodia, the small bodies of protoplasm are enabled to move from place to place, crawling over the objects to whose surfaces they cling at a rate which can only be measured under the microscope. *Amœbæ* are able to traverse a distance of ½ mm. in a minute.

In this manner the white blood corpuscles during inflammation are able to pass through the walls of the capillaries and of the smaller vessels, and the lymph corpuscles make their way as wandering cells into the connective tissue spaces, such as the interlamellar spaces of the cornea, where the resistance to be overcome is not great, or they force their way between epithelial cells, and so reach the surface of an epithelial membrane.

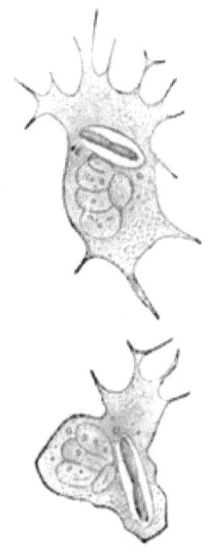

FIG. 37.—A Leucocyte of the *Frog* containing a *Bacterium* which is undergoing the process of digestion. The *Bacterium* has been stained with vesuvine. The two figures represent two successive changes of shape in the same cell. (After Metschnikoff, Fig. 54.)

This extension and retraction of pseudopodia is most marked in a small *Amœba* (Fig. 38), which was described as far back as 1755 by Roesel von Rosenhof, who on account of its energetic changes of form called it the small *Proteus*.

Somewhat different movements take place in Myxomycetes, and in Thalamophora, Heliozoa, and Radiolaria.

The plasmodia of some species of Myxomycetes, such as the *Æthalium septicum*, often spread

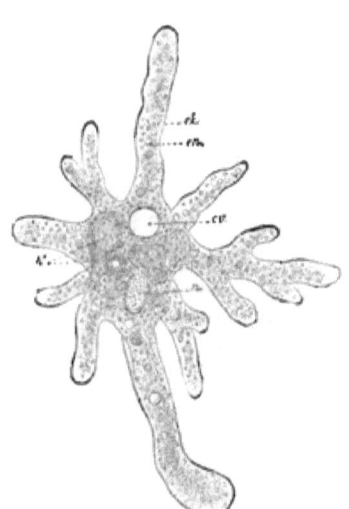

FIG. 38.—*Amœba proteus* (after Leidy; from R. Hertwig, Fig. 16): n nucleus; cv contractile vacuole; N food vacuoles; en endoplasm; ek ectoplasm.

themselves out over the object upon which they rest, in large masses about the size of a fist. In order to make a suitable preparation for observation of such a plasmodium, it is best to hold a moistened slide near to its edge in an oblique position, and to cause a stream of water by means of a special contrivance to flow slowly down the slide. The plasmodia of the *Æthalium* possess the property of moving in a direction opposite to that of the stream of water (rheotropism); hence they protrude innumerable pseudopodia, and by this means crawl up on to the moistened slide, where they spread themselves out, and, by uniting neighbouring pseudopodia together by means of transverse branches, they form a delicate transparent network (Fig. 39). When this network is examined with a high power, it can be seen to exhibit two kinds of movements.

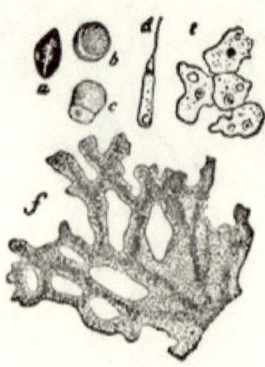

FIG. 39.—*Chondrioderma difforme* (after Strasburger). Part of a fairly old plasmodium. *a* Dry spore; *b* the same, swollen up in water; *c* spore, the contents of which are exuding; *d* zoospore; *f* mœboid forms produced by the transformation of zoospores, which are commencing to unite together to form a plasmodium. (In *d* and *e* the nuclei and contractile vacuoles may be distinguished.)

At first the granular protoplasm which is present in the threads and strands, where it is surrounded by a thin peripheral layer of hyaline protoplasm, is seen to have a quick, flowing movement, which is chiefly observable because of the movement of the small granules, and which resembles the circulation of the blood in the vessels of a living animal. There is no distinct boundary line between the motile endoplasm and the non-motile ectoplasm, for the granules at the edge of the stream move much more slowly than those in the centre; indeed, sometimes they may keep quite still for a time, to be later on again caught up by the stream and carried along with it. In the thinner threads there is always only one stream flowing longitudinally, but in the thicker branches there are often two flowing along side by side in opposite directions. "In the flat membrane-like extensions" which are developed here and there in the network, "there are generally a large number of branched streams flowing either in the same or in different directions; not infrequently we find streams flowing along side by side in opposite directions." Further, the rate of movement may vary in different places, or it may gradually alter; it may be so great that under a powerful lens the granules appear to travel so fast that the eye can scarcely follow them; on the other hand, it may be so small that the granules scarcely appear to change their place.

The second kind of movement consists of a change of form in the individual threads and in the network as a whole. As in the *Amœba*, processes are protruded and withdrawn from various places, a mass of homogeneous protoplasm being first protruded, into which the granular protoplasm flows later on. Occasionally, when the streaming movements are very powerful, it appears as though the granular endoplasm is pressed forcibly into the newly formed processes. By this means the plasmodium can, like the *Amœba*, crawl slowly along over a surface in a given direction; new processes are continually being protruded from the one edge, towards which the endoplasm chiefly streams, whilst others are withdrawn from the opposite one.

Gromia oviformis (Fig. 40) is a classical object amongst the Reticularia, for the study of protoplasmic movements (see p. 29). If the little organism has not been disturbed in any way, a large number of long fine threads may be distinguished stretching out from the protoplasm, which has made its way out of the capsule, and spreading themselves out radially in every direction into the water; here and there lateral branches are given off, and occasionally all the threads are united together into a network by such branches. Even the most delicate of these threads exhibit movements. As Max Schultze (I. 29) aptly describes it, " a gliding, a flowing of the granules which are imbedded in the thread substance," may be seen with a high power; " they move along the thread, more or less quickly, either towards its periphery or in the other direction; frequently streams flowing in both directions may be seen at the same time even in the finest threads. When granules are moving in opposite directions, they either simply pass by each other, or else move round one another for a time, until after a short pause they either both go on in their original directions, or one takes the other along with it. All the granules in a thread do not move along at the same rate; hence sometimes one may overtake another, either passing it or being stopped by it." Many evidently pass along the outermost surface of the thread, beyond which they can be plainly seen to project. Frequently other larger masses of substance, such as spindle-

70 THE CELL

shaped swellings or lateral accumulations in a thread, may be seen to move in a similar manner. Even foreign bodies which adhere to the thread substance, and have been taken in by it, are seen to join in this movement, the rate of which may attain to ·02 mm. per second. Where several threads overlap each other granules may be seen passing from one into the other. At such places broad flat surfaces may be produced by the heaping up of the thread substance.

A special kind of protoplasmic movement is described by Engelmann (III. 5, 7) under the name of *gliding movement* (Glitschbewegung). It has been observed chiefly in Diatoms and Oscillaria. In the former the protoplasm is surrounded by a siliceous shell, in the latter by a cellulose membrane. However, outside this covering there is an exceedingly delicate layer of hyaloplasm, quite free from granules, which cannot be seen in the living ob-

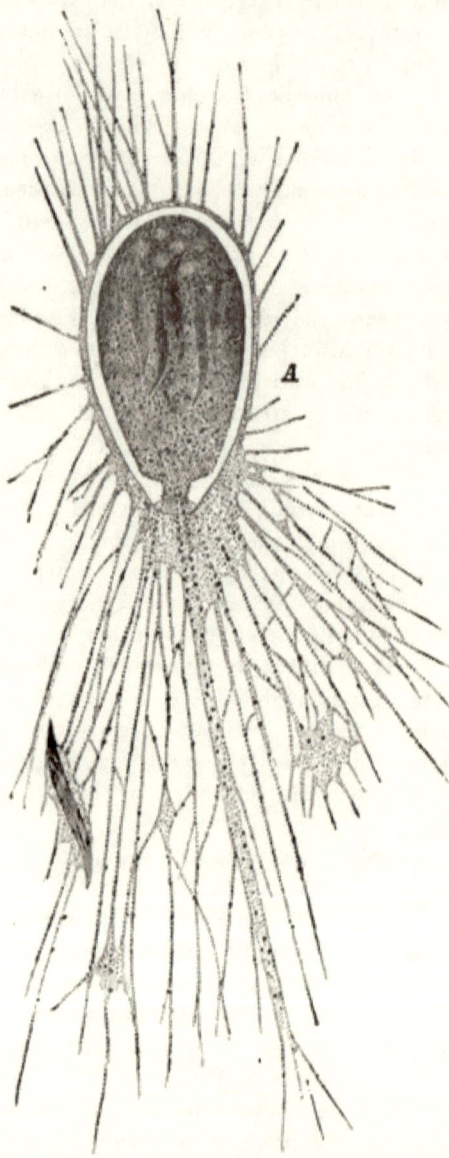

FIG. 40.—*Gromia oviformis.* (After M. Schultze.)

ject, but which may sometimes be demonstrated by means of reagents. Hence, since this layer moves in a certain direction over the siliceous shell, or cellulose membrane, the small organisms can "move in a gliding or creeping fashion over a solid surface" (Engelmann).

b. **The movements of Protoplasm inside the Cell Membrane.** This kind of movement is chiefly seen in the vegetable kingdom, and as a rule is best observed in the cells of herbaceous plants rather than in those of shrubs and trees. According to de Vries (III. 25), these movements are never totally absent in any plant-cell, but frequently they are so slow as to escape direct observation. They are best seen in vascular tissues, and in those where materials have been stored up, and further at such times when considerable quantities of plastic substances are being transported in order to supply the material necessary for the continuation of growth, for local accumulations, and for special needs (de Vries). Hence this movement of the protoplasm appears to be directly of importance during the conveyance of materials from one part of the plant to another. More rarely it may be seen in the lower organisms, and in the animal kingdom, as in *Noctiluca* in the vesicular cells in the centre of the tentacles of *Cœlenterata*, etc.

Two kinds of movements may be distinguished in plants, *Rotation* and *Circulation*.

These movements of *rotation* were first observed in 1774 by Bonaventura Corti (I. 8); after that they were lost sight of for a time, but were re-discovered by Treviranus. The most suitable objects for studying them are afforded us by the Characeæ; root-hairs of the *Hydrocharis morsus ranæ*, and of *Trianea bogotensis*, leaves of *Vallisneria spiralis*, etc., are also very convenient for observations. In the large cells of the Characeæ, the protoplasm, as has already been described on p. 33, is spread out as a thick cohesive layer upon the inner surface of the cellulose membrane, surrounding the large quantity of cell-sap like a closed sac. In this lining two distinct layers of protoplasm can always be distinguished: an outer one, touching the cellulose membrane, and an inner one, in contact with the cell-sap. The former is always motionless; in Hydrocharis it is very thin, in Characeæ it is somewhat thicker, and it also contains a greater number of chlorophyll grains, which remain motionless. This immotile layer gradually passes over into the inner motile one, which in

Chara contains no chlorophyll corpuscles, but only nuclei and granules. The protoplasm of the inner layer, which, compared to that of the outer layer, appears to be richer in water, exhibits rotatory streaming movements, which take place in the following manner. The current passes up along the longitudinal wall of an elongated cell, then, turning round past a transverse wall, flows down the opposite longitudinal side, until, curving round again at the second transverse wall, it reaches the starting point, when the cycle recommences. *Between the upward and downward streams there is a more or less broad neutral strip where the protoplasm is at rest*, and where as a rule it is reduced to a very thin layer. In *Nitella* there are no chlorophyll corpuscles along this neutral strip in the outer layer.

A connecting link between the rotatory movement and true circulation is afforded us by the so-called "*fountain-like rotation*" (Klebs III. 14). This, which as a rule but rarely occurs, is found in young endosperm cells of *Ceratophyllum*, in young wood vessels of the leaf-stem of *Ricinus*, etc., etc. Here the protoplasm, in addition to spreading itself out in a thick layer over the inner surface of the cellulose wall, stretches itself in the form of a thick central strand along the longitudinal sap-cavity of the cell. Under these circumstances a single stream flows along this central strand, spreading itself out in all directions like a fountain upon the transverse wall, upon which it impinges; then streaming down the sides of the cell, it collects again at the opposite transverse wall, where it re-enters the main axial stream.

The motion which is described as *circulation* is observed in those plant and animal cells in which the protoplasm spreads itself out, both as a thin layer beneath the membrane, and also in the form of more or less delicate threads, which traverse the sap-cavity and are united together to form a net-like structure.

The objects which have been most examined are the staminal hairs of the various kinds of *Tradescantia*, and young hairs of the stinging nettle, and of pumpkin shoots.

The phenomenon of circulation resembles that observed in the protoplasmic nets of Myxomycetes, and of the delicate pseudopodia of the Rhizopoda. Circulation consists of two kinds of movements. In the first place attention must be drawn to the streaming movements of the granules. In the thinnest threads they move more or less quickly over the surface of the walls in one direction, whilst in broader bands several separate streams may circulate quite close together, sometimes in the same, sometimes in opposite directions. The nucleus, as well as the chlorophyll and starch grains, which lie embedded in the protoplasm, are carried slowly along by the current. Similarly in this case the most external hyaline layer of protoplasm, which is in contact with the cellulose membrane, is, comparatively speaking, at rest. In the second place, the whole body of protoplasm itself slowly moves along, in consequence of which it changes its form. Broad bands become narrowed, and may after a time disappear, delicate threads increase in size, and new processes are formed, just as new pseudopodia are protruded to the exterior by Myxomycetes and Rhizopoda. Large masses of protoplasm become heaped up here and there upon the layer lining the cell-wall, whilst at other places the coating becomes thinner.

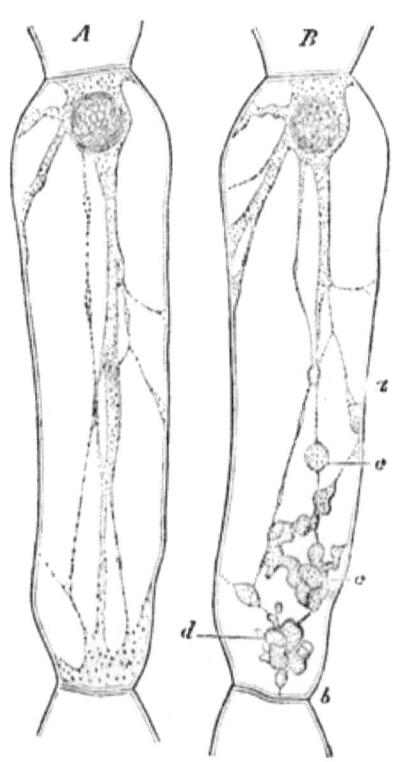

FIG. 41.—*A B*, cells of a staminal hair of *Tradescantia virginica*. *A* Undisturbed streaming movements of protoplasm. *B* Protoplasm which has run together into ball-like masses in consequence of irritation: *a* cell-wall, *b* transverse wall of two cells; *c d* protoplasm which has massed itself together into small balls. (After Kühne; from Verworn, Fig. 13.)

c. **Theories concerning Protoplasmic Movements.** Attempts have lately been made by various investigators, Quincke (III. 17), Bütschli (II. 7B), Berthold (III. 2), and others, to compare these protoplasmic movements with those exhibited by a mixture of inorganic substances, and thus to explain them.

Quincke has carefully investigated the movements which occur at the areas of contact of various fluids. He placed in a glass containing water a drop of a mixture of almond oil and chloroform, the specific gravity of which is slightly greater than that of water, and then, by means of a fine capillary tube, he caused a drop of 2 per cent. solution of soda to approach the globule of oil. This latter then exhibited changes in shape, which are similar to those observed with the microscope in certain *Amœbæ*. The explanation of this is that the soda solution gradually spreads itself out over the surface of the oil, forming a soap.

Quincke is of opinion that the protoplasmic movements are analogous to these. In the plasmolysis of plant cells, the protoplasm frequently breaks up into two or more balls, which spread themselves out, and then either re-unite, or remain separated from one another by an even surface, just as two soap bubbles of equal size which are placed in contact may touch each other, without uniting. In consequence of these appearances he is of opinion that, considering the physical properties of delicate solid or fluid lamellæ, the protoplasm must be surrounded by a very delicate *fluid* membrane, just as in the soap bubble the air is enclosed in a thin skin layer of soap solution. "The substance of the membrane surrounding the protoplasm," as Quincke proceeds to state, "must be a fluid which forms drops in water. Since of all the substances known in nature oil is the only one which possesses this property, the membrane must consist of an oil, that is to say of a fluid fat. The thickness of this layer may be most minute, less than ·0001 mm., and hence it is not perceptible even with the microscope." Through the action of the albumen upon this oil, a substance is produced upon the areas of contact, which is soluble in water, and spreads itself out just like the soap produced by the combination of soda and oil. Hence it is called albuminous soap.

Thus Quincke considers the cause of the protoplasmic movements to be a periodic spreading out of albuminous soap upon the inner surface of the envelope of oil surrounding the protoplasm. This soap, in being continually re-formed on the area of contact as fast as it is dissolved and diffused throughout the surrounding fluid, remains constant in quantity; thus, since the presence of oxygen is necessary in this chemical process, the fact is explained, that, in its absence, the protoplasmic movements are arrested, and similarly their cessation at extreme temperatures may be ascribed to chemico-physical conditions.

Bütschli, being stimulated by these investigations of Quincke, has undertaken some interesting experiments based on the assumption of his foam or emulsion theory of protoplasm, and these, as it appears to him, throw light upon the cause of the protoplasmic movements. He prepared frothy structures of oil in various ways. The most delicate and instructive masses were obtained by mixing a few drops of olive oil, which had been kept for some time in a warm chamber, with some finely powdered K_2CO_3, until a viscous mass was produced; a small drop of this mixture was then introduced into water. The emulsion which is produced in this manner is milky white in appearance, and consists of minute vacuoles, filled with the solution of soap, which is formed at the same time: it may be cleared by adding to it a few drops of dilute glycerine. By this means active streaming movements are produced, which, in a successful preparation, last for at least six days, and which are certainly surprisingly like the protoplasmic movements of an *Amœba*. "From one place on the edge the current flows through the axis of the drop; it then streams away from the edge down both sides, in order to unite again, gradually to form the axial current again. Here and there a blunt process is protruded and withdrawn, and so on; indeed, individual drops may exhibit fairly active locomotive powers for a time." Bütschli, in accordance with Quincke's experiments, explains these phenomena of movement in the following manner: "On some place on the surface some of the delicate chambers of the froth structure burst, and thus the soap solution at this region is able to reach the surface of the drop, which is composed of a very thin lamella of oil. The necessary consequence of this is a diminution of surface-tension at this spot, and hence a slight bulging and out-streaming occur. Both of these induce a flow of foam-substance from the interior to this spot. A few more meshes may be broken down by this current, and so on, the result being that a streaming, once induced, is persistent unless considerable obstacles present themselves." Bütschli is quite convinced that the streaming movements seen in these saponified fat drops are identical in all essentials with amœboid protoplasmic movements.

These experiments of Quincke and Bütschli are of the greatest interest, for they prove that very complex phenomena of movement may be induced by means of comparatively simple methods. On the other hand, various objections may be raised against their deduction, that in protoplasmic movements similar processes occur. Even the hypothesis, that the protoplasmic substance is enveloped by a delicate lamella of oily substance, is exceedingly doubtful. For if we only take into account the single fact that protoplasm is composed of a great number of chemical substances, which, during the metabolic processes upon which its life depends, are continually undergoing chemico-physical changes, we cannot but think that conditions much more complex in their nature must be necessary for its movements, than those required for a moving drop of saponified oil, and, indeed, the complexity of these conditions must be proportionate to the immense difference in the complexity of the chemical composition and organisation of the two substances in question [cf. statements already made on this subject on p. 22; and *Die Bewegung der lebendigen Substanz* by Verworn (III. 24)]. Further, *all* the protoplasmic movements—the streaming movements, the radial arrangement round attraction centres, the movements of cilia and flagella, and muscular contraction—together constitute a single group of correlated phenomena which demand a common explanation. This, however, is not afforded us by the experiments of either Quincke or Bütschli. The movements, induced by them in a mixture of substances, bear the same relation to the movements of living bodies, as the structure of the artificial cell produced by Traube does to the structure of the living cell.

Fig. 42, which is taken from a paper by Verworn (III. 24), shows what very different appearances, closely resembling the various kinds of pseudopodic extensions, may be produced by the simple spreading out of a drop of oil upon a watery solution; *a–d* is a drop of salad oil which has spread itself out upon soda solutions of different degrees of concentration; in *a* it has assumed the form of *Amœba guttula*, in *b* and *c* of *Amœba proteus*, and in *d* of a plasmodium of a Myxomycete. Figs. *e* and *f*, which represent drops of almond oil, resemble the formation of pseudopodia in Heliozoa and Radiolaria, whilst *g* is taken from Lehmann's *Molecular Physics*, and represents a drop of creasote in water, in which it has assumed a form resembling a typical *Actinosphærium* (Verworn III. 24, p. 47).

Other attempts to explain the protoplasmic movements (Engelmann III. 6, Hofmeister II. 20, Sachs) lead us into the domain of theories upon the molecular structure of organised bodies, since the cause of the movements is supposed to lie in the changes of form of the most minute particles. A discussion of Verworn's latest attempt (III. 24) would lead us too far in another direction.

THE VITAL PROPERTIES OF THE CELL

Once for all, it must be admitted that none of the hypotheses which have, up till now, been propounded, are able to furnish us with a satisfactory conception of the causes and mechanical conditions of the plasmic movements, and that, therefore, we must confine ourselves to a simple description of observed conditions. This, however, is not to be wondered at, when we consider what

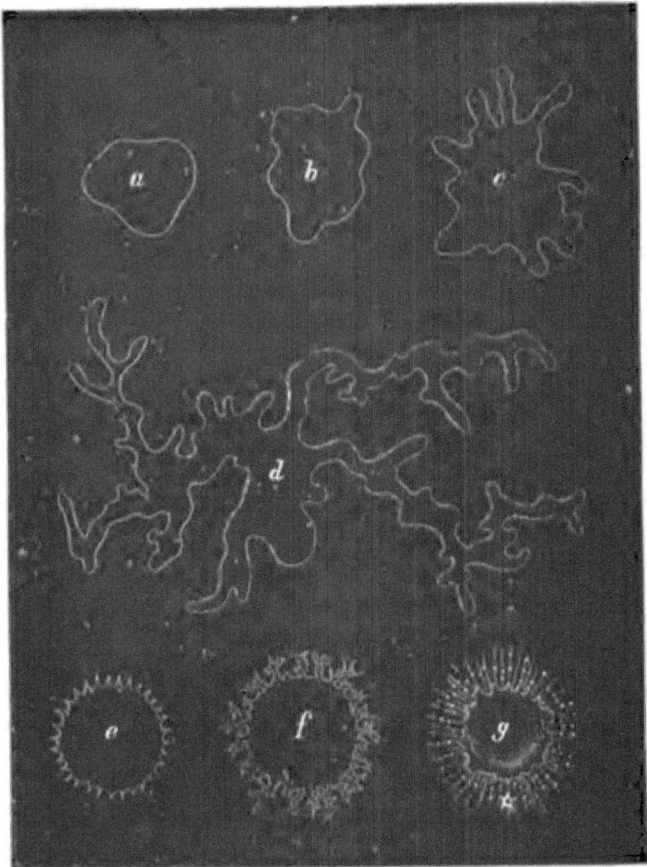

FIG. 42.—Different appearances assumed by drops of oil, which have spread themselves out. (After Verworn, Fig. 11.)

a number of different opinions are held with regard to the ultimate structure of protoplasm itself (see pp. 18-26), and this must of course affect the explanations tendered of its movements.

II. **Movements of Flagella and Cilia.** Unicellular organisms, by means of their flagella and cilia, are able to move from place to place much more rapidly than can be effected by means of pseudopodia. Flagella and cilia are delicate hair-like processes, which extend in greater or less numbers from the surface of the cell. They are composed of a homogeneous, non-granular substance, and in this respect resemble short, thin pseudopodia, when these consist of hyaloplasm alone. However, they differ from pseudopodia in two respects: firstly, they move in a different and more energetic fashion, and secondly, they are not transitory, but permanent organs, being neither protruded nor withdrawn. Fundamentally, however, the movements of flagella and pseudopodia are identical in kind, as is shown by the observations made by de Bary (I. 2) on swarmspores of Myxomycetes, and by Haeckel, Engelmann, R. Hertwig (III. 12B), and others on Rhizopoda.

Many of the lower organisms reproduce themselves by means of small spores, which resemble *Amœbæ* in their appearance and in their mode of movement (Fig. 43). After a time such spores usually protrude two thread-like pseudopodia (Fig. 43 a), which exhibit slow oscillatory movements, and develop into flagella, whilst the remainder of the body withdraws all its other pseudopodia, and so becomes spherical in shape. As the movements become stronger, the spore travels more and more rapidly, by means of its two flagella, through the water (Fig. 43 b); thus a "swarmspore" has developed out of the little amœboid cell.

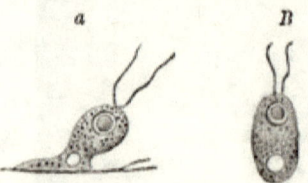

FIG. 43.—*Microgromia socialis*. An amœboid cell (a) which has been produced by division, and has wandered from the colony; and which, having withdrawn all its pseudopodia, with the exception of two, which have developed into flagella, becomes transformed into a swarmspore (b). (From *Hertwig*, Pl. 6, d and e.)

It may be safely deduced from these discoveries that *flagella are developed from delicate protoplasmic processes, which become especially contractile, and in consequence differ somewhat in their properties from the remaining protoplasm.* Hence they may be considered as constituting a special *plasmic product or cell-organ*, composed of contractile substance.

Flagella and cilia always arise directly from the body of the cell. If the cell is enveloped by a membrane, they protrude through pores in it. At their bases they are always somewhat thickened, frequently starting from the surface of the protoplasm as small button-like protuberances, whilst at their free ends they gradually become reduced to fine points.

Ciliary organs may occur in large or small numbers. In the latter case, when only from one to four are present, and when they are generally longer and more powerful, they are called *flagella*; in the former case, they cover the whole surface of the cell in large numbers, thousands being frequently present, they are then smaller and more delicate, and are called *cilia*.

a. **Cells with Flagella.** Flagella occur either at the anterior or posterior end of the body, producing a correspondingly different movement in the body. In the first case the flagella travel forwards, dragging the body along after them; in the second they propel it from behind. The former mode of locomotion has

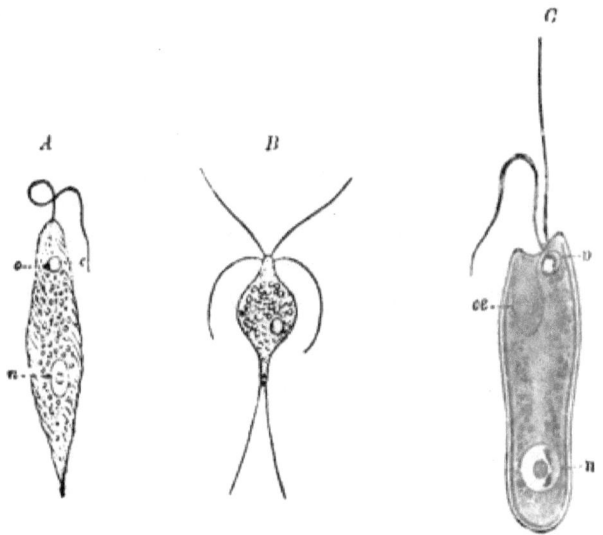

FIG. 44.—*A Euglena viridis* (after Stein): *n* nucleus; *c* contractile vacuole; *o* pigment-spot. *B Hexamitus inflatus* (after Stein). *C Chilomonas paramæcium* (after Bütschli): *oe* cytostome; *v* contractile vacuole; *n* nucleus. (From Hertwig, Figs. 130-132.)

been chiefly observed in Flagellata and kindred organisms (Fig. 44 *A*, *B*, *C*), in many kinds of Bacteria (Fig. 33 *b*), in antherozoids (Mosses, Ferns, Equisitaceæ), and in swarmspores, under which name the reproduction bodies of many Algæ and Fungi are included; the latter method of locomotion occurs in the spermatozoa of most animals (Fig. 45).

The ciliary organs of unicellular organisms have a double

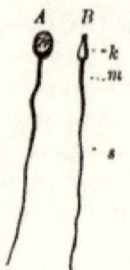

Fig. 15.—Mature human spermatozoon from two points of view. It is composed of *k* head; *m* middle portion; and *s* tail.

function to perform. Firstly, they have to keep the cell body afloat by means of their activity, since its specific gravity is somewhat greater than that of the surrounding medium. This is proved by the fact that dead swarmspores and spermatozoa sink to the bottom of the vessel. Secondly, they have to propel the body in a certain direction by means of their movements.

Nägeli (III. 16) has made most careful observations upon the *mechanism of the movements* of the motile cells of plants. According to this investigator, the oscillations of the flagella impart a two-fold movement of the body—a forward, and at the same time a rotatory movement. Hence the resultant motion resembles that of a ball shot out of a rifle. Such motions may be divided into three types:—

"Many motile cells travel forwards in a straight or somewhat curved line, the anterior and posterior ends of their axes remaining exactly in the same direction; these swim steadily forward, without deviating from a fixed path. With others it may be distinctly seen that they describe a straight, or somewhat bent spiral, in which one revolution round the axis always corresponds to one turn of the spiral (a given side of the cell always facing outwards), whilst the axis of the cell is parallel to that of the spiral. Finally, there are other cells whose anterior ends describe spirals, whilst their posterior ends proceed in a straight line, or in a spiral of smaller diameter. The nature of the second and third of these movements can only be distinguished if they are very slow. If they are rather quicker, only a kind of wavering can be made out, which, especially in the third, is of a peculiar character."

The direction in which the motile cells rotate about their longitudinal axis generally remains constant for each kind, species, or family; many rotate from south to west (*Ulothrix*), others from south to east (antherozoids of Ferns), others are somewhat uncertain in their rotations, turning now from south to east, and now from south to west (*Gonium*). If motile cells strike against any object, they cease for a time their forward movements, but continue to rotate about their longitudinal axes; then, "as a rule, they commence to retreat, their posterior ends being in advance, and to rotate themselves in an opposite direction." This backward movement never lasts for long, and is always slower than the forward one; however, the cell soon returns to its normal mode of progression, which usually takes place in a somewhat oblique direction.

In consequence of his investigations, Nägeli is of opinion that if zoospores and spermatozoa be quite regular in form, if their substance be evenly distributed throughout their mass, and further, if the medium be quite homogeneous, they must travel in a perfectly straight line, and hence that all deviations from this straight line, both as regards rotation round the axis and forward progression, must be ascribed either to the circumstance that they are not symmetrical in form, and that their centres of gravity are not in the centres of their bodies; or to the fact that the frictional opposition which they encounter is not equal in every direction.

By means of flagella a far greater speed is attainable than by means of pseudopodia. According to Nägeli, zoospores usually proceed at the rate of one foot per hour; the quickest, however, take only a quarter of an hour to traverse the same space; whilst a man, at ordinary speed, traverses a distance of rather more than half his length in a second, a swarmspore in the same time covers a distance of nearly three times its own diameter. However, although the rate at which they move appears, when they are seen under the microscope, to be very great, we must take into account the fact that the distance is also magnified, and that in consequence they appear to move much more rapidly than they do in reality. As a matter of fact, their movements are exceedingly slow. "Without magnification, even if the organisms could be plainly seen, no movement could be perceived on account of its slowness."

Spermatozoa (Fig. 45) may be distinguished from the zoospores of plants by their possessing one single thread-like flagellum, situated at the posterior end of the body. The spermatozoon, being propelled by it, advances by means of snake-like movements, resembling those of many fishes. In some cases the structure is more complicated, a delicate *contractile or undulating membrane*, which may be compared to the edge of a fish's fin, being present. This is especially well developed on the posterior end of the large spermatozoa of the Salamander and the Triton (Fig. 46).

82 THE CELL

If this undulating membrane be examined with a very high power of the microscope, waves are seen to travel continually over its surface, passing from the front to the back. "These," as Hensen explains, "are caused by each successive transverse portion passing one after the other from one extreme position (Fig. 47) to the other. For instance, if at the initial period a portion of the edge, which is seen from above, occupies position I to I^1 (Fig. 47), it is seen at the end of the first quarter of the period to have assumed position II to II^1, or, which amounts to the same thing, position II^1 to II^2. At the end of the second quarter the portion II^1 to II^2 is in the position III to III^1 or, which is the same, III^1 to III^2. At the end of the third quarter III^1 to III^2 has passed into the position IV to IV^1, whilst at the end of the whole period it has again taken up position I to I^1. The movements follow after each other with a certain degree of force and speed; it remains now to be seen how a forward motion results from them. Any one point on the surface of the undulating border (Fig. 47) moves, as is indicated by the arrow, from δ to γ with the force $x = a\gamma$. This force can be resolved into its two components $a\beta$ and $\beta\gamma$. The force $a\beta$ is exerted in the direction of the border, compressing it, and apparently producing no further effect. Force $\beta\gamma$ may be again split up into $\gamma\delta$ and $\gamma\epsilon$. $\gamma\epsilon$ exerts a direct backward pressure on the water, and hence, in consequence of the resistance of the water, propels the body in a forward direction. Force $\gamma\delta$ would cause the body to rotate on its own axis; but opposed to it is the opposite force, which is developed at all the places where the arrows point in an opposite direction (as for instance over D). Further the same force $\gamma\epsilon$ is present in Fig. D as in Fig. C, only the shaded portions of Fig. A develop the forces which are opposed to $\gamma\epsilon$. It is seen, however, that the size of the surfaces in question, and hence

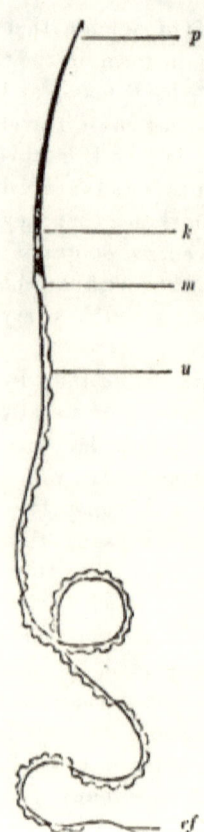

Fig. 48.—Spermatozoon of *Salamandra maculata*: *k* head; *m* middle portion; *cf* tail; *sp* anterior end; *u* undulating membrane.

of their force components, is invariably of minor importance" (Hensen III. 11).

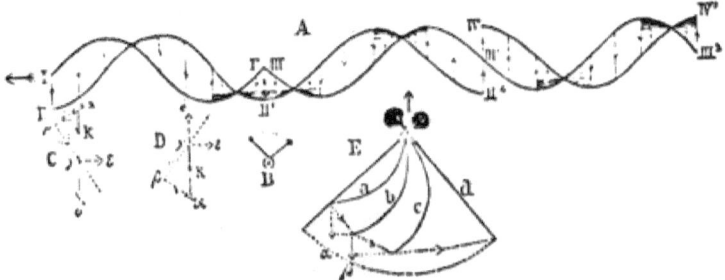

Fig. 47.—Explanation of the mechanism of the movements of spermatozoa (after Hensen, Fig. 22). *A* The four phases of position assumed by the border of the flagellum when an undulation passes over it. *I* to I^1, the first; *II* to II^1 to II^2, the second; *III* to III^1 to III^2, the third; *IV* to IV^1, the fourth stage of the bending of the border in a longitudinal undulation. *B* Section of the thread-like tail and membrane, in its two positions of greater elongation. *C* and *D* resolution of forces. *E* Movement of an ordinary spermatozoon; *a b c* various phases of this movement.

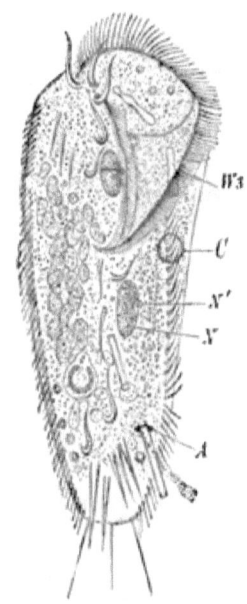

Fig. 48.—*Stylonychia mytilus* (after Stein; from Claus' *Zoology*) seen from the ventral surface. W^s Adoral zone of cilia; *C* contractile vacuole; *N* nucleus; N^1 nucleolus; *A* anus.

b. **Cells with numerous Cilia.** The Infusoria are chiefly to be distinguished from other unicellular organisms by *the large number of cilia* they possess, on which account they are called *Ciliata* (Fig. 48). Cilia are much smaller than flagella, being, as a rule, about ·1 to ·3 μ thick, and about 15 μ long. They may number many thousands. For example, it has been calculated, that the *Paramœcium aurelia* possesses approximately 2,500. As for the *Balantidium elongatum*, which is parasitic in the Frog, and which is very thickly ciliated, Bütschli (III. 3) is of opinion that it has nearly ten thousand cilia; these are generally arranged in several longitudinal rows, which either encircle the body in spirals, or are confined to a certain portion of its surface.

In addition to the cilia, many Infusoria possess special large organs of locomotion, cirri, and undulating membranes. The former may be distinguished from cilia by their greater thickness and length, and by the fact that they are somewhat wide at the base, whilst they taper off to a fine point (Fig. 48). Further, like other special contractile tissues (muscular fibres), they exhibit a fibrillar differentiation, so that they may be split up into many delicate fibrils (Bütschli). These cirri occur with especial frequency in hypotrichous Infusoria, being situated chiefly around the mouth. The undulating membranes also terminate at the mouth cavity. They are locomotive organs which have been developed superficially; they may frequently be seen to be distinctly marked with delicate striæ extending from the base to the free edge, and hence they, like the cirri, must possess a fibrillar structure.

Infusoria have various methods of locomotion. As a rule the body, when it moves freely through the water, revolves about its longitudinal axis. It has the power of changing the direction in which it travels; the rate at which the cilia move may suddenly be altered, being either slowed or quickened; the body may even keep still for a short time, without any apparent external cause. Hence various kinds of movements take place, suggesting the idea of volition. In addition, it is remarkable that the cilia, often thousands in number, of one and the same individual, always act together in a strictly co-ordinate fashion. "They do not only always oscillate at the same rate, and with the same amplitude of beat (rhythm), but they always strike the water in the same direction, and in the same order" (Verworn). This co-ordination is carried out to such an extent, that two individuals which have been produced by the division of a parent cell always exhibit uniform and synchronous movements as long as they are united by a bridge of protoplasm. Hence it follows, that although the cilia possess the power of spontaneous contraction, yet their working together is regulated by stimuli from the protoplasmic body itself.

The ectoplasm seems to play an especially important part in the transference of these stimuli, as is shown by an experiment made by Verworn (IV. 40). He made a slight incision with a lancet in *Spirostomum ambiguum* (Fig. 49) and in *Stentor cœruleus* in the ectoplasm supporting the rows of cilia. "Under these circumstances it could be plainly seen that the ciliary waves did not cross the area of the incision, but were confined to the one side, and could not be seen on the other." Occasionally also he observed that the mean position through which the cilia oscillated was different for a time in one half of the rows of cilia from that seen on the other side.

III. The Contractile Vacuoles, or Vesicles, of Unicellular Organisms.

Contractile vacuoles occur very frequently in Amœbæ, Reticularia, Flagellata (Figs. 7, 43, 44), and Ciliata (Fig. 50 *cv*). In the last, where they have been most accurately examined, there is generally only one single vacuole in the whole body; occasionally two are present (Fig. 50), or rarely a few more; they are always situated just below the surface of the body, under the ectoplasm. They may be easily distinguished from the other fluid vacuoles, of which large numbers may be distributed throughout the body, by the fact, that at regular intervals they discharge their contents to the exterior, and then gradually fill up again. Hence they temporarily disappear (Fig. 50 *cv*) to reappear again in a short time (cv^1).

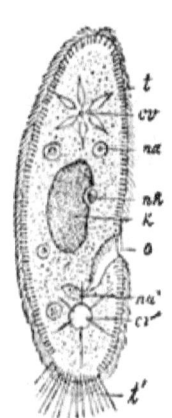

FIG. 50.—*Paramœcium caudatum* semi-diagrammatic (R. Hertwig, Zoologie, Fig. 139): *K* nucleus; *nk* secondary nucleus; *o* mouth aperture (cytostome); na^1 food vacuole in process of formation; *na* food vacuole; *cv* contractile vacuole, contracted; cv^1 the same contractile vacuole, distended; *t* trichocysts, t^1 the same with their threads ejected.

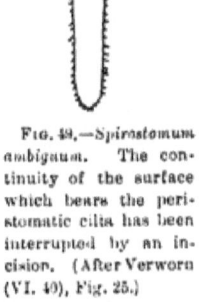

FIG. 49.—*Spirostomum ambiguum*. The continuity of the surface which bears the peristomatic cilia has been interrupted by an incision. (After Verworn (VI. 40), Fig. 25.)

The evacuation takes place through one or more special pores, which can be observed on the surface of the infusorian immediately over the vacuole. "Each pore appears as a rule as a minute circle, the border of which is dark, but which is bright in the centre; this brightness of the centre is due to the refractive power of the pellicular and alveolar layer. Sometimes each pore is connected to the vacuole by means of a fine excretory tube. In addition, it is not uncommon to find special conducting canals (1, 2, or more) regularly arranged in its neighbourhood. In

Paramæcium aurelia and *Paramæcium caudatum* (Fig. 50), there is a system of conducting canals, which have been known for a long time, and have been worked at more than any others; from each of the two dorsal vacuoles about eight to ten fairly straight tubes radiate; their course may be traced almost all over the whole body. However, the two systems remain independent throughout their whole extent." They are thickest in the neighbourhood of the vacuoles, becoming gradually finer distally.

The *Paramæcium* affords us an excellent subject for a closer study of the working of this peculiar apparatus. When both the contractile vacuoles have attained their greatest size, their whole contents are suddenly and energetically ejected to the exterior through their efferent canals and pores, so that for a time the vacuole cavities quite disappear. This condition, as with the heart, is termed the *systole*, whilst the period during which the vacuoles become again filled with fluid, and hence distended and visible, is called the *diastole*.

They become filled in the following manner: Even before the systole has commenced, the above-described conducting canals have collected fluid from the endoplasm of the body of the infusorian; this fluid probably is charged with carbonic acid and other decomposition products. According to Schwalbe (III. 21) the process occurs in consequence of " the condition of pressure of the fluid in the animal's body, this pressure being due to the ever-increasing amount of water which is continually being taken in by the mouth." The conducting canals can be easily seen, at this time being full of water. They become swollen in the neighbourhood of the contractile vacuole, which is now fully distended, so that they look like a circle of rosette-shaped vacuoles surrounding it; these have been called formative vacuoles by Bütschli. In consequence of their being in this condition, the contractile vacuole cannot, during its systole, discharge its contents back through them, but only forwards to the exterior. As soon as the diastole again occurs, the distended formative vacuoles empty themselves into the contractile vacuole, which in consequence becomes visible again; it then gradually distends itself until it reaches its maximum size. Hence at the commencement of the diastole the emptied formative vacuoles disappear for a time; however, they continue to collect fluid from the parenchyma of the body until the commencement of the next systole.

When several vacuoles are present they generally empty themselves in turn, with the result that the water is discharged as regularly as possible. The frequency with which these evacuations take place varies considerably in different species. According to the observation of Schwalbe (III. 21) the following law may be stated: that the smaller the vacuoles are, the more frequently are they emptied. For instance, in *Chilodon cucullulus* they contract about 13 to 14 times in two minutes, in *Paramæcium aurelia*, only 10 or 11 times in the same period, whilst in *Vorticella microstoma*, only once or twice. In *Stentor* and *Spirostorum* the contractions occur less frequently still. Of all the above-mentioned animals, the two last have the largest contractile vacuoles, next comes *Vorticella*, then *Paramæcium aurelia*, and lastly *Chilodon cucullulus*, whose vacuoles are only half as large in diameter as those of *Paramæcium*, where the diameter is about ·0127 mm.; in *Vorticella* it is ·0236 mm (Schwalbe).

The interval which elapses between the two evacuations is very regular at the same temperature; it is, however, considerably affected if the temperature is raised or lowered (Rossbach III. 19, Maupas). For instance, with *Euplotes charon*, the interval between the contractions is 61 seconds; at 30° Celsius, it has diminished to 23 seconds (Rossbach); that is to say, the frequency has become nearly trebled.

The amount of water which in this manner passes through the animal is extremely great. According to the computations of Maupas, *Paramæcium aurelia*, for example, evacuates, in 46 minutes at 27° Celsius, a volume of water equal to its own volume.

From the above-mentioned observations, it may be concluded that *contractile vacuoles are not merely simple variable drops of water in the plasma, but that they are permanent morphological differentiations in the body of the Protozoon; that is to say, true cell organs, which appear to perform an important function in the carrying on of breathing and excretion.* The energy with which the vacuole discharges its contents, so that it completely disappears, indicates that its walls, which consist of hyaline substance resembling the flagellum substance, must be contractile to an exceptional degree, and by means of this property are to be distinguished from the endoplasm of the infusorian body. It must, however, be admitted that no special membrane, clearly defined from the remainder of the body mass, can be seen microscopically, just as with smooth muscle fibres the contractile substance and the protoplasm are not sharply defined from one another, and further as flagella pass over imperceptibly at their base into the protoplasm of the cell.

Therefore I agree with Schwalbe (III. 21) and with Engelmann, that the vacuoles possess contractile walls although they are not clearly defined from the rest of the protoplasm. In addition, it is well known that delicate membranes are often imperceptible with the microscope although they are undoubtedly present. In many plant cells it is impossible to see the so-called primordial utricle as long as it adheres closely to the cellulose membrane; its existence, however, cannot be doubted, as its presence can be proved by plasmolysing it.

In this opinion, however, I find myself in opposition to Bütschli (III. 3). He considers that the contractile vesicle is simply a drop of water in the plasma. "Each vacuole after evacuation ceases to exist as such. The one that takes its place is a new formation, a newly created drop, which in its turn only exists until it has discharged itself." In his opinion they are due to the flowing together of several formative vacuoles, which separate out as small drops in the plasma, where they increase in size until, by breaking down the partition walls, they coalesce. However, the existence of the conducting and afferent canals, described by Bütschli himself, the fact that the number of vacuoles present remains constant, and the circumstance that during the diastole the vacuole is seen to occupy the same position as during the systole, and moreover, that the frequency of contraction bears a fixed relation to changes of temperature, all appear to me to support the former view, and to be opposed to Bütschli's theory. The fact that at the end of the systole the vacuole, having evacuated its contents, is for a moment invisible, does not seem to weigh much against the theory of its constancy, especially if one considers that even large lymph spaces and capillary blood vessels in vertebrates elude perception in an uninjected condition.

IV. **Changes in the Cell during passive movement.** In order to complete the subject of the movements of protoplasm, it is necessary to consider finally the changes of form which, to a certain extent, the cell may experience in consequence of *passive movements*. Under these circumstances, the cell is in the same condition as a muscle which, being excited by an external stimulus, becomes extended and then contracted again.

In this manner the cells of an animal body may become considerably altered in form, in adapting themselves to all the changes of shape which an individual organ experiences as a consequence of muscular action or of distension through a collection of fluid or nutriment. Thread-like epithelial cells have to become cylindrical, and cylindrical ones to become flat, when the surface increases in size through the distension of an organ, whilst, on the other hand, the reverse takes place when the whole organ, including its surface, decreases in size.

How powerful and sudden may be the changes of form which the protoplasm of a cell, in consequence of passive movement, may experience without damage to its delicate structure, can be best seen in Cœlenterata, in which extended portions of the body, like palpocils, may suddenly shorten by about a tenth or more of their length, in consequence of sudden energetic muscular contraction (Ill. 12 a). The form which an epithelial cell assumes varies very considerably, according as to whether it has been taken from a portion of a body which is moderately or strongly contracted, as may be seen by comparing Fig. 51 A, B. The former was taken from the tentacle of an Actinia, which was only moderately contracted,

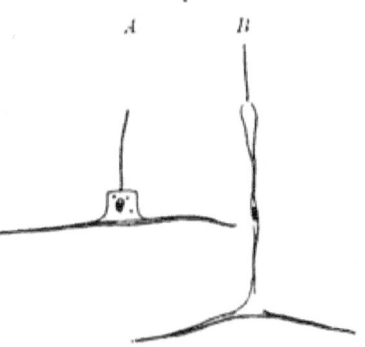

Fig. 51.—Muscular epithelial cell from the endodermal surface of the tentacle of an Actinia (Sagartia parasitica) (after O. and R. Hertwig, Pl. VI., Fig. 1); from Hatschek, Fig. 10a): A extended condition of tentacle; B strongly contracted condition of same.

since by means of chemical reagents it had been rendered nonsensitive before it was killed; the second was derived from the tentacle of another individual which had contracted strongly in death.

Literature III.

1. DE BARY. *Die Mycetozoen. Zeitschrift f. wissenschaftl. Zoologie.* Bd. 10. 1860.
2. G. BERTHOLD. *Studien über Protoplasmamechanik.* Leipzig. 1886.
3. BÜTSCHLI. *Protozoen. First Volume of Bronn's "Classen und Ordnungen des Thierreichs."* 1889.
4. ALEX. ECKER. *Zur Lehre vom Bau u. Leben der contractilen Substanz der niedersten Thiere. Zeitschrift f. wissenschaftl. Zoologie.* Bd. I. 1849.
5. ENGELMANN. *Protoplasm and Ciliary Movement,* trans. by Bourne from Hermann's *"Handbuch der Physiologie."* Bd. I. *Quar. Jour. Mic. Soc.* 1880.

6. ENGELMANN. *Contractilität und Doppelbrechung.* Archiv. f. die gesammte Physiologie. Bd. XI.
   See also E. A. SCHÄFER. *On the Structure of Amœboid Protoplasm, etc., with a Suggestion as to the Mechanism of Ciliary Action.* Proc. Roy. Soc. 1891.
   J. CLARK. *Protoplasmic Movements and their Relation to Oxygen Pressure.* Proc. Roy. Soc. 1889.
7. ENGELMANN. *Ueber die Bewegungen der Oscillarien und Diatomen.* Pflügers Archiv. Bd. XIX.
8. ENGELMANN. *Ueber die Flimmerbewegung.* Jenaische Zeitschrift f. Medicin und Naturwissenschaft. Bd. IV. 1868.
9. FROMMANN. *Beobachtungen über Structur und Bewegungserscheinungen des Protoplasmas der Pflanzenzelle.* Jena. 1880.
10. FROMMANN. *Ueber neuere Erklärungsversuche d. Protoplasmaströmungen u. über Schaumstructuren Bütschli's.* Anatom. Anzeiger. 1890.
11. HENSEN. *Physiologie der Zeugung.* Handbuch der Physiologie. Bd. IV. 1881.
12A. O. and R. HERTWIG. *Die Actinien.* Jena. 1879.
12B. RICHARD HERTWIG. *Ueber Mikrogromia socialis, eine Colonie bildende Monothalamie des süssen Wassers.* Archiv. f. mikroskop. Anat. Bd. X. 1874.
13. JÜRGENSEN. *Ueber die in den Zellen der Vallisneria spiralis stattfindenden Bewegungserscheinungen.* Studien des Physiol. Instituts zu Breslau. 1861. Heft I.
14. KLEBS. *Form und Wesen der Pflanzlichen Protoplasmabewegung.* Biologisches Centralblatt. Bd. I.
15. KOLLMANN. *Ueber thierisches Protoplasma.* Biol. Centralblatt. Bd. II.
16. C. NÄGELI. *Die Bewegung im Pflanzenreiche.* Beiträge zur wissenschaftlichen Botanik. Heft II. 1860.
    NÄGELI. *Rechts und links. Ortsbewegungen der Pflanzenzellen und ihre Theile.*
17. G. QUINCKE. *Ueber periodische Ausbreitung an Flüssigkeitsoberflächen u. dadurch hervorgerufene Bewegungserscheinungen.* Sitzungsber. der Akademie der Wissenschaften zu Berlin. 1888.
18. PURKINJE u. VALENTIN. *De phaenomeno generali et fundamentali motus vibratorii continui.* 1835.
19. ROSSBACH. *Die rhythmischen Bewegungserscheinungen der einfachsten Organismen und ihr Verhalten gegen physikalische Agentien u Arzneimittel.* Arbeiten a. dem. zool. zoot. Institut zu Würzburg. 1874.
20. SACHS. *Experimentalphysiologie der Pflanzen.* Leipzig. 1865.
21. SCHWALBE. *Ueber die contractilen Behälter der Infusorien.* Archiv. für mikroskopische Anatomie. Bd. II.
22. VELTEN. *Einwirkung strömender Elektricität auf die Bewegung des Protoplasmas, etc.* Sitzungsber. d. Wiener Akademie. 1876. Bd. 73.
23. VERWORN. *Studien zur Physiologie der Flimmerbewegung.* Pflügers Archiv. Bd. 48. 1890.
24. VERWORN. *Die Bewegung der lebendigen Substanz.* Jena. 1892.
25. DE VRIES. *Ueber die Bedeutung der Circulation und der Rotation des Protoplasmas für den Stofftransport in der Pflanze.* Botanische Zeitung. 1885.

CHAPTER IV.

THE VITAL PROPERTIES OF THE CELL (*continued*).

Phenomena of Stimulation. The most remarkable property of protoplasm is its power of reacting to stimuli:—its *Irritability*.[1]

By this is understood, as Sachs (IV. 32a) expresses it, "the power possessed by living organisms alone of reacting to the most various external stimuli in one way or another." It is chiefly through this irritability that living objects can be distinguished from non-living ones, and in consequence the earlier natural philosophers considered that it was the expression of a special *vital force* which was only to be seen in organised nature.

Modern science has discarded the *theory of vitalism* (vitalismus); instead of explaining irritability by means of a special vital force, it is considered to be a very complicated chemico-physical phenomenon, differing only in degree from other chemico-physical phenomena of inanimate nature. That is to say, the external stimuli come into contact with a substance very complex in structure, an organism, which is an exceedingly complicated material system, and in consequence they give rise to a series of very complex phenomena.

However, care must be taken in accepting this mechanical conception not to fall into the very common mistake of trying to explain vital processes as being due *directly to mechanical causes*, in consequence of their analogy to many phenomena seen in

---

[1] Claude Bernard (IV. 1a), in his lectures on vital phenomena, arrives at the same conclusion, his opinion being based on a number of considerations: "Arrivés au terme de nos études, nous voyons qu'elles nous imposent une conclusion très générale, fruit de l'expérience, c'est, à savoir, qu entre les deux écoles qui font des phénonèmes vitaux quelque chose d'absolument distinct des phénonèmes physico-chimiques ou quelque chose de tout à fait identique à eux il y a place pour une troisième doctrine, celle du vitalisme physique, qui tient compte de ce qu'il y a de spécial dans les manifestations de la vie et de ce qu'il y a de conforme à l'action des forces générales : l'élément ultime du phénomène est physique; l'arrangement est vital!"

1. Any change in external conditions is a stimulation.

inanimate objects. It must never be forgotten that there is no substance in inanimate nature which remotely approaches the living cell for complexity of structure, and that hence the reactions of such a substance are of necessity correspondingly complex in character.

The field of the phenomena of irritability is exceedingly wide, since it embraces all the correlations which take place between the organism and the outer world. The stimuli which act upon us from without are innumerable. For the sake of clearness, we will consider them under five heads: (1) thermal stimuli, (2) light stimuli, (3) electrical stimuli, (4) mechanical stimuli, (5) the almost infinite variety of chemical stimuli.

The manner in which an organism responds to one of these stimuli is called its reaction. This may vary very considerably with different individuals even when they are exposed to the same stimulus. It depends entirely upon the structure of the organism, or upon its finer properties, although these may not be perceptible to us. Different organisms, to use a simile of Sachs (IV. 32a), may in this respect be compared with variously constructed machines, which, when set in motion by the same external force, heat, produce different useful effects according to their internal structures. *Similarly, the same stimulus may produce quite different effects in different organisms, according to their specific structure.*

We shall see later on that many protoplasmic bodies are to a certain extent attracted, whilst others are repelled, by light; a similar difference will be seen when the action of chemical reagents, etc., on protoplasm is studied. The terms *positive and negative heliotropism, positive and negative chemotropism, galvanotropism*, and *geotropism* are used to describe these varying effects.

Another phenomenon, in some respects the exact opposite of the ones described above, must also be explained by the varying specific structure of the stimulated substance; the term *specific energy* has been used to describe this phenomenon. Whilst, as described above, we see that protoplasmic bodies, differing in structure, react in various ways to the same stimulus, we find, on the other hand, that similar effects are produced upon the same protoplasmic body by very different stimuli, such as light, electricity, or mechanical injury.

A muscle cell responds to all kinds of stimuli by contracting, a gland cell by secreting; an optic nerve can only experience the sensation of light, whether stimulated by light waves, electricity, or pressure. Similarly, as Sachs has pointed out, plant cells also are furnished with their specific energies. Tendrils and roots bend themselves in a manner peculiar to themselves, whether stimulated by light, gravitation, pressure, or electricity. *The effect of a stimulus bears the specific stamp, so to speak, of the special structure of the stimulated substance, or, in other words, irritability is a fundamental property of living protoplasm, but it manifests itself in specific actions according to the specific structure of the protoplasm under the influence of the external world.*

The same idea is expressed by Claude Bernard (IV. 1a) in the following words: "La sensibilité, considerée comme propriété du système nerveux, n'a rien d'essentiel ou de spécifiquement distinct; c'est l'irritabilité spéciale au nerf, comme la propriété de contraction est l'irritabilité spéciale au muscle, comme la propriété de sécrétion est l'irritabilité spéciale à l'élément glandulaire. Ainsi, ces propriétés sur lesquelles on fondait la distinction des plantes et animaux ne touchent pas à leur vie même, mais seulement aux mécanismes par lesquels cette vie s'exerce. Au fond tous ces mécanismes sont soumis à une condition générale et commune, l'irritabilité."

In speaking generally of irritability, another peculiar phenomenon deserves especial attention, namely the *transmission or conduction of stimuli*. If a small portion of the surface of a protoplasmic body is stimulated, the effect produced is not limited to this point alone, but extends to far outlying ones. Hence the changes produced by the stimulus at the point of contact must be more or less quickly shared by the rest of the body. Stimuli, as a rule, are more quickly transmitted in animal than in vegetable bodies; in human nerves, for example, the rate is 34 metres per second; it is always slower in plant protoplasm.

We imagine that the substance which is capable of receiving stimuli forms a system of exceedingly elastic particles in a condition of unstable equilibrium. In such a system it is sufficient for one of the particles to receive a slight shock, in order to set all the others in motion, since each transmits its movement to another. This theory explains the phenomenon, that exceedingly great effects are often produced by very slight stimuli, just as a small spark, by setting on fire a single grain of powder, may cause a powder magazine to explode.

Finally, another peculiarity of organic matter is its capacity of returning more or less completely to its original condition, after a period, varying in length, of rest or recuperation has elapsed since the cause of irritation was removed. I say advisedly more or less completely, for very often the organic substance is permanently altered in its structure and reacting powers by the application, for a considerable period, of a stimulus, or by the repeated action of the same stimulus. The phenomena thus produced are spoken of as the *after-effects of stimulation*.

As a rule, we are not in a position to determine whether or no a body can be stimulated, that is to say, whether it reacts to changes in its environment, since *most of the effects due to stimulation are imperceptible to us*. Sometimes the protoplasm responds by exhibiting movements, or by striking changes of form; but, as has been just remarked, such phenomena constitute only a small and limited portion of the results produced, although naturally they are the most important to the investigator, since they are apparent to his perception. In consequence, in the following pages, we will chiefly consider the way in which protoplasm responds, by means of movements, to the stimuli, which have been grouped into the above five classes. I have therefore decided to commence my considerations of the vital properties of the elementary organism with contractility.

I. **Thermal Stimuli.** One of the essential conditions for the vital activity of protoplasm is the temperature of its environment. This temperature can only vary between certain fixed limits; if it oversteps either of these, the protoplasm invariably dies immediately. These limits, it is true, are not the same for all protoplasmic bodies; some are able to withstand extremes of temperature better than others.

The maximum temperature for plants and animals is generally about 40° C. Exposure for a few minutes to such a temperature suffices to cause the protoplasm to swell up and become coagulated, and thereby its irritable structure and its life are destroyed. If an *Amœba* is placed in water at 40°, it dies immediately; it draws in its pseudopodia and "converts itself into a globular vesicle, whose sharply defined double contour encloses a large, turbid mass which, by transmitted light, looks brownish in colour" (Kühne IV. 15). The same temperature causes "death from heat" in *Æthalium septicum*, coagulation being induced. In *Actinophrys*, however, instantaneous death occurs at a temperature of 45°, whilst the cells of *Tradescantia* and *Vallisneria* are only killed by a temperature of 47–48° C. (Max Schultze I. 29).

The protoplasm of organisms which live in hot springs is able to sustain much higher temperatures. Cohn found specimens of *Leptothrix* and *Oscillaria* in the Karlsbad springs at 53° C., whilst Ehrenberg observed *Alyx* in the warm springs of Ischia.

But even in these cases we have not arrived at the extreme limit of heat which can be sustained for a time by living substance. For endogenous spores of *Bacilli*, which are furnished with unusually resistent envelopes, remain capable of germination after they have been heated for a short time in a liquid at a temperature of 100°. Many even can endure 105-130° (de Bary IV. 5*b*, p. 4). It is only after a substance has been exposed to the action of dry heat of 140°, for a period of three hours that we can assume with certainty that all life has been completely destroyed in it.

It is even more difficult to determine the lower limit at which "death from cold" occurs. As a rule, temperatures below 0° are less injurious to protoplasm than high ones. It is true that if the eggs of *Echinodermata*, which are about to divide, are placed in a freezing mixture at a temperature of from 2° to 3° C., the process of division is temporarily arrested (IV. 12); but division recommences and proceeds in a normal fashion when the eggs are slowly warmed, even if they have been kept in the freezing mixture for a quarter of an hour. Indeed, the greater number of the eggs are found to be uninjured even if they have been kept at this temperature for two hours. Plant-cells may be frozen until ice crystals develop in the sap, and yet, after they have been thawed, they exhibit the streaming movements of protoplasm (IV. 15).

Sudden exposure to temperatures below zero produces striking changes of form in the protoplasm of plants; however, it reverts to its normal condition on being thawed. When Kühne (IV. 15) examined in water cells of *Tradescantia*, which had been kept for a little more than five minutes in a freezing mixture at 14° C., he found, in the place of the ordinary protoplasmic net, a large number of isolated, round drops and globules. These, after a few seconds, began to show active movements, and in a few minutes commenced to join themselves one to another, and thus to gradually become reconstructed into a network, in which active streaming movements soon commenced to take place.

Kühne describes in the following words another experiment:—
"If a preparation of *Tradescantia* cells is kept for at least one hour in a space which is maintained by means of ice at a tempera-

ture of 0°, the protoplasm is found to exhibit an inclination to break up into separate drops. Even where the network still persists, it is composed of extremely fine threads, which are studded here and there with large globules and drops; several other globules float about freely in the cell fluid, in which, without moving much from place to place, they revolve about their own axes with active, jerking movements. After a few minutes, the free globules are seen to unite themselves to the delicate threads, or to amalgamate themselves with some of the globules hanging on to the threads, until the appearance of the streaming protoplasmic network is quite restored."

In plants, as a rule, their power of resistance to cold is inversely in proportion to the amount of water they contain; seeds which have dried in the air, and winter-buds, the cells of which consist almost entirely of pure protoplasm, can withstand intense cold, whilst young leaves, with their sap-containing cells, are killed even by frosty nights. However, the power of resistance to cold varies according to the specific organisation of different plants, or rather of their cells, as is proved by daily experience (Sachs IV. 32B).

Micro-organisms are able to resist exceedingly low temperatures. Frisch has discovered that the spores, and indeed the vegetative cells of the *Anthrax bacillus* do not lose their capacity of development by being cooled down in a liquid to a temperature of $-110°$ C., from which they were extracted after it had been thawed.

Before reaching the above-mentioned extreme temperatures, at which death by heat or cold is produced, phenomena known as heat rigor or heat tetanus, and cold rigor, occur; when the protoplasm is in either of these conditions, all the attributes which show it to be alive, especially those of movement, are arrested so long as the temperature in question is maintained; but when this is either raised or lowered, as the case may be, after a period of rest, they again manifest themselves.

Cold rigor generally occurs at a temperature of about $0°$ C., whilst heat rigor sets in at a temperature only a few degrees lower than that at which immediate death results; in both cases the protoplasmic movements become gradually slower and slower, until at last they quite cease. *Amœbæ, Reticularia*, and white blood corpuscles draw in their pseudopodia and become converted into globular masses. Most plant cells assume the appearance described above by Kühne. If the temperature is either slowly raised or lowered, as the case may be, the vital appearances gradually become normal. It is true that if the condition of rigor produced by cold is maintained for a considerable time, death may ensue, although cold is better withstood, and for a longer time, than heat. When the protoplasm dies it becomes coagulated and turbid, whilst commencing to swell up and to decompose. At the temperatures lying between these extremes, the vital processes are performed in a manner which varies in intensity with the degree of temperature. This is especially true of the movements which take place at different speeds, increasing in rate up to a certain point, as the temperature rises, until they reach a certain fixed maximum speed. This occurs at the so-called *optimum temperature*, which is always several degrees below that at which heat rigor is produced. As the temperature passes this limit, the protoplasmic movements are seen to slacken, until at last rigor sets in.

White blood corpuscles have been much used in studying the effects produced by heat; for this purpose Max Schultze's warm stage, or Sachs' warm cells, are most suitable. In a fresh drop of blood the corpuscles are seen to be motionless and globular in form. If the drop is warmed—the necessary precautions being of course observed—the corpuscles gradually commence to extend pseudopodia, and to move about. As the temperature approaches the optimum for the time being, these changes of shape become more rapid. In *Myxomycetes*, *Rhizopoda*, and plant cells, the effect produced by an access of heat is exhibited by an increase of rapidity of the streaming movements of the granules. Thus, according to the measurements of Max Schultze (I. 29), the granules in the hair-cells of *Urtica* and *Tradescantia* travel at ordinary temperatures at a rate of ·004-·005 mm. per second, whilst if the temperature is raised to 35° C., their speed is increased to ·009 mm. per second. In *Vallisneria* the rate of circulation may be increased to ·015 mm., and in a species of *Chara* even to ·04 mm. per second. The difference between the slow and accelerated movements may be so great that whilst with the former the length of a foot is traversed in fifty hours, with the latter the same distance may be covered in half an hour.

Nägeli (III. 16) has expressed the acceleration produced by an accession of heat in the granular streaming movements in the cells of *Nitella* by the following figures: in order to traverse a distance of ·1 mm. the granules require 60 seconds at 1° C.; 24 seconds at 5° C.; 8 seconds at 10° C.; 5 seconds at 15° C.; 3·6 seconds at

20° C.; 2·4 seconds at 26° C.; 1·5 seconds at 31° C.; and ·65 seconds at 37° C. From these figures it is apparent that "each consecutive degree of temperature produces a corresponding slight acceleration" (Nägeli, Velten).

Finally, it is necessary to mention the remarkable behaviour of protoplasm towards sudden great fluctuations of temperature, and also towards partial or uneven heating.

Fluctuations of temperature may be either positive or negative, that is to say, they may be caused by a raising or a lowering of temperature. The consequence of a violent thermal stimulation is a temporary cessation of all movements. However, after a time, the motion recommences at a rate corresponding to the temperature (Dutrochet, Hofmeister, de Vries). The accuracy of these observations has been questioned by Velten (IV. 38). According to his experiments, fluctuations of temperature between the necessary limits produce neither a cessation nor a slackening of the protoplasmic movements, which, on the contrary, immediately proceed at a rate corresponding to the temperature which has been attained.

Stahl (IV. 35), in his experiments upon the plasmodia of *Myxomycetes*, has made some very interesting discoveries concerning the effects produced by *partial heating*. If a portion of such a plasmodium, which has spread its network out over an even surface, be cooled, the protoplasm is seen to travel gradually from the cooler to the warmer part, so that the one portion of the network is seen to shrink up, whilst the other becomes swollen. The experiment may be conducted in the following manner: Two beakers, one filled with water at 7°, and the other with water at 30°, are placed quite close to one another; a wetted strip of paper over which a plasmodium has spread itself is then placed over their contingent edges, so that one of its ends dips into each beaker; the temperature of the water in the beakers is not allowed to vary. After a time the plasmodium, by stretching out and drawing in its protoplasmic thread, succeeds in creeping over to the medium which is best adapted to it.

No doubt most free-living protoplasmic bodies move somewhat in this fashion, for as a rule their movements are regulated by expediency, that is to say, they take place in order that the life of the organism may be maintained. For instance, flowers of tan sink down during the autumn to a depth of several feet into the warmer layers of the tan, in order to pass the winter there.

Then during the spring, as the temperature rises, they move in an opposite direction, ascending to the warmer superficial layers.

II. **Light Stimuli.** In many cases light, like heat, acts as a stimulus to animal and plant protoplasm. It induces characteristic changes of form in individual cells, and causes movements in fixed directions in free-living unicellular organisms. Botanists have obtained especially interesting results in this department.

The plasmodia of *Æthalium septicum* only spread themselves out on the surface of the tan in the dark; in the presence of light they sink down below the surface. If a small pencil of light is allowed to fall upon a plasmodium which has spread its network upon a glass slide, the protoplasm is immediately seen to stream away from the illuminated portion, and to collect in the parts which are in shadow (Barenezki, Stahl IV. 35).

Pelomyxa palustris, an organism resembling the *Amœba*, is actively motile in shadow, extending and protruding broad pseudopodia. If a fairly powerful ray of light impinges upon it, it suddenly draws in all its pseudopodia, and transforms itself into a globular body. Only after it has rested quietly in the shade for a time does it gradually recommence its amœboid movements. "If, on the other hand, daylight is admitted gradually during a period of rather less than a quarter of an hour, no effects of stimulation are to be perceived; this is also the case when, after a prolonged illumination, the light is suddenly withdrawn" (Engelmann IV. 6 b).

The star-shaped pigment cells of many invertebrates and vertebrates, which have been described under the name of *chromatophores* (IV. 3, 29, 30, 33), react very actively to light; they are the cause of the changes of colour so often seen in many Fishes, Amphibians, Reptiles, and Cephalopods. For example, the skin of a Frog assumes a lighter shade of colour when under the influence of light. This is due to the fact that the light causes the black pigment cells, which extend their numerous processes through the thick skin, to contract up into small black points. In addition, as they become less prominent, the green and yellow pigment cells, which do not contract, become more easily seen.

Further, the *pigment cells of the retina* become considerably altered in form under the influence of light, both in vertebrates (Boll) and in invertebrates, for instance in the eyes of *Cephalopoda* (Rawitz IV. 31).

It is a well-known fact that many unicellular organisms which propel themselves by means of cilia or flagella, such as *Flagellata*, *Ciliata*, the swarm-spores of *Algæ*, etc., prefer to collect either on that side of the cultivation dish which is nearest the window, or on the one which is away from it.

This may be easily proved by means of a simple experiment described by Nägeli (III. 16). A piece of glass tubing three feet in length is filled with water containing green swarm-spores of *Algæ* (tetraspores), and is placed perpendicularly. Then, if the upper part of the tube is covered with black paper, and light is allowed to fall upon the lower portion, it is seen after a few hours that all the spores have collected in this lower portion, leaving the upper part colourless. If now the upper portion is uncovered, and the paper is transferred to the lower part, all the swarm-spores ascend the tube, and collect on the surface of the water.

Euglena viridis is exceedingly sensitive to light (Fig. 44 A, IV. 8). If a drop of water containing *Euglenæ* is placed upon a slide, and only a small portion of it is illuminated, all the individuals collect in this area, which, to quote an expression of Engelmann's, acts like a trap. This organism is especially interesting, because the perception of light is restricted to a definite portion of the body. Each *Euglena* consists of two portions, a large posterior one containing chlorophyll, and a colourless anterior, flagella-bearing one, in which there is a red pigment spot. Now, it is only when this anterior portion comes into contact with light, or is placed in shadow, that the organism is seen to react by altering the direction of its movements (Engelmann). Hence, in this case, a certain part of the body functions to a certain extent as an eye.

Stahl (IV. 34) and Strasburger (IV. 37) have investigated most fully the action of light upon *swarm-spores*. The former sums up his results in the following words:—"Light effects an alteration in the direction of the movements of swarm-spores by causing them to make their longitudinal axes coincide approximately with the light. The colourless flagellated end may be directed either towards or away from the source of light. Either position may become exchanged for the other under otherwise unaltered external conditions, and, indeed, this occurs at very different degrees of light intensity. The intensity has the greatest influence over relative positions. When the light is very intense, the anterior end is directed away from the source; when it is less strong, the swarm-spores move towards the light."

This sensitiveness towards light varies considerably both in different species and in individual members of the same species; indeed, even in the same individual, considerable differences may be seen under different external conditions. This varying power of reaction in swarm-spores has been called phototonus or light-tone by Strasburger.

Swarm-spores of the *Botrydium* and *Ulothrix*, which react somewhat differently under the influence of light, are very suitable for experiments on this subject.

If some swarm-spores of *Botrydium* are placed in a drop of water upon a coverglass, and are kept in shadow, they spread themselves out evenly in the water. If a light is allowed to fall on them, they are seen to immediately direct their anterior ends towards the source of light, and to hurry in fairly parallel paths towards it. After a short time, at most from one and a half to two minutes, almost all of them have collected at the illuminated side of the drop, which, for the sake of brevity, Strasburger has named the positive edge, to distinguish it from the opposite or negative edge. Here they are seen to intermingle and to conjugate in large numbers. If the slide is now turned round through an angle of 180°, all the spores which are still capable of movement immediately forsake the edge of the drop, which is now turned away from the light, and hasten back towards the light. If the microscope is fitted with a rotating stage, it is possible by turning the latter to make the swarm-spores continually keep changing their course. They always travel in a straight line towards the light.

Ulothrix zoospores behave in a somewhat different manner. "These also travel quickly, and in approximately straight paths towards the positive edge of the drop; however, as a rule, they do not all move in this manner; on the contrary, it is generally the case that a larger or smaller number of individuals in each preparation are seen to move rapidly in the opposite direction, that is to say, towards the negative edge. A most peculiar spectacle is thus produced, for the spores, since they go in opposite directions, appear to travel at double speed as they pass each other. If the preparation is turned through an angle of 180°, the spores which had collected on the side which was positive are seen to hasten to the other edge, whilst the others, which were collected on the side which was negative, travel in the opposite direction, and having arrived at their destination, they commence to move about amongst themselves, keeping more or less close to the edge of the drop, according to the condition of the preparation. Continually, individual spores are seen to suddenly forsake the side, either positive or negative, at which they were stationed, and to hurry through the drop to the opposite one. Such an exchange is continually taking place between the two sides. Indeed, it frequently occurs that certain individuals, which have just left one side and arrived at the other, hasten back to the one from which they originally came. Others become arrested in the middle of their course, and then return to their starting-point, in order eventually to oscillate backwards and forwards for a considerable time like a pendulum."

The following experiment, described by Strasburger, shows how sensitively and quickly the zoospores react to light:—" If a piece of paper is placed between the microscope and the source of light, just as the zoospores are on their way from one edge of the drop to the other, they immediately turn to one side, many rotating in a circle; this, however, only lasts for a moment, after which they continue to move in the same direction as before (interruption movements)." *Strasburger* (IV. 37) *has named those zoospores which hasten towards the source of light light-seeking* (*photophylic*), *and those which travel from it light-avoiding* (*photophobic*).

As has been already remarked, the way in which the spores collect at one or other side of the drop, thus indicating their special *kind of phototonus*, depends upon *external circumstances*, such as the intensity of the light, the temperature, the aëration of the water, and their condition of development.

It is possible to entice spores, which under intense illumination have collected on the negative side, to come over to the other side. The intensity of the light must be gradually diminished in proportion to their phototonus by introducing one, two, three or more screens of ground glass between the preparation and the source of light. The same result may be more easily obtained by moving the microscope slowly away from the window, and thus rendering the illumination less intense.

The *temperature* of the environment often has a considerable influence upon the degree of sensitiveness to light which is evinced by many zoospores. When the temperature is raised they become, so to speak, attuned to a greater degree of sensitiveness; whilst, at the same time, their movements are rendered more active: the reverse is the case when the temperature is lowered. In the first case they also become more photophylic (light-seeking), and in the latter more photophobic (light-avoiding).

"In addition, zoospores alter as regards their phototonus during the course of their development, for they appear to be able to withstand greater intensity when they are young than when they are old."

As is shown by the experiments of Cohn, Strasburger, and others, not all the rays of the spectrum are able to exert an influence upon the direction of the movements of the spores, *it being only those which are strongly refracted (blue, indigo and violet)* that produce stimulation.

If a vessel containing a deep-coloured solution of ammoniated copper oxide, which only transmits blue or violet rays, be placed between the source of light and the preparation, the spores are seen to react just as if they came in contact with ordinary white light; on the other hand, they do not react at all to light which has passed through bichromate of potassium solution, through the yellow vapour of a sodium flame, or through ruby-red glass.

Another very important and complex manifestation of the effects due to light is seen in the *movements of the chlorophyll corpuscles in plant cells.* The light acts as a stimulus to protoplasm, which contains chlorophyll, causing the latter to collect by means of slow movements in suitable places within the cellulose membrane.

The most suitable object for the study of these phenomena is the *Alga, Mesocarpus,* upon which Stahl (IV. 34) has made some most convincing observations.

In the cylindrical cells, which are united together to form long threads, a narrow band of chlorophyll is extended longitudinally along the middle of the vacuole, which is thus divided into two equal parts; the ends of this band pass over into the protoplasmic lining of the wall. Now this chlorophyll band changes its position according to the direction of the impinging light. If it is exposed directly from above or below to *weak daylight*, it turns its surface towards the observer. If, however, on the contrary, it is arranged so that only such rays as are parallel to the stage of the microscope are allowed to reach the preparation from one side, the green plates are seen to turn about through an angle of $90°$, so that they take up an exactly vertical position, assuming now an appearance of dark green longitudinal stripes, stretching themselves through the otherwise transparent cell. The band is able to assume every possible intermediate position in its endeavour to place its surface at right angles to the impinging light. On a warm summer's day this change of position is effected in a very few minutes, being brought about by the active movements which the protoplasm makes inside the cell membrane.

The effect produced varies in this case also, as with the zoospores, according to the *intensity of the light*. Whilst diffuse daylight has the effect described above, direct sunlight brings about a quite opposite result, for in this case the chlorophyll bands turn one of their edges to the sun. Hence we can educe the following: "Light exerts an influence upon the position of the chlorophyll bands of *Mesocarpus*. If the light is fairly weak, the bands turn themselves at right angles to the path of the rays; if, however, it is intense, they place themselves in the same direction as the rays." Stahl calls the first arrangement *surface position*, and the second, *profile position*.

If illuminated intensely for a considerable period, the whole band contracts to form a dark green vermiform body; it is, however, under favourable conditions capable of resuming its original form.

The purpose of all these various movements of the protoplasm under the influence of light is, on the one hand, to bring the chlorophyll bands into a favourable position for the exercise of their functions; and, on the other, to protect them from the injurious action of a too powerful illumination.

Further, the plant-cells which contain *chlorophyll granules*, and which are connected to form tissues, are also subjected to the influence of light, as is so plainly seen in *Mesocarpus*. Only in this case the phenomena are somewhat more complex (Fig. 52).

Sachs was the first to notice that the colour of leaves is lighter when they are exposed to direct sunlight, than when they are in shadow, or when the light is less intense. In consequence of this discovery, Sachs was able to produce light pictures upon leaves, by partially covering them with strips of paper, and exposing them to intense light (IV. 32a); after a certain time the strips of paper were removed, and it was then seen that the portions which they covered appeared as dark-green stripes upon a light-green background.

This phenomenon may be explained by the law which was laid down in the case of *Mesocarpus*; this has been proved by the investigation of Stahl (IV. 34), which he conducted on the lines laid down by Famintzin, Frank, and Borodin. When the illumination is faint, or when the leaves are in shadow, the protoplasm moves so that the chlorophyll granules are arranged upon those external surfaces of the cells which are turned towards the light (Fig. 52*A*), having completely forsaken the side-walls. On the other hand, the protoplasm, under the influence of direct sunlight, streams away towards the side-walls, until the external surface is quite free from chlorophyll granules, that is to say, in the first

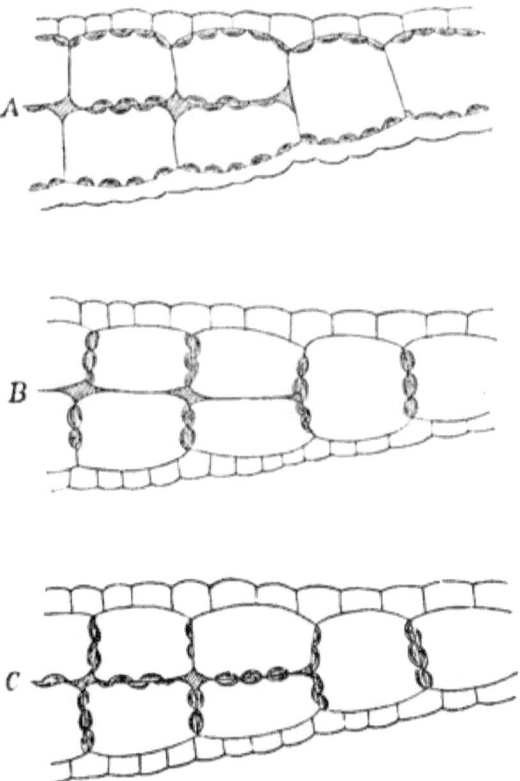

Fig. 52. — Transverse section through the leaf of *Lemna trisulca* (after Stahl): *A* surface position position assumed in diffused sunlight; *B* arrangement of chlorophyll granules under the influence of intense light; *C* position assumed by chlorophyll granules in the dark.

case, the whole chlorophyll-bearing substance, as in *Mesocarpus*, assumes a surface position towards the impinging light, and in the second, a profile position; hence the varying colour of the leaves.

In addition, *the chlorophyll granules themselves, when under the influence of intense light, alter their shape, becoming smaller and more globular.*

All these occurrences serve to accomplish the same end: "Chlorophyll granules protect themselves by turning on their axes (*Mesocarpus*), by migration, or by altering their shapes from intense illumination." "If the illumination is weak, the largest surfaces are turned towards the light, in order that as much of it may be received as possible. The behaviour is exactly the opposite when the light is strong, a smaller surface being then exposed to the light."

III. **Electrical Stimuli.** As has been shown by the experiments of Max Schultze (I. 29), of Kühne (IV. 15), of Engelmann, and of Verworn (IV. 39), electrical currents, both constant and induced, act as stimuli upon protoplasm, when they flow directly through it.

If some *staminal hairs of Tradescantia* (Fig. 53) are placed between non-polarisable electrodes which are close together, and are then stimulated by means of weak induction shocks, the granular streaming movements can be seen to have been influenced in that portion of the protoplasmic net through which the current flowed. Irregular masses and globules develop upon the protoplasmic threads; these separate off at the thinnest places, and become absorbed into neighbouring threads. After a short period of rest, the movements recommence, the

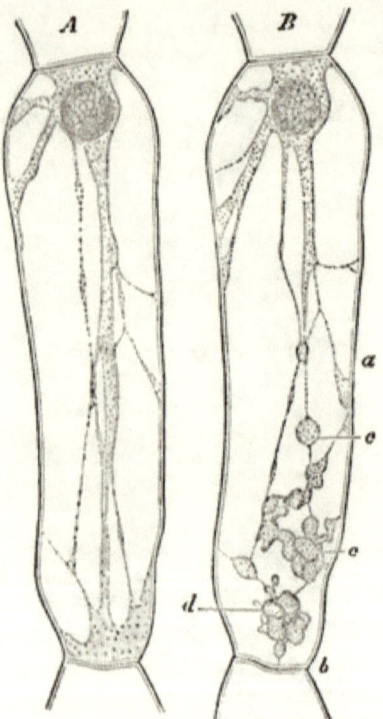

Fig. 53.—*A, B* cell of a staminal hair of *Tradescantia virginica*. *A* Normal condition of protoplasm before it has been disturbed. *B* The protoplasm, in consequence of stimulation, has massed itself into balls; *a* cell-wall; *b* transverse wall of two cells; *c, d* balls of protoplasm. (After Kühne; from Verworn, Fig. 13.)

masses and globules being gradually taken up by the neighbouring streams of protoplasm, carried along by them, and finally split up. If strong shocks are repeatedly administered, so that the whole cell is affected, a return to the normal condition is impossible, for the protoplasmic body, by becoming partially coagulated, has been transformed into turbid flakes and masses.

In *Amoeba* and *white blood corpuscles* the streaming motions of the granules and the crawling movements of the whole cell are both arrested for a time by slight induction shocks; after a while they are resumed and proceed in a normal fashion. If stronger induction shocks are administered, the result is that the pseudopodia are quickly withdrawn, and the body contracts up into a ball; finally, very strong shocks cause the bursting and consequent destruction of the contracted spherical body.

If the induction current is applied for a considerable time to one of the lower unicellular organisms, it can be gradually destroyed bit by bit, and thus diminished in size. In *Actinosphaerium* the process is as follows: the pseudopodia, which are parallel to the current, soon exhibit varicosities; they are gradually completely withdrawn, whilst the protoplasm becomes massed together to form little balls and spindles (Fig. 54); then at this place the surface of the body becomes gradually destroyed by a process resembling to a certain extent a kind of melting down, during which the vacuoles, which are contained in the protoplasm, burst. On the other hand, those pseudopodia which are at right angles to the current are unaffected. When the stimulus is removed, the body, which has thus been reduced to about a half or a third of its original size, gradually recovers, and reproduces the parts which have been destroyed.

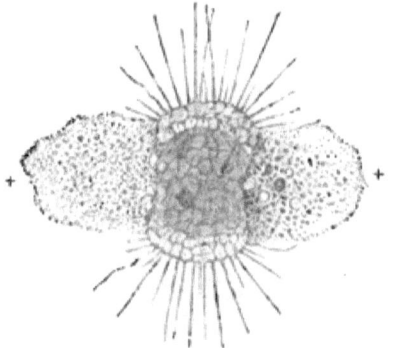

FIG. 54.—*Actinosphaerium Eichhornii*, action of an interrupted current. Progressive destruction of protoplasm is equal at both poles. (After Verworn, Tab. 1, Fig. 5.)

The action of the constant current upon the *Actinosphaerium* (Fig. 55), *Actinophrys*, *Pelomyxa*, and *Myxomycetes*, is similar to this. When the circuit is closed, an excitation occurs at the positive pole or anode (in Fig. 55, +) which is manifested by the retraction of the pseudopodia, and, if the stimulus lasts long, by the destruction of the protoplasm at the place where the current enters. When communication is broken, the destructive process at the anode immediately ceases, whilst, on the other hand, a transitory contraction occurs at the surface which is turned towards the cathode.

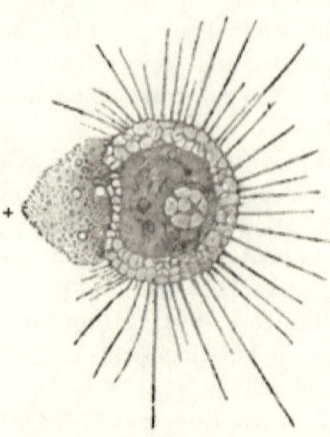

Fig. 55.—*Actinosphærium Eichhornii*, between the poles of a constant current. A short time after the closing of the current, granular destruction of the protoplasm commences at the anode (+). At the cathode the pseudopodia have become normal again. (After Verworn, Tab. 1, Fig. 2.)

Perhaps even more interesting and important than these processes are the **phenomena produced by Galvanotropism**, which have been observed by Verworn in a number of unicellular organisms (IV. 39, 40).

Many organisms, in consequence of the influence of the constant current, are caused to move in certain fixed directions, just as they move when stimulated by a ray of light (heliotropism). "If a drop, containing as many *Paramœcia aurelia* as possible, is placed upon a slide between two non-polarisable electrodes, and the constant galvanic circuit is closed, it is seen that the *Paramœcia* immediately leave the anode in a mass, and hurry in a dense swarm to the cathode, where they collect in great numbers. After a few seconds the rest of the drop becomes completely free from Protista, whilst at the cathode there is a dense seething crowd of them. Here they remain as long as the current persists. When connection is broken, the whole swarm immediately forsakes the cathode to swim back in the direction of the anode. However, they do not all collect at the anode, part of them remaining scattered about in the drop; at first they do not come near to the cathode, but after a time they gradually approach it, until finally all the Protista are again evenly distributed throughout the drop."

If pointed electrodes are employed, the *Paramœcia* swarm inwards to form a galvanic figure around the cathode (Fig. 56 A).

An appearance similar to that produced when iron filings are attracted by a magnet is seen. "Under the circumstances," as

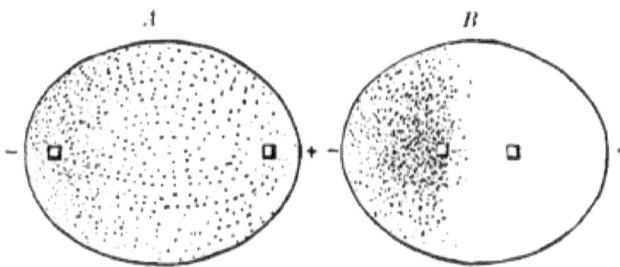

Fig. 56.—On completing the circuit of the constant current all the *Paramæcia* in a drop of water swim within the curve of the electric current towards the negative pole (*A*), until after a time they collect on the other side of the pole (*B*). (After Verworn IV. 10, Fig. 20.)

Verworn remarks, "it may be observed that after all the *Paramæcia* have wandered over to the negative pole, the largest collection is formed behind, that is to say—reckoning from the positive pole—on the other side of the negative pole, and that only a few remain on this side of the pole (Fig. 56 *B*). When the connection is broken the Protista swim back again, in the manner described above, towards the positive pole, keeping at first, just as before, well within the curve of the electric current, until gradually the movement, and with it the division into groups, becomes irregular again."

In the same manner, a number of other Ciliata (such as *Stentor, Colpoda, Halteria, Coleps, Urocentum*) and Flagellata (such as *Trachelomonas, Peridinium*) are galvanotropic.

Amœba react in a similar manner. At the first moment after the circuit of the constant current has been completed a cessation of the streaming movements of the granules occurs; very soon, however, the hyaline pseudopodia are suddenly protruded from the end which is turned towards the cathode, and, whilst the remainder of the body substance flows in the same direction, and keeps continually stretching out new pseudopodia, the *Amœba* creeps towards the cathode. When the current is reversed it is seen that the granular streaming movements are also immediately reversed, and the *Amœba* commences to creep in the opposite direction.

The movement towards the cathode may be called *negative galvanotropism*. As there exist both negative and positive heliotropism and thermotropism, so we occasionally find isolated instances of *positive galvanotropism*. It has been observed by Verworn in *Opalina ranarum*, and in a few Bacteria and Flagellata such as *Cryptomonas* and *Chilomonas*. When the circuit is completed the above-named species travel towards the anode instead of towards the cathode, and collect there. If Ciliata and Flagellata are present side by side in one drop, they are seen under the influence of the constant current to hasten in opposite directions, so that finally two distinct groups are to be seen, the Flagellata being at the anode, and the Ciliata at the cathode. If the current is now reversed they advance like two hostile armies upon one another, until they assemble again at the opposite poles. Each time the current was made it produced in a few seconds a distinct sorting out of the crowd of Infusoria, which were otherwise in inextricable confusion.

IV. **Mechanical Stimuli.** Pressure, violent shaking, crushing, all these act as stimuli to protoplasm. Weak mechanical stimulations only produce an effect upon the point of contact; strong stimuli affect a larger area and produce a more rapid and more powerful effect than weak ones. If a cell of a *Tradescantia* or *Chara* or the plasmodium of an *Æthalium* be violently shaken, or pressed upon at one place, the granular movement is temporarily arrested, whilst swellings and knots may even appear on the protoplasmic threads, such as are produced by the electrical current. Hence it frequently occurs, that in preparing the slide for observation all the protoplasmic movements may be brought to a standstill, simply by putting on the coverglass. They gradually return after a period of rest.

Amœbæ and white blood corpuscles withdraw their pseudopodia and assume a globular form when they are violently shaken. Reticularia, which have extended their long processes, often withdraw them with so much energy that the ends which were attached to the slide are torn off (Verworn). A localised stimulus can be produced at a given point with a fine needle. If the stimulus is weak the effect is confined to this point, a varicosity being formed and a shortening of the pseudopodium being produced. Strong and repeated stimuli cause neighbouring pseudopodia, which were not directly touched, to contract (Fig. 57 *B*).

If an Infusorian or other small animal comes in contact with an outstretched pseudopodium, it is firmly grasped by it, and becomes surrounded by the protoplasm. As the pseudopodium

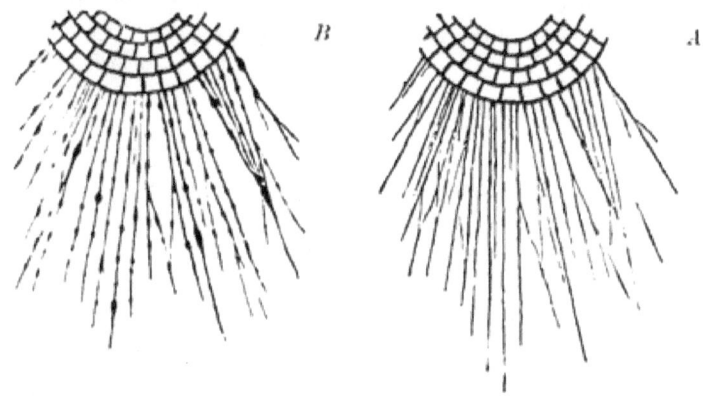

FIG. 57.—*Orbitolites.* A portion of the surface with its pseudopodia: *A* undisturbed; *B* the whole has been stimulated by repeated shaking. (After Verworn III. 24, Fig. 7.) This is of importance to Rhizopoda in absorbing food.

gradually shortens itself, a motion in which the neighbouring threads eventually participate, the Infusorian is gradually drawn into the centre of the protoplasmic mass, where it undergoes digestion.

V. **Chemical Stimuli.** A living cell is able to a certain extent to adapt itself to chemical changes in its environment. For this, however, one thing is most important, namely that the changes should be made gradually, not suddenly.

Æthalium plasmodia flourish in a 2 per cent. solution of grape-sugar, if the latter is added in gradually increasing quantities to the water (IV. 35). If they were to be transferred straight from pure water into this chemically different environment, the sudden change would result in their death; this would also occur if they were to be suddenly placed back into pure water from the 2 per cent. sugar solution. It is evident that the protoplasm needs time to adapt itself to its altered condition, probably by increasing or diminishing the amount of water it contains.

Marine Amœbæ and Reticularia remain alive after the water which contains them, in consequence of being in an open vessel, has evaporated so much that it contains 10 per cent. of salt. Fresh water Amœbæ can gradually accustom themselves to a 4 per cent. solution of common salt, whereas, if they are suddenly immersed in a 1 per cent. solution, they contract into balls, and in time become broken up into glistening droplets. During the process of adaptation to a new chemical environment, the individual cells may undergo greater or less changes in their structure and vital properties. When such changes are apparent to us, we speak of the *effects of chemical stimulation*. These phenomena, which are so exceedingly numerous, may vary considerably, according as to whether the whole, or only part, of the cell-body is affected by the stimulus.

a. **First group of experiments. Chemical stimuli which affect the whole of the body.** In order to throw light upon this first group of phenomena, the *behaviour of protoplasm towards certain gases*, which are grouped under the common name of anæsthetics, must be investigated.

The protoplasmic movements of a plant cell soon become arrested, if, instead of being put into water, it is placed in a drop of olive oil, by which means the air is excluded (IV. 15). After the oil has been removed, the movements are seen to gradually recommence.

The streaming movements may in a similar manner be slackened and finally completely stopped, if the air is replaced by carbon dioxide or hydrogen. For these experiments special slides with gas chambers have been constructed through which a current of carbon dioxide or hydrogen may be conducted. If the plant cell is kept from 45 minutes to an hour in a current of carbon dioxide, the movements are as a rule completely stopped; when hydrogen is used, a longer time must be allowed (III. 5). This protoplasmic paralysis may, if it has not been allowed to last too long, be removed by the addition of oxygen. "Apparently living protoplasm unites chemically with the oxygen of its environment. The definite oxygenated compound thus produced, of which under ordinary conditions a considerable amount must be assumed to exist in every protoplasmic body, is continually broken down during the movements, whilst carbon dioxide is probably given off" (Engelmann III. 5). Hence the removal of oxygen has a paralysing effect upon the irritability, and indeed upon all the vital activities of the protoplasm.

Such anæsthetics, as chloroform, morphia, chloral-hydrate, etc., have a marked influence upon the vital activities of the cell. These substances do not affect the nervous system alone, as is frequently believed, but all the protoplasm of the body. The difference is only a matter of degree; the irritability of the nerve-cells is more quickly lowered and finally destroyed than that of the protoplasm of other cells. Further, when narcotics are employed medicinally, the attempt is made to act upon the nervous system alone, for if all the elementary cells were affected, a cessation of the vital processes would result, and death might ensue. However, the following examples will prove clearly that the irritability of animal and vegetable protoplasm may be temporarily destroyed without permanent harm.

The sensitive plant, or *Mimosa pudica*, is very easily affected by mechanical stimulation. When a leaflet is shaken a little, it immediately closes itself up, and forsaking its upright position, droops downwards. In addition, it forms an example of the rapid manner in which a stimulus is conducted in plants, in which, since no nerves are present, it must be simply transmitted by each protoplasmic cell quickly conveying the impulse to its neighbour. In consequence of this, if the stimulus is sufficiently strong, not only do the leaves which were directly touched close up, but also those on the same branch, and eventually even the whole plant, are affected. In consequence of the stimulation, certain mechanical arrangements, not suitable for present discussion, come into play.

In order to study the effect of anæsthetics, a sensitive plant, in a condition of normal irritability, should be placed under a belljar, and when the leaves are fully extended, a sponge soaked with chloroform or ether should be inserted (Claude Bernard IV. 1). After about half an hour it is seen that the chloroform or ether vapour has caused the protoplasm to lose all its irritability. When the bell-jar is removed, the leaves, which are spread out as usual, may be touched, or even severely crushed or cut, without any reaction being produced; the result is the same as that produced on one of the higher animals provided with nerves. And yet, if proper precautions have been taken, it is found that the protoplasm has not been killed, for after the sensitive plant has been for a short time in the fresh air, the narcosis gradually disappears; at first, individual leaves gradually close up when they are roughly handled, until finally complete irritability is restored.

Ova and *spermatozoa* may be subjected to the action of narcotics in a similar manner. When Richard Hertwig and myself (IV. 12a) placed the actively motile spermatozoa of a sea-urchin in a ·5 per cent. solution of chloral-hydrate in sea water, we found that after five minutes, their motions were completely arrested; however, these soon recommenced, after the chloral solution had been diluted with pure sea water. Further, those spermatozoa which had been temporarily paralysed in this manner united with ova when they were brought to them, almost as quickly as fresh spermatozoa. When they were kept for half an hour in the chloral solution, a more marked paralysis was produced, which persisted for a long time after the noxious agent had been removed. It was not until some few minutes had elapsed that certain individual isolated spermatozoa commenced to exhibit snake-like movements, which gradually became more active. Even when they were brought into the neighbourhood of ova, it was observed, that after ten minutes none of these were fertilised, although several spermatozoa had attached themselves to their surfaces, and had bored their way in. But even in this case fructification and the subsequent normal division of the eggs took place finally.

Similarly, egg-cells become affected, as regards their irritability, by a ·2 to ·5 per cent. solution of chloral hydrate or of some similar drug; this may be recognised by the abnormal manner in which, after the seminal fluid has been added, the process of fertilisation takes place. For whilst under ordinary circumstances only one single spermatozoon penetrates into the ovum, with the result that a firm yolk membrane is immediately formed, which prevents the entrance of other spermatozoa, in *chloralised eggs multiple fertilisation* takes place. It has been proved that, according to the intensity of the action of the chloral, that is to say, the stronger the solution, and the longer it is allowed to act, the greater is the number of spermatozoa which make their way into the ovum before the formation of the yolk and membrane. Evidently the effect of this chemical reagent is to lower the power of reaction of the egg plasma, so that the stimulus which is produced by the entrance of one spermatozoon is now no longer sufficient, but the ovum must be stimulated by the entrance of two, three, or even more spermatozoa, before it is sufficiently excited to form a membrane.

Finally, another example will show that the *chemical processes of the cell may also be hindered by anæsthetics*. As is well known, the yeast fungi (*Saccharomyces cerevisiæ*) produce alcoholic fermentation in a solution of sugar, and during this process bubbles of carbon dioxide rise through the fluid. When Claude Bernard (IV. 1) added chloroform or ether to the solution of sugar, before adding the yeast, no fermentation took place, although in other respects the circumstances were favourable. But when the yeast, after having been filtered out from the chloroform solution, and rinsed with clean water, was placed in pure sugar solution, he found that fermentation soon occurred; hence the yeast had recovered its power of converting sugar into alcohol and carbon dioxide, this power having, by the action of the chloroform and ether, been temporarily suspended.

In a similar manner the functions which the chlorophyll performs in plants, and the dependent process of giving off oxygen in the sunlight, may be arrested by means of chloroform (Claude Bernard).

b. **Second Group of Experiments. Chemical Stimuli which come into contact with the cell-body at one spot only.** Very interesting and varying phenomena are produced when chemical substances, instead of coming into contact with the body all round, only impinge upon it, at a definite fixed point. Such stimuli may produce changes in form, and movements in a definite direction, which phenomena have been classed under the name of *Chemotropism (Chemotaxis)*.

Chemotropic movements may be directed towards the stimulating source, or, on the contrary, away from it. In the first case the chemical substance is said to *attract*, and in the second to *repel*, the protoplasmic body. This depends partly upon the chemical nature of the substance, partly upon the individual properties of the special kind of plasma, and, finally, upon the degree of condensation of the chemical substance. A substance, which when dilute may attract, may repel when the solution is strong. Here, as with strong and weak light, special differences are present. Just as heliotropism may be positive or negative, so may chemotropism be positive or negative.

We will first examine the action of gases, and next that of solutions; at the same time we will become acquainted with a very ingenious method of investigation, for which we must especially thank the botanist Pfeffer (IV. 26).

1. **Gases.** Oxygen has great attractive powers for freely moving cells, as has been shown by the experiments of Stahl, Engelmann, and Verworn.

Stahl has made experiments upon the plasmodia of *Ethalium septicum* (IV. 35). He half filled a glass cylinder with thoroughly boiled water, which, in order to exclude the air, he covered with a very thin layer of oil. He then took a strip of filter paper, over which a plasmodium had extended itself, and placed it along the side of the cylinder in such a manner that one half of it was immersed in the water. The strands of protoplasm, which were placed in the non-oxygenated water, were seen to grow gradually thinner, until after a time all the protoplasm had crept up above the layer of oil, which, except in excluding the air, had no deleterious effect upon it, to the upper portion of the cylinder, where it could come into contact with the oxygen of the air. Another method of performing the same experiment is to place a plasmodium in a cylinder which is quite full of thoroughly boiled water; to close the opening with a perforated cork, and then to place the cylinder upside down in a plate of fresh water. Very soon the plasmodium is seen to have wandered through the small hole in the cork into the medium which contains oxygen.

Engelmann (IV. 7) has made some very interesting experiments upon the directing influence exerted by oxygen upon the movements of bacteria. He shows *that many species of bacteria may be used as a very delicate test for minute quantities of oxygen.* If into a fluid which contains certain bacteria, a small alga or diatom is introduced it is seen after a short time to be surrounded with a dense envelope of bacteria, which have been attracted by the oxygen set free by the action of its chlorophyll.

Verworn (IV. 40) saw a diatom quite enclosed by a wall of motionless *Spirochætæ* whilst the rest of the preparation was quite free from them (Fig. 58). Suddenly the diatom moved a short distance away, getting out of the crowd of Bacteria. The *Spirochætæ*, so suddenly left in the lurch by the producer of oxygen, remained quiet for a second, but soon commenced to move about quickly, and to swim after the diatom in dense masses. After a minute or two they had nearly all reassembled round about it, after which they remained motionless as before.

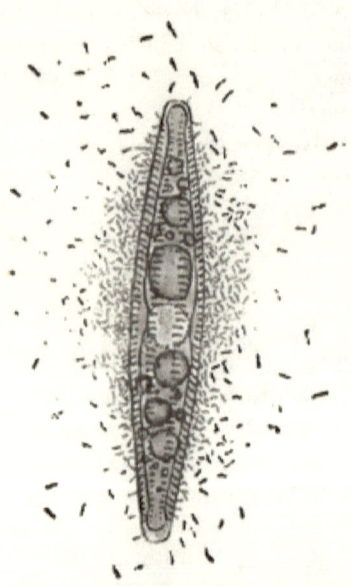

Fig. 58.—A large diatom (*Pinnularia*) surrounded by a large number of *Spirochætæ plicatilis*. (After Verworn IV. 40, Fig. 14.)

This attractive power possessed by oxygen explains the fact that in microscopic preparations almost all Bacteria, Flagellata, and Ciliata are found collected together round the edges, or round any air bubbles which may be present in the water.

Verworn describes a most instructive experiment (IV. 40). A large number of *Paramoecia* are placed in a test-tube, which is filled with water, poor in oxygen. The test-tube is then reversed and placed under mercury. Very soon the movements of the cilia commence to slacken, in consequence of the lack of oxygen. If now a bubble of pure oxygen is introduced through the mercury into the test-tube, it will be seen after a few seconds to be surrounded by a thick white envelope of *Paramoecia*, "which, driven by their thirst for oxygen, throw themselves energetically upon the bubble of this gas."

2. **Liquids.** Stahl and Pfeffer have made systematic experiments upon the stimulating action of fluid substances.

Stahl (IV. 35) has again made great use of flowers of tan. Upon this organism even pure water has a stimulating effect, a phenomenon described by Stahl as *positive and negative hydrotropism*. If a plasmodium is evenly spread out over a strip of damp filter paper, it is seen, as soon as the paper commences to dry, that the plasmodium makes its way to the dampest parts. If, whilst the drying process is going on, a slide covered with gelatine is held perpendicularly at about two mm. distance above the paper, a few branches are seen to extend themselves upwards towards the gelatine, attracted by the water vapour it gives off, until finally they reach it and spread themselves out upon it possibly, during the course of a few hours, the whole plasmodium may transfer itself to the damper surface. When Myxomycetes are about to fructify, negative instead of positive hydrotropism takes place. Under these conditions the plasmodia seek the driest portions of the environment, and withdraw themselves from any damp gelatine or moistened filter paper which may be brought into their neighbourhood.

These phenomena of hydrotropism are easily explained by the fact that protoplasm contains a certain quantity of imbibition water, which may fluctuate up to a certain extent, and may even increase or decrease during the development of the cell-body. The more saturated the protoplasm is with water, the more active as a rule are its movements. During the vegetative period the plasmodium of the *Æthalium* tends to increase its supply of water, and hence it moves towards the source of water; when the reproductive period commences, it shuns moisture, because, at the time when spores are being formed, it diminishes its water supply.

Many chemical substances attract, whilst others repel plasmodia. If a net of *Æthalium*, which has spread itself out upon a moist substratum, is brought into contact with a ball of filter paper, which is saturated with an infusion of tan, individual strands of plasma immediately commence to creep towards the nutrient medium. After a few hours all the spaces in the paper ball are filled up with the slime fungus.

In order to study negative chemotropism, a crystal of common salt or of saltpetre, or a drop of glycerine, may be brought to the edge of the piece of damp filter paper upon which the slime fungus has spread itself out. It can then be seen how, as the concentrated solution of salt or of glycerine gradually creeps along the filter paper, the protoplasm shrinks away from the source of stimulation in ever-widening circles.

Hence the naked plasmodia, which are so easily destroyed, possess the marvellous property, on the one hand, of avoiding harmful substances, and, on the other, of searching all through the medium in which they are, for substances which are of value to them for purposes of nutrition, and of absorbing them. "For instance, if one of the numerous branches of a plasmodium, by chance comes across a place which is rich in nutriment, an influx of plasma immediately occurs to this favourable spot."

Pfeffer has very accurately examined the chemotropism of small, freely motile cells, such as spermatozoa, Bacteria, Flagellata, and Ciliata, in some pioneering investigations that he has made, and by this means has discovered a very simple and ingenious method of investigation.

He takes some fine glass capillary tubes from 4 to 12 mm. long; one end of each tube is closed, whilst at the other there is an opening varying in inside diameter from ·03 to ·15 mm., according to the size of the organism to be examined. He fills these tubes for about a half or a third of their length with the stimulating substance, there being a space filled with air at the closed end.

In order to explain their use, we may quote the following experiment. Pfeffer has discovered that malic acid has a strong affinity for the antherozoids of Ferns, and that probably it is on this account that it is secreted normally by the archegonia. A capillary tube is filled with ·01 per cent. of malic acid, and after its surface has been most scrupulously cleansed, is reversed and carefully placed in a drop of water containing a large number of Fern antherozoids. With a magnifying power of 100 to 200 diameters, it can be seen that some antherozoids immediately begin to make their way towards the opening of the tube, from which the malic acid commences to diffuse itself throughout the water. They soon force their way right into the tube itself, until after five or ten minutes several hundreds of them have collected there. After a short time there are only a few left outside of the tube.

If experiments are made with solutions of malic acid of varying strengths, a law similar to that of the effect produced by various degrees of heat upon protoplasmic streaming movements may be deduced. *Beyond a certain minimum concentration (about ·001 per cent.) which may be considered to constitute the stimulative starting point, every increase in concentration produces a corresponding increased effect, until a certain fixed point is reached, when the optimum or maximum result is produced;* if the concentration is increased above this point the attraction of the malic acid for the antherozoids decreases, until finally the positive chemotropism is converted into negative chemotropism.

Hence a very strong solution produces an exactly opposite effect to that produced by a weak one, the antherozoids being repelled instead of attracted. How small a quantity of malic acid is necessary to produce a result may be seen from the fact that in a capillary tube which contains a ·001 per cent. solution only ·0000000284 milligramme, or $\frac{1}{350000000}$ of a milligramme, of malic acid is present.

As has been already stated, if the chemical stimulus is to produce movements in a certain direction, it must only be strongly applied at one point, or at any rate from one side. This is the case in the above experiment, for as the malic acid becomes diffused through the opening in the surrounding water, the antherozoids, passing through the opening and making their way up the tube, come into contact with solutions gradually increasing in strength. The diffusion causes an unequal distribution of the stimulus about the bodies of the antherozoids: "thus varying with its varying degrees of concentration, the malic acid exerts a stimulus which causes a movement in a fixed direction."

The antherozoids, as might be expected, are distributed evenly throughout a homogeneous solution, yet even under these conditions a specific stimulative effect is exerted upon them. This, however, can only be perceived indirectly, and can only be explained by the supposition that the attitude, so to speak, of the antherozoids towards malic acid has experienced some modification. Pfeffer is able in this case to demonstrate a relation similar to that expressed by the Weber-Fechner law for the mental perceptions of man: "Whilst the stimulus increases in geometrical progression, the perception or reaction increases in arithmetical progression."

This ratio, which in many respects is very important, can be observed in the behaviour of antherozoids towards malic acid.

To the fluid, containing the fern antherozoids, some malic acid is added in such a quantity that when the two are well mixed together a solution of ·0005 per cent. is produced. If now a capillary tube containing a solution of ·001 per cent. is inserted, attractive influence, as was the case when the antherozoids were in pure water, can be perceived. The tube must now contain a ·015 per cent. solution in order to produce an effect, and if the water, in which the antherozoids are, contains ·05 per cent. of malic acid, the solution in the tube must be 1·5 per cent. in strength. Or more generally expressed, *the solution in the tube must be thirty times as strong as that from which the antherozoids are to be attracted. The sensitiveness to stimuli, or the stimulation tone of the antherozoids, is affected, if they are present in a liquid which contains a certain proportional amount of the substance which is to act as the stimulus.* Thus it is possible in an artificial way to render them non-sensitive towards weak solutions of malic acid, which under ordinary circumstances constitute excellent stimuli, whilst on the other hand they may be made susceptible to attraction from strong concentrations of malic acid, which would repel antherozoids accustomed to living in pure water.

Individual cell bodies behave very variously towards chemical substances, just as they do towards light. Malic acid, which exerts such a powerful attraction upon fern antherozoids, does not affect those of Feather-moss at all. For these, however, a 1 per cent. solution of cane sugar acts as a stimulus, whilst on the other hand neither of these substances has any effect on Liverwort or *Characeæ*.

A 1 per cent. solution of meat extract or of Asparagin exerts a strong attraction upon *Bacterium termo*, *Spirillum undula*, and many other unicellular organisms. Even after a short period, varying from two to five minutes, a distinct plug of bacteria is seen to have collected at the mouth of a capillary tube, which has been placed in a drop of water containing these micro-organisms.

On account of the different ways in which various cell bodies react towards different chemical stimuli, the method, which Pfeffer has perfected and used with various reagents, may be employed, not only to attract one individual organism sensitive to one special reagent, but also to separate different species which are mixed together, as has also been done by means of galvanotropism or heliotropism. Glass tubes provided with suitable attractive material, and inserted in fluids, may be used as traps for *Bacteria* or *Infusoria*.

Further, it follows from the above-mentioned experiments, that organisms which are specially sensitive towards a given chemical substance may be used as reagents to indicate the presence of this stimulating substance. Thus, according to Engelmann (IV. 7), certain Schizomycetes form an excellent test for oxygen, of which such a minute portion as one trillionth of a milligramme is sufficient to attract them.

Not every substance which attracts an organism is useful to it as food, or is even innocuous to it; many, such as sodium salicylate, saltpetre, strychnine, or morphia, even cause the immediate death of the organisms which they have enticed. However, as a rule the substances which are hurtful to protoplasm generally repel it; this is the case with most acid and alkaline solutions. Even 2 per cent. solutions of citric acid and sodium carbonate exert a distinctly repellent influence.

Hence, within the above-mentioned limitations, the general rule may be stated that organisms are, through positive chemotropism, enabled to seek suitable nutriment, whilst in consequence of negative chemotropism they avoid hurtful substances.

These phenomena of chemotropism are of the greatest importance in understanding many processes in the bodies of man and of other vertebrates. Here also there are cells which react to chemical stimuli by changes of shape, and movements in special directions. These cells are the white blood corpuscles and lymph cells (leucocytes or wandering cells).

The chemical irritability of leucocytes has been established as a fact by the experiments of Leber (IV. 17a, b); Massart and Bordet (IV. 20, 21); Steinhaus (IV. 36); Gabritschevsky (IV. 10); and Buchner (IV. 2). If, in accordance with Pfeffer's method, fine capillary tubes, filled with small quantities of some "irritating substance," are introduced into the anterior chamber of the eye or the lymph sac of a frog, they become filled in a short time with leucocytes, whilst tubes filled with distilled water exert no attractive power upon the leucocytes. When introduced into the subcutaneous connective tissue the tubes cause the outwandering of the leucocytes from the neighbouring capillary vessels (diapedesis), and under certain conditions produce suppuration.

Amongst substances which will set up inflammation, many micro-organisms and their metabolic products are in the first rank. Thus, Leber found during his experiments that an extract of *Staphylococcus pyogenes* proved very effectual as an inflammatory agent. Hence the study of chemotropism is of the greatest importance in the investigation of the diseases produced by the presence of pathogenetic micro-organisms. Accurate knowledge of the former will no doubt explain many apparently contradictory phenomena, which are met with in the study of infectious diseases.

It may be taken for granted at the outset, that if leucocytes can be stimulated by means of chemical substances produced by micro-organisms, such stimulation can only occur in accordance with laws similar to those which have been established generally with regard to cells. Positive and negative chemotropism—excitation, and the variations which may occur in it owing to the even distribution of the existing agent—the effects of stimulation—all these must be taken into account.

Hence the behaviour of the leucocytes towards the stimulating substance assumes the form of a complicated process, which may vary very considerably according to the special conditions. For the metabolic products excreted by micro-organisms may, according to their nature and state of concentration, exert an attractive or repellent influence. In addition, the effect produced may vary according as to whether these products are restricted to the region where they are produced, and from which they attack the leucocytes, or whether they are in addition evenly distributed throughout the blood. For in the latter case the presence of the bacterial products in the blood will modify the way in which the leucocytes react towards those which are collected in considerable quantities near the diseased spot; and as was the case with the antherozoids and malic acid (pp. 118-120), the result will depend upon the relative proportions of the stimulating substance which is present in each region.

The numerous possibilities may be grouped under two heads.

First group.—The metabolic products are evenly distributed or approximately so throughout the blood and the diseased tissues. Since under these conditions there can be no special point of stimulation, it stands to reason that the leucocytes cannot wander away from the diseased spot.

Second group.—The collections of products are unequal in concentration, and further, the difference in their concentration is sufficient to give rise to an effective stimulation. Two alternatives may occur. Either the higher degree of concentration is present at the seat of the disease, or in the blood-vessels. In the first case only will the leucocytes collect around the affected tissue.

The consideration of these relative conditions appears to me to explain many interesting phenomena, which have been observed by certain French investigators, Roger, Charrin, Bouchard (IV. 1b), etc., during their various experiments with the catabolic products of the *Bacillus pyocyaneus*, of the *Anthrax bacillus*, etc.; and by Koch in his observations upon the action of *Tuberculin*. I have endeavoured to explain such phenomena in a short popular paper: "Ueber die physiologische Grundlage der Tuberculin wirkung, eine Theorie der Wirkungsweise bacillärer Stoffwechselproducte" (IV. 13), to which I refer the reader for information with regard to physiological experiments and the explanation of the special phenomena of disease.

Literature IV.

1a. CLAUDE BERNARD. *Leçons sur les phénomènes de la vie commune aux animaux et aux végétaux.*

1b. BOUCHARD. *Théorie de l'infection. Verhandl. des X. intern. med. Congresses zu Berlin.* Bd. I. 1891.

2. BUCHNER. *Die chemische Reizbarkeit der Leukocyten und deren Beziehung zur Entzündung und Eiterung. Berliner klinische Wochenschn.* 1890.

3. BRÜCKE. *Untersuchungen über den Farbenwechsel des afrikan. Chamaeleons. Denkschrift d. math. nakurw., Classe der Akad. d. Wissensch.* Bd. IV. 1851.

T. LAUDER BRUNTON. *Action of Drugs on Protoplasm. Pharmacology Therapeutics and Materia Medica.* London.

4. BUNGE. *Vitalismus und Mechanismus.*

5a. DE BARY. *Vorlesungen über Bacterien.* 1885.

5D. DEHNECKE. *Einige Beobachtungen über den Einfluss der Preparationsmethode auf die Bewegungen des Protoplasmas der Pflanzenzellen. Flora* 1881.

6A. ENGELMANN. *Beiträge zur Physiologie des Protoplasmas-Pflügers Archiv. Bd. II.* 1869.

6B. ENGELMANN. *Ueber Reizung contractilen Protoplasmas durch plötzliche Beleuchtung. Pflügers Archiv. Bd. XIX.*

7. ENGELMANN. *Neue Methode zur Untersuchung der Sauerstoffausscheidung pflanzlicher u. thierischer Organismen. Pflügers Archiv. Bd. XXV.*

8. ENGELMANN. *Ueber Licht u. Farbenperception niederster Organismen. Pflügers Archiv. Bd. XXIX.* 1882.

9. ENGELMANN. *Bacterium photometricum. Ein Beiträg zur vergleichenden Physiologie des Licht und Farbensinnes. Pflügers Archiv. Bd. XXX.*

10. GABRITCHEVSKY. *Sur les propriétés chimiotactiques des leucocytes. Annales de l'Institut Pasteur* 1890.

11. RICHARD HERTWIG. *Erythropsis agilis, eine neue Protozoe. Morph. Jahrb. Bd. X.*

12A. OSCAR u. RICHARD HERTWIG. *Ueber den Befruchtungs und Theilungsvorgang des thierischen Eies unter dem Einfluss äusserer Agentien.* 1887.

12B. OSCAR u. RICHARD HERTWIG. *Experimentelle Studien am thierischen Ei vor, während und nach der Befruchtung.* 1890.

13. OSCAR HERTWIG. *Ueber die physiologische Grundlage der Tuberculinwirkung. Eine Theorie der Wirkungsweise bacillärer Stoffwechselproducte. Jena.* 1891.

14. KLEBS. *Beiträge zur Physiologie der Pflanzenzelle. Untersuch aus dem botanischen Institut zu Tübingen. Bd. II. p. 489.*

15. W. KÜHNE. *Untersuchungen über das Protoplasma und die Contractilität.* 1864.

16. KÜNSTLER. *Les yeux des infusoires flagellifères. Journ. Mic. Paris. 10th year.*

17A. LEBER. *Ueber die Entstehung der Entzündung und die Wirkung der entzündungserregenden Schädlichkeiten. Fortschritte der Medicin,* 1888, p. 460.

17B. LEBER. *Die Enstehung der Entzündung und die Wirkung der entzündungserregenden Schädlichkeiten. Leipzig.* 1891.

18. J. LOEB. *Der Heliotropismus der Thiere und seine Uebereinstimmung mit dem Heliotropismus der Pflanzen. Würzburg.* 1890.

19. J. LOEB. *Weitere Untersuchungen über den Heliotropismus der Thiere. Pflügers Archiv. Bd. XLVII.* 1890.

20. J. MASSART et BORDET. *Recherches sur l'irritabilité des leucocytes et sur l'intervention de cette irritabilité dans la nutrition des cellules et dans l'inflammation. Journ. de la Soc. R. des Sciences médicales et naturelles de Bruxelles.* 1890.

21. J. MASSART et BORDET. *Annales de l'Institut Pasteur.* 1891.

22. METCHNIKOFF. *Lectures on the Comparative Pathology of Inflammation,* trans. by F. A. and E. H. Starling. 1893.

23. W. PFEFFER. *Handbuch der Pflanzenphysiologie. Bd. I.* 1881.

24. W. Pfeffer. *Locomotorische Richtungsbewegungen durch chemische Reize.* Untersuch. aus d. botan. Institut zu Tübingen. Bd. I.
25. W. Pfeffer. *Zur Kenntniss der Contactreize.* Untersuch. aus dem botan. Institut zu Tübingen. Bd. I.
26. W. Pfeffer. *Ueber chemotactische Bewegungen von Bakterien, Flagellaten und Volvocineen.* Untersuch. aus d. botan. Institut zu Tübingen. Bd. II.
27. George Pouchet. *D'un œil véritable chez les Protozoaires.* C. R. soc. Biol. No. 36.
28. George Pouchet. *Du rôle des nerfs dans les changements de coloration des poissons.* Journ. de l'anat. et de la phys. 1872.
29. George Pouchet. *Note sur l'influence de l'ablation des yeux sur la coloration de certaines espèces animales.* Journ. de l'anat. et de la phys. T. X. 1874.
30. F. A. Pouchet. *Sur la mutabilité de la coloration des reinettes et sur la structure de leur peau.* Compt. rend. T. 25.

F. E. Beddard. *Animal Colouration.* London.

Poulton.

31. Rawitz. *Zur Physiologie der Cephalopodenretina.* Archiv f. Anat. u. Physiologie. 1891.

A. Ruffer. *On the Phagocytes of the Alimentary Canal.* Quar. Journ. Mic. Soc. 1891.

A. Ruffer. *Immunity against Microbes.* Quar. Journ. Mic. Soc. 1891.

32a. Sachs. *Lectures on the Physiology of Plants,* trans. by Marshall Ward. Oxford. 1887.

32b. Sachs. *Handbuch der Experimentalphysiologie der Pflanzen.* 1865.

Sachs. *Text-book of Botany,* trans. by Bennet and Dyer. 1875.

33. Seidlitz. *Beiträge zur Descendenztheorie.* Leipzig. 1876.
34. Stahl. *Ueber den Einfluss von Richtung u. Stärke der Beleuchtung auf einige Bewegungserscheinungen im Pflanzenreich.* Botan. Zeitung. 1880.
35. Stahl. *Zur Biologie der Myxomyceten.* Botan. Zeitung. 1884.
36. Steinhaus. *Die Aetiologie der acuten Eiterungen.* Leipzig. 1889.
37. Strasburger. *Wirkung des Lichts und der Wärme auf die Schwärmsporen.* Jena. 1878.

S. Vines. *Lectures on the Physiology of Plants.* Cambridge. 1886.

38. Velten. *Einwirkung der Temperatur auf die Protoplasmabewegungen.* Flora, 1876.
39. Verworn. *Die polare Erregung der Protisten durch den galvanischen Strom.* Pflügers Archiv. Bd. XLV. u. XLVI.
40. Verworn. *Psycho-physiologische Protisten-Studien.* Jena. 1889.

CHAPTER V.

THE VITAL PROPERTIES OF THE CELL (*continued*).

Metabolism and Formative Activity.

GENERAL CHARACTERISTICS. Each living cell exhibits the phenomena of metabolism; it absorbs nutrient material, which it elaborates, retaining certain portions of it within its body, whilst it rejects others; it resembles a small chemical laboratory, for the most varying chemical processes are almost continually taking place in it, by means of which substances of complex molecular structure are on the one hand being formed, and on the other are being broken down again. The more intense is the vitality of the cell, the more considerable are these processes of destruction and reconstruction, the latter keeping pace with the former. In the chemistry of the cell these two principal phenomena must be clearly kept apart, namely the phenomena of progressive and retrogressive metabolism, or, as Claude Bernard (IV. 1a) expresses it, "les phénomènes de destruction et de création organique, de décomposition et de composition."

During its destruction the living substance, as a result of its own decomposition, passes through a series of intermediate stages of more simple chemical combinations, the precise nature of which is at present unknown. Carbon dioxide and water are the simplest final products of this decomposition. Tension (potential energy) is converted into active vital force (kinetic energy). Intra-molecular heat becomes free, and represents the living force, which is the essential condition for the production of work in the cell body. The fact that the slightest shock often suffices to call forth great changes and to cause work to be done shows that vital substances are exceedingly unstable in composition: as Pflüger (V. 25, 26) remarks: "Are not the forces which act in a ray of light truly inconceivably small? and yet they produce most marked effects upon the retina and the brain. How infinitesimal are the forces which serve to excite the nerves; how extremely minute the amount of certain poisons which suffices to kill a large living animal."

In the reconstruction of living substance, or in progressive metabolism, new material is taken up from outside, to replace that which has been used up; these substances become incorporated and transformed into new chemical combinations. During the execution of this work, more or less heat is rendered latent, and is converted into potential energy; this latent heat is derived partly from the intramolecular heat, which is released by the process of decomposition, partly, and in the case of plants chiefly, from the vivifying heat of the sun's rays, by means of which a large amount of kinetic energy is conveyed to the organic world, and is converted in the protoplasmic body into potential energy. The substances taken up from outside, and the heat rays from the sun, supply in the last instance the material and energy required for the carrying on of the vital processes of alternate decomposition and reconstruction.

According to Pflüger's definition,—" The vital force is the intramolecular heat. The highly unstable molecules of albumen, which are built up in the cell substance, and which become decomposed through a splitting up of the molecules—carbon dioxide, water, and nitrogenous bodies being chiefly formed —becoming continually regenerated and rearranged."

In spite of the great variety of metabolic processes which occur in a single individual, there is a series of fundamental processes, which are common to all organic bodies, and which take place in the lowest unicellular organisms, as well as in the bodies of plants and animals. Thus the unity of the entire organic kingdom is exhibited in these fundamental processes of metabolism, just as in the phenomena of movement and of reaction to stimuli.

Up to this point they may be included in the general anatomy and physiology of the cell. This uniformity is especially noteworthy in the following three points:—

1. Each cell, whether plant or animal, respires, that is to say, it is essential to it, to take up oxygen from its environment, by means of which it oxidises the carbo-hydrates and albuminous substances of its own body, and produces as end products carbon dioxide and water.

2. In both organic kingdoms to a large extent, corresponding substances make their appearance during metabolism, such as pepsin, diastase, myosin, xanthin, sarcin, sugar, inosit, dextrin, glycogen, lactic acid, formic acid, acetic acid, and butyric acid.

3. In both kingdoms a great many identical, or at any rate very similar, processes occur, by means of which complex chemical combinations are produced. These, however, differ essentially from the synthetical methods employed by chemists for the production of different organic compounds. In the chemistry of the cell, whether plant or animal, ferments play an important part (diastase, pepsin, trypsin, etc.). By the term ferment is understood an organic substance, produced by the living cell, of which an exceedingly minute quantity is sufficient to bring about a considerable chemical effect, and which, without being itself, to any appreciable extent, consumed, is able to produce characteristic chemical changes both in carbo-hydrates and albuminous bodies.

"Le chimisme du laboratoire est exécuté à l'aide d'agents et d'appareils que le chimistre a créés, et le chimisme de l'être vivant est exécuté à l'aide d'agents et d'appareils que l'organisme a créés" (Claude Bernard IV. 1a).

In the following pages we will consider the individual phenomena of metabolism, chiefly from a morphological point of view, without entering more fully into the chemical processes, which for the most part are very complicated, and as yet to a great extent obscure. During the course of metabolism three stages may be recognised: the absorption of new material, the consequent transformation effected in the interior of the protoplasm, and the excretion of waste products. We will first consider together the first and third of these stages, and later on the second by itself.

1. **Absorption and Excretion.** All cells absorb gases, and also substances in a fluid or dissolved, and hence diffusible, condition; finally many cells can make use of solid substances as food. These three series of phenomena must be considered apart.

1. **The Absorption and Excretion of Gaseous Material.** Protoplasm can absorb the most various kinds of substances in a gaseous condition (oxygen, nitrogen, hydrogen, carbon dioxide, carbon monoxide, nitrous oxide, ammonia, chloroform, ether, and a large number of similar substances).

Amongst these substances, oxygen and carbon dioxide are the only ones of general importance in metabolism, and of these oxygen is the more important.

Without the absorption of oxygen, that is to say without respiration, life cannot continue. With very few exceptions (anaërobic *Bacteria*, etc.) the *respiration of oxygen* is a fundamental characteristic of the whole of organic nature, being absolutely necessary for the continuance of the metabolic processes upon which life depends, and through which by the oxidising of complex molecular compounds the vital forces must be produced. As a rule the lack of oxygen very quickly arrests the functions of the cell (its irritability, powers of movement, etc.): and finally death of necessity ensues.

Some of the fermentation organisms, the fission and pullulating fungi, appear to form an exception to this fundamental process of respiration. For they are able to grow and multiply in a suitable nutrient fluid when completely shut off from oxygen. In this case, however, the oxygen necessary for the oxidation processes in the protoplasm is obtained through the decomposition of the fermenting substance. Similarly intestinal parasites are able to exist in an environment comparatively free from oxygen by splitting up of compounds of which a superfluity is supplied to them (Bunge V. 2).

What is the part played by the oxygen after it has been taken up by the cell?

It was formerly believed that the oxygen directly oxidised the living material, so that, as it was figuratively expressed, a process of combustion was called forth, as the result of which heat was given off. However, there seems to be little doubt but that the forces which result in the combination of the oxygen originate in the vital substance itself. In this mixture of special albuminous bodies, and their derivatives, which goes under the name of protoplasm, and in which, moreover, fats and carbohydrates are stored up, important molecular re-arrangements and re-groupings of atoms, often the result of very minute exciting causes, take place; amongst these, decomposition and dissociation occur. "Under these circumstances many decomposition products continually develop an affinity for free oxygen (oxidative decomposition), and it is in this way that oxygen takes part in the process of metabolism" (Pflüger V. 25, 26). Hence in consequence of respiration, and at the cost of the organic substance, combinations rich in oxygen are produced; and finally, through the repeated dissociation and oxidation of these substances, carbon dioxide and water, the most important final products of the destructive processes of living substance during respiration are produced.

This is true for every animal and every plant cell.

If plant cells (staminal hairs of *Tradescantia*, cells of *Characeæ*), in which active streaming protoplasmic movements are taking place, are immersed in a drop of pure olive oil, the movements, in consequence of the exclusion of the oxygen, soon commence to slacken, and finally quite cease. The same occurs when plant-cells are introduced into an atmosphere consisting exclusively of carbon dioxide or of hydrogen, or of a mixture of the two. At first the functions of the protoplasm are only arrested, and if the olive oil, carbon dioxide, or hydrogen, be soon removed, the irritability and movements return gradually after a period of rest. If however the cells are deprived of oxygen for a considerable time, their functions become paralysed, until finally death, accompanied by the turbidity, coagulation, and decomposition of the protoplasm, ensues.

In a similar manner each animal cell respires. If a hen's egg, which has been incubated, and which, being in an early stage of development, consists simply of small cells, is placed in an atmosphere of carbon dioxide, or if its porous shell is so saturated with oil that no interchange of gases can take place between the embryo and the outer air, the egg dies in a few hours.

The oxygen which is absorbed by man through the lungs serves to satisfy the need of oxygen evinced by all the cells contained in the various tissues of our bodies. This last process is designated in animal physiology internal or tissue respiration, in contradistinction to the taking in of oxygen or lung respiration.

In the whole organic kingdom, respiration is united with the excretion of carbon dioxide and with the production of heat. The following is a simple chemical law: "A certain amount of heat is evolved during respiration, just as it is produced in every other case when carbon and hydrogen are oxidised into carbon dioxide and water" (Sachs IV. 32a). Plant cells expire carbon dioxide and evolve heat, just like animal cells.

The formation of heat is most easily demonstrated in portions of plants which are growing rapidly; such as in germinating seeds. It can be especially well detected in the flowers of *Aroideæ*. These become heated to as much as 15° C. above the temperature of their surroundings.

The living cell itself is able, by means of its respiration, to regulate the amount of oxygen which it consumes. This depends simply upon the degree of its functional activity, to which the decomposition of organic substance is proportionate. An unfertilised egg-cell and a resting plant seed breathe in very minute quantities of oxygen; however, after the egg-cell has been fertilised, and division is proceeding rapidly, or when the plant seed germinates, the amount of oxygen which is absorbed increases. This absorption of oxygen is one of the functions of active living protoplasm (Sachs). Thus the following is easily explained, that the absorption of oxygen by the living cell "is, within certain wide limits, quite independent of the gaseous tension of the oxygen" (Pflüger).

One important phenomenon must be described before closing this chapter on respiration. Even when oxygen is absent the cells are able to excrete carbon dioxide and evolve heat for a longer or shorter time. If germinating plants are introduced into a Torricellian vacuum, they continue to exhale a normal quantity of carbon dioxide for about an hour, after which the quantity gradually decreases.

According to Pflüger's experiments, Frogs can live for several hours in a bell-jar which is free from oxygen and filled with nitrogen, during which time they exhale a considerable quantity of carbon dioxide.

Both these experiments prove, that for a time, without direct access to oxygen, but simply through the decomposition of organic substances, carbon and oxygen atoms may unite together in the cell to form carbon dioxide.

This process is termed *intramolecular respiration*. As long as this persists, the cell lives, and remains irritable and capable of performing its functions, although with continually decreasing energy, by using up a portion of the oxygen contained in combination in its substance. However, when oxygen is withheld for a considerable time, death invariably ensues.

Upon these phenomena of intramolecular respiration the proposition already mentioned rests: "that the first impulse to the chemical processes of respiration is not given by the oxygen which enters from without, but that first and primarily a decomposition of albumen molecules resulting in the formation of carbon dioxide takes place inside the protoplasm, and that hence the incoming oxygen effects a *restitutio in integrum*."

In fermentation processes, during which the ferments grow, multiply, and evolve carbon dioxide, without having access to oxygen, we see an instance which resembles intramolecular respiration; to this Pfeffer (V. 22) has called especial attention.

Whilst the absorption of oxygen and the giving up of carbon dioxide indicate the beginning and end of a series of complicated processes which belong chiefly to retrogressive or destructive metabolism (catabolism), the absorption and elaboration of carbon dioxide in the cell afford us an insight into the opposite process, progressive metabolism (anabolism), or the reproduction of organic substance. This process, in contradistinction to respiration, is termed assimilation.

Respiration of oxygen and assimilation of carbon dioxide are in every respect opposite processes. The former is a fundamental phenomenon common to nearly the whole organic kingdom, the latter is confined to the vegetable kingdom alone, and even here occurs *only in such cells as contain chlorophyll or xanthophyll in their protoplasm.* The respiration of oxygen conduces to oxidation decomposition processes, whilst on the contrary the assimilation of carbon dioxide causes the reduction of the latter, and the synthetic formation of complex molecular organic substances. These are carbo-hydrates, especially starches, which are found deposited in the form of small granules in the green portions of plants (chlorophyll corpuscles and chlorophyll bands).

The individual stages of the synthetic processes which take place in the plant-cell during the assimilation of carbon dioxide are as yet unknown. Only so much may be said : carbon dioxide and water form the initial material for the synthesis ; further, as a result of the reduction of the carbon dioxide and water, oxygen is evolved, and is given off largely in the form of a gas. This transformation can only take place in protoplasm when chlorophyll is present ; but it is possible that other chemical substances are also concerned in the process. Finally, carbon dioxide assimilation can only occur under the influence of light. Heat is necessary in order to liberate the oxygen from the molecules of carbon dioxide and water. In this point also carbon dioxide assimilation and oxygen respiration are opposed : in the latter heat is evolved through oxidation, which is a process of combustion, and vital force is set free ; in the former heat is used up in reducing the carbon dioxide, and as potential heat is rendered latent in the assimilation products. The heat required for this process is afforded by the sun's rays.

If an aquatic plant is introduced into water containing carbon dioxide, and is placed in the sunlight, innumerable small bubbles of gas are soon seen to rise ; if these are collected in a bell-jar, they can be shown by chemical analysis to consist chiefly of oxygen. The amount of oxygen exhaled is in proportion to the carbon dioxide which is simultaneously absorbed out of the water, and the carbon of which is elaborated into carbo-hydrates. It has already been mentioned in a previous chapter (p. 103), that the living protoplasm, which is sensitive to light, endeavours to bring the chlorophyll corpuscles into favourable positions for receiving the direct powerful rays of light.

The process of assimilation proceeds in such an energetic manner under the influence of sunlight that, in comparison to it, the respiration of oxygen and the exhalation of carbon dioxide, which are absolutely essential for the maintenance of the vital processes, are placed quite in the background, so much so, indeed, that in former times they were quite overlooked. But in plants which are placed in the dark, the expiration of oxygen and, to an equal degree, the absorption of carbon dioxide are immediately arrested, whilst respiration continues in precisely the same manner as when the plants were in the light. The gas now given off is seen to be carbon dioxide, the quantity of which, however, is much less than that of the oxygen in the preceding experiment.

Claude Bernard (IV. 1a) has drawn attention to a very interesting difference existing between the respiration of oxygen and the assimilation of carbon dioxide in plants. He narcotised water-plants by means of chloroform or ether, and then found that they no longer gave off oxygen in direct sunlight. Thus the function of the chlorophyll, the capacity of forming starch by synthesis from carbon dioxide and water, is absolutely suspended during narcosis, just as the irritability and power of motion are arrested in the protoplasm. This capacity returns when the plants are transferred into pure water. But it is still more remarkable that respiration, including the exhalation of carbon dioxide, is uninterrupted during narcosis. This difference may be probably traced back to the fact that respiration, and the decomposition in connection with it, stand in a much closer relationship to the whole vital economy, and hence can only be quite extinguished with the life of the cell itself. But long before this occurs, the functions of the cell are paralysed during narcosis, and with them the chlorophyll function.

2. **The Absorption and Excretion of Fluid Substances.** Most of the substances concerned in metabolism are taken up by the organism in a fluid condition. Unicellular and aquatic plants extract

them from the fluid by which they are surrounded, whilst terrestrial plants take them up with their roots from the soil, which is saturated with moisture. The cells of the higher animals nourish themselves by absorbing substances held in solution in fluid media, which must first, by means of complicated processes, be introduced by them into their bodies. These fluid media are the chyme of the intestinal canal, blood, chyle, and lymph. They play the same part in the economy of the animal cell as the water and moisture of the earth do in that of the lower organisms and of plants.

In opposition to the antiquated physiological view that the principal metabolic processes take place in the fluids of the body, too much stress cannot be laid upon the following proposition,—that the cells are the site of the absorption, excretion, and transformation of material; the fluids only function in conveying the nutrient material in a fluid condition to the cells, and in carrying away the waste products.

Between the cell and its surrounding medium, there exist the most complicated physical and chemical conditions of interchange. Their investigation is a most difficult undertaking, and can only be entered into here to a very limited extent.

Each cell adapts itself most closely in its organisation to the surrounding medium, any considerable variation in the concentration or composition of which causes its death. However, in many cases, great alterations may be permanently endured, provided that the consecutive stages are allowed to merge slowly and gradually into one another, so that the cell has time to adapt itself to its new conditions.

As has been already mentioned in the chapter on chemical stimuli (p. 111), fresh-water *Amœbæ* are able to accustom themselves to living in salt water, whilst marine animals can adapt themselves to the presence of a greater or less percentage of salt in the water surrounding them. Apparently they adapt themselves by adjusting the fluid they contain to the surrounding medium. It is on this account that when the changes are made suddenly, death immediately ensues, the protoplasm either swelling up, or shrinking and coagulating.

Since in Vertebrates the cells which are bathed in the tissue-fluids exist under such extremely complex conditions, it is difficult to keep small portions of tissue alive, even for a short time, when once they have been separated from the rest of the body; for even the tissue-fluids become quickly altered as soon as they are separated from the living body. Hence, in examining a tissue outside of the body, blood serum, aqueous humour, amniotic fluid, iodised serum, or artificially prepared mixtures resembling these fluids, only function, to a certain extent, as indifferent, supplementary fluids. As a matter of course, they cannot at all supply the natural conditions for the cell.

In endeavouring to understand the relationship which exists between the cell and the fluid which bathes it, care must be taken at the outset to avoid the idea that the former is simply saturated by the latter. Such a conception is wholly fallacious; on the contrary, each cell is an independent unity which selects certain substances from the mixture of fluids surrounding it, and absorbs a varying quantity of them, whilst others it quite rejects. In all these respects different cells behave very differently: in a word, the cells, to a certain extent, make a selection from the substances offered them.

Such selective powers, often very different in character, may be easily demonstrated by the following:—

Amongst the lowest unicellular organisms there are some which possess silicious skeletons, whilst others construct theirs out of carbonate of lime. Hence they exhibit quite opposite powers of selection towards these two substances, both of which occur in small quantities in solution in water, and by this means very important effects have been produced in the formation of chalk, and of the geological strata, consisting of silicious shells. Similarly, different plants, which thrive side by side under similar conditions and in the same water, take up from it very different salts, and these in very varying quantities. The relative proportions which occur may be easily computed by drying and burning the plants, and then reckoning out the proportion which the ash bears to the whole of the dried substance, and further the proportion the separate constituents of the ash bear to the pure ash.

The ashes of several kinds of Fucus which were collected on the west coast of Scotland were examined, and the results obtained were tabulated by Pfeffer (V. 23) in his *Plant Physiology*.

	Fucus vesiculosus.	Fucus nodosus.	Fucus serratus.	Laminaria digitata.
Pure ash . per cent.	13·89	14·51	13·89	18·64
K_2O . . . ,,	15·23	10·07	4·51	22·40
Na_2O . . ,,	24·54	26·59	31·37	24·09
CaO . . ,,	9·78	12·80	16·36	11·86
MgO . . ,,	7·16	10·93	11·66	7·44
Fe_2O_3 . . ,,	·33	·29	·34	·62
P_2O_5 . . ,,	1·36	1·52	4·40	2·56
SO_3 . . ,,	28·16	26·69	21·06	13·26
SiO_2 . . ,,	1·35	1·20	·43	1·56
Cl . . . ,,	15·24	12·24	11·39	17·23
I . . . ,,	·31	·46	1·13	3·08

Marine plants show most clearly, in what very unequal proportions, they absorb from the multitude of salts offered them in seawater, the ones which are necessary to them. For instance, they only store up very small quantities of common salt, of which about 3 per cent. is present in the water, whilst, on the contrary, they take up relatively large amounts of potassium, magnesium, and calcium salts, of which there are only traces. And in a similar manner, the analysis of the ashes of different land-plants which have flourished side by side in the same earth yields very different results.

Investigation of the metabolism occurring in the animal body leads to the same conclusion. Only certain cells have the tendency to take possession of the lime-salts, which are present in almost inappreciable amounts in the fluids of the body, and to deposit them in the osseous tissues; other groups of cells, such as those in the kidneys, take up the substances from the blood, and excrete them in the form of urine; others store up fat, etc., etc.

The factors concerned in this absorption and non-absorption of matter are at present quite beyond our comprehension. It is curious that the need which is evinced by the economy of a cell for a certain substance does not always imply that this will be taken up. Cells may absorb materials which are either directly hurtful or completely useless to them. In this respect the very different ways in which living plant cells take up aniline dyes are very instructive (Pfeffer V. 22b).

Although solutions of methylene blue, methyl violet, cyanin, Bismark brown, fuchsine and safranin, are absorbed, those of nigrosin, aniline blue, methyl blue, eosin, and congo-red, are not.

As to whether a given substance will be absorbed or not can, according to Pfeffer, who has carefully studied the subject, only be decided empirically.

The substances excreted by cells also vary. Just as with absorption, excretion depends upon the special individual properties of the living cell body. The red or blue-coloured petals of phanerogamic flowers do not allow the concentrated solution of colouring matter which they contain to become diffused into the surrounding water as long as they are alive. However, as soon as the cells die, the colouring matter commences to pass through the cell-wall.

In order to really understand all these complicated phenomena, it would be necessary to possess an exhaustive knowledge of the chemistry and physics of the cell. For the property, which I have designated above as the power of selection, must in the last instance be traced back to the chemical affinities of the very numerous substances which, being formed during the process of metabolism, are present for a time in the cell. The same thing, doubtless, occurs here as with the absorption of oxygen and carbon dioxide, which can only take place when, through metabolic processes, substances with chemical affinities for them are set free. It is on this account that no carbon dioxide is taken up by plants in the dark, although it is immediately absorbed, if, under the influence of direct sunlight, the chemical process for which it is necessary is started.

The same thing occurs when living cells absorb aniline dyes. *Azolla*, *Spirogyra*, the root-hairs of *Lemna*, etc., gradually draw up into themselves so much colouring matter out of a very weak solution of methylene blue, that they acquire a deep blue colouration, such as is seen in a 1 per cent. solution. The methylene blue does not stain the protoplasm itself, but simply passes through it, thus forming in the cell sap a solution of ever-increasing strength. Hence the death of the cell, which would inevitably occur if the poisonous methylene blue were to be collected in such quantities in the protoplasm itself, does not ensue. This storing up in the cell sap is caused by the presence in it of substances which, with the aniline dye, form compounds, which osmose with difficulty. Pfeffer considers that the tannin which is so frequently found in plant cells is a substance of this nature. This tannin, with the aniline colour, forms compounds which are sometimes insoluble, and hence are precipitated in the cell sap (methylene blue, methyl violet), and sometimes are more or less soluble (fuchsine, methyl orange, tropæolin).

Further, animals afford us good examples of this storing up in living cells. Fertilised eggs of *Echinoidea* acquire a more or less intense blue colouration, if they are placed for a short time in a very dilute solution of methylene blue (Hertwig IV. 12b). A small accumulation of colouring matter does not arrest the process of segmentation, which still continues, although somewhat slowly, in a normal fashion, and in some cases may go on even until the gastrula is formed. Here the colouring matter is chiefly deposited in the endoderm cells, which points to the conclusion that it is by the agency of the yolk material that the accumulation takes place. Living Frog and Triton larvæ become of an intense blue colour if they are left for from five to eight days in a weak solution of methylene blue. In this case the colouring matter combines with the granules in the cells (Oscar Schultze V. 44). After remaining for days in pure water they commence to become colourless again. If indigo-carmine is injected directly into the blood of a mammal, it is soon taken up both by the liver-cells and by the epithelium of the convoluted tubules of the kidney, and then is excreted either into the biliary ducts, or into the kidney tubules (Heidenhain V. 42). If methylene blue is injected into the blood, it combines with the substance of the nerve fibres, imparting to them a dark blue colouration (Ehrlich V. 41). Alizarin is stored up in the ground substance of the bones.

Next to the chemical affinities, which exist between the particles of matter within the cell and those outside of it, the study of the physical processes of osmosis is of the greatest importance for the comprehension of the absorption and rejection of matter. We must here observe whether the membrane, when present, is more or less permeable. As a rule it is much more permeable to dissolved substances than is the protoplasmic substance itself. This latter is separated from the exterior by a peripheral layer (*cf.* p. 15), which, according to Pfeffer, plays a most important part in the process of osmosis. If some substance in solution is to be taken up into the protoplasm, it must first be imbibed by the peripheral layer; that is to say, its molecules must become deposited between the plasmic particles, and from there be transferred to the interior. Further, a substance in solution can, even if it be not actually absorbed, produce an osmotic action by exerting an attraction upon the water contained in the cell, and by thus inducing a flow of water towards the exterior. "Essentially osmosis consists in this, that two fluids simultaneously pass through a membrane in opposite directions; with regard to an endosmotic equivalent (a term expressing the proportionate interchange, upon which there is frequently too much stress laid), this cannot be spoken of in such cases where only water is diosmosed through a membrane" (Pfeffer V. 23).

On account of their fragility and small size, experiments upon osmosis can only be made in animal cells with great difficulty. Hence the osmotic processes have been investigated chiefly by botanists in plant cells, which are much more suitable, and our

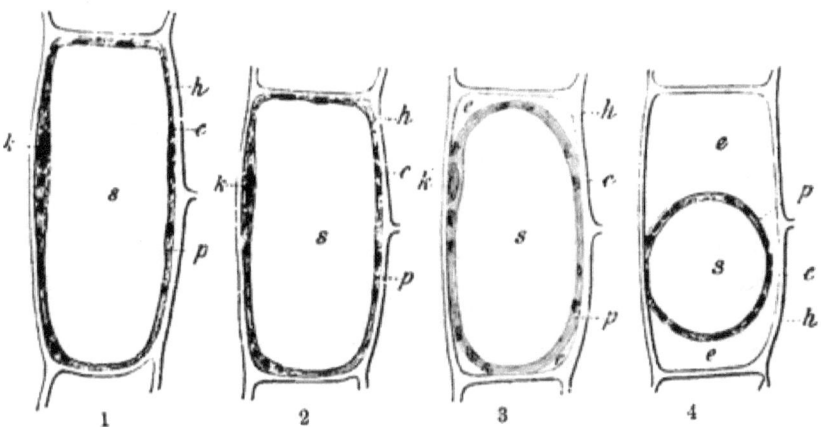

Fig. 59.—1, A young, at most half-grown, cell from the cortical parenchyma of the flower peduncle of *Cephalaria leucantha*. 2. The same cell immersed in a 4 per cent. solution. 3. The same cell in a 6 per cent. solution. 4. The same cell in a 10 per cent. solution (Nos. 1 and 4 are taken from nature, Nos. 2 and 3 are diagrammatic; all in optical longitudinal section). h Peripheral layer; p protoplasmic coating of wall; k nucleus; c chlorophyll granules; s cell sap; e salt solution which has penetrated into the interior. After de Vries (V. 36).

knowledge has been especially advanced by the following experiments.

If plant cells containing a large sap space are placed in a 5 to 20 per cent. solution of a suitable salt, or of sugar or glucose (Fig. 59), they are seen to diminish somewhat in size from having given up water from the interior to the exterior; in consequence, as this process of water abstraction proceeds, the protoplasmic coating becomes separated from the cellulose membrane, which, on account of its greater firmness, is unable to shrink any more (de Vries V. 36).

Thus the salt or sugar solution must make its way through the cellulose membrane, after which it continues to abstract more water from the protoplasm, which shrinks more and more according to the concentration of the solution, so as to occupy a smaller and smaller space. The sap which it encloses becomes correspondingly more concentrated. In spite of these changes, which are grouped together under the same *plasmolysis*, the protoplasm may remain alive for weeks, and exhibit its usual streaming movements; it may even surround itself with a new peripheral layer, although it remains in its contracted condition.

Two conclusions may be deduced from the process of plasmolysis : (1) that the cellulose membrane is pervious to the salt solutions which were used ; (2) "that the amount of dissolved salt which diosmoses through the peripheral layer is not worth mentioning, for if a considerable quantity penetrated into the protoplasm, or into the cell sap, an increase in the quantity of the substances setting up osmosis would be produced within the protoplasmic membrane, and thus an increase in the volume of the protoplasmic body would result " (Pfeffer).

If the cells which have become flaccid through plasmolysis are carefully removed and placed in pure water, the reverse process occurs. The sugar solution which was enclosed within the cellulose membrane becomes diffused into the water. In consequence, the peripheral protoplasm layer becomes distended, because its cell sap is now richer in osmotolytic substances than its environment, and so water is caused to flow in the opposite direction. This distension gradually increases, as the water becomes absorbed, until the peripheral layer of protoplasm comes into close contact with the cellulose membrane, and until finally the cell has dilated to its original size.

Other experiments have shown that the sap contained in the plant cell is under a considerable pressure, often of several atmospheres. This produces the natural turgescence of certain portions of plants. The cause is,—that powerfully osmotolytic substances are present in the cell sap, such as saltpetre, vegetable acids, and their potassium salts, which have a strong affinity for water (Pfeffer V. 23 ; de Vries V. 36).

Therefore under these conditions the protoplasmic coating containing the cell sap may be compared to a very elastic thin-walled bladder, which is filled with a concentrated salt solution. If such a bladder is put into pure water, the solution attracts the water, and so produces a current, the result being that the bladder swells up in consequence of the increased pressure of its contents, and its wall grows thinner and thinner. The distension of the bladder only ceases when the external and internal liquids are in osmotic equilibrium. Thus the protoplasmic coating of many plant-cells would be very much distended in consequence of the internal pressure (turgor) were it not that a limit is set to its distension by the less elastic cellulose membrane.

Equilibrium between the cell-sap and the surrounding fluid might be established, if the osmotic substances were to become diffused into the water, so as to remove the cause of the internal pressure. However, this is prevented by the properties of the living plasmic membrane. As the plasmic membrane, if the expression may be allowed, decides whether a body may be admitted into the interior of the cell or no, similarly it has the important power of retaining in the cell-sap dissolved substances which otherwise would be washed out by the water bathing the cell; of this property mention has already been made, and an instance cited (Pfeffer V. 23).

That, in fact, the cell-sap exists under a pressure greater than that of its environment, for instance, that the pressure in aquatic plants is greater than that of the surrounding water, may be easily proved by some simple experiments, as has been shown by Nägeli (V. 16). If a cell of *Spirogyra* be opened by an incision, so that part of its contents flows out, the transverse walls of the two neighbouring cells bulge out towards the cavity of the injured one. Hence the pressure in the uninjured cells must be greater than that in the injured one, the tension of which has sunk down to the level of that of the surrounding water.

3. **Absorption of Solid Bodies.** Cells, which either are not surrounded by a special membrane, or possess apertures in their membranes, are able to take solid bodies up into their protoplasm, and to digest them. Thus Rhizopoda capture other small unicellular organisms with which their widely outstretched pseudopodia come into contact (Figs. 10, 60). The pseudopodia which have seized the foreign body contract, and so gradually draw it into the mass of the protoplasm; here the nutrient substances are extracted, whilst the indigestible remains, such as skeletal structures, are after a time ejected to the exterior. Even solid substances, which possess but small nutritive value, are taken up. If carmine or cinnabar granules are introduced into the water,

the Rhizopoda eagerly seize upon them, so that after a short time their whole bodies are quite filled with them.

Infusoria (Fig. 50) eat Flagellata, unicellular Algæ and Bacteria, conveying them into their endoplasm through an opening in their cuticle which functions as a mouth. Here a vacuole filled with fluid forms itself round each foreign body, which undergoes digestion.

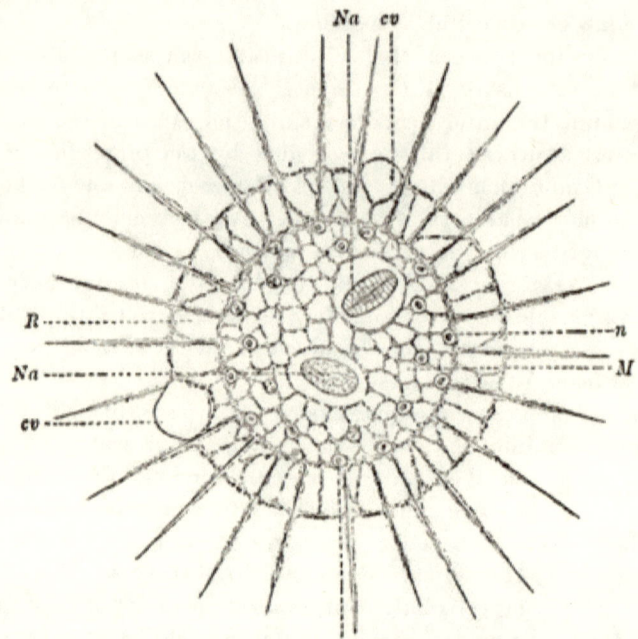

FIG. 60.—*Actinosphærium Eichhorni* (after R. Hertwig, Zool., Fig. 117): M medullary substance with nuclei (n); R cortical substance with contractile vacuoles (cv); Na nutrient material.

In a similar manner to that shown by unicellular organisms, many tissue cells of Metazoa devour solid substances offered to them, and digest them.

Intracellular digestion, as it has been termed by Metchnikoff (V. 12), occurs very frequently in Invertebrates; it may be best demonstrated by means of feeding experiments with easily recognisable substances, such as granules of colouring matter, globules of milk, spores of fungi, etc. In some Cœlenterata the ectoderm as well as the endoderm takes up foreign bodies. The tentacular ends of *Actinia* may load themselves with carmine granules, which may also be found distributed throughout the whole endoderm of *Actinia* larvæ after suitable feeding.

But white blood corpuscles, lymph cells and the migratory cells of the mesoblast, in both Vertebrates and Invertebrates, afford us the best material for observation, in consequence of their power of absorbing and digesting solid bodies. This important fact was first observed by Haeckel (V. 4a), who injected a mollusc (*Tethys*) with indigo, and found after a short time that indigo granules were present inside the blood corpuscles.

Metchnikoff (V. 12) has further investigated the phenomenon most thoroughly. He found that if powdered carmine were injected under the skin of another species of mollusc (the transparent *Phyllirhoë*), the smaller granules were eaten up by some of the migratory cells, while the larger ones attracted a number of other migratory cells around them, which surrounded them like an envelope, and fused themselves together to form a plasmodium or multinucleated giant cell.

That the same thing occurs in Vertebrates may be easily proved by injecting some carmine into the dorsal lymph sac of a Frog, and, after a short time has elapsed, removing some drops of lymph, and examining them with the microscope. Further, the eating process can be directly followed under the microscope if powdered carmine or a little milk be added to some fresh drops of lymph or of blood which have been carefully drawn off, certain precautions having been observed. If the blood has been taken from man or some other mammal, the preparation must be carefully heated on Max Schultze's warm stage until it has attained a temperature of $30-35°$ Celsius (V. 43). The white blood corpuscles now commence to show amœboid movements; they seize with their pseudopodia the carmine granules, or milk globules with which they come in contact, and draw them into their bodies. On this account Metchnikoff designates them as *phagocytes*, and the whole process as *phagocytosis*.

This capacity of the amœboid elements of the animal to take up solid substances is of great physiological importance; for herein the organism possesses a means of ridding itself of foreign and noxious organic particles which are present in its tissues. There are three different conditions of the body, partly normal and partly pathological, when the phagocytes exercise this function.

Firstly, during the process of development in many Invertebrates and also in Vertebrates, certain larval organs lose their importance, and undergo fatty degeneration. Thus, during the metamorphosis of *Echinoderm* larvæ and of *Nemertines*, certain portions disappear; and, similarly, the young Frog during its development loses its conspicuous tail, which acted as a rudder. In all these cases the cells of these degenerating organs undergo a fatty metamorphosis, die and disintegrate. In the meantime a large number of migratory cells or phagocytes have collected in their neighbourhood, and these commence to devour and digest the degenerated tissue, as can be plainly seen during life in transparent marine animals.

Secondly, just as during the normal processes of development, the phagocytes occupy themselves in reabsorbing particles, the death or disintegration of which has been brought about either by normal or pathological conditions. Red blood corpuscles become destroyed after they have circulated in the blood for a certain time. In splenic blood their remains have been seen in the bodies of white corpuscles, which here again fulfil their function of getting rid of dead material. When in consequence of a wound an effusion of blood occurs in the tissue, and thousands of blood corpuscles and elementary particles are destroyed, the migratory cells again set to work, and produce reabsorption and healing.

Thirdly, and lastly, the phagocytes during infectious diseases constitute a body-guard to the organism, in opposing the spread of the micro-organisms in the blood and tissues.

Metchnikoff has rendered great service in drawing attention to this circumstance (V. 13-15, IV. 22). He succeeded in showing that the Cocci of erysipelas, the Spirilla of relapsing fever, and the Bacilli of anthrax were eaten up by the wandering cells, and thus rendered harmless (Fig. 61). The micro-organisms, of which as many as from ten to twenty may be present in one cell, after a certain time show distinct signs of degeneration. If the micro-organisms are present in the blood, they are destroyed, especially in the spleen, liver, and red bone marrow. If they succeed in settling down in some place in the tissue, the body endeavours to get rid of the intruders by collecting as the result of inflammatory processes a large number of migratory cells to the spot.

As Metchnikoff expresses it, between micro-organisms and phagocytes an active war is raging. This is settled in favour of one or other party, resulting, as the case may be, in the recovery or death of the affected animal.

The power possessed by migratory cells of destroying certain species of micro-organisms appears to vary considerably in different animals, and to depend largely upon the most varying conditions; for instance, chemical stimuli play an especially important part, as has been already mentioned on p. 121 (negative and positive chemotropism; Hertwig IV. 13). Apparently it is upon this that the greater or less immunity of organisms from many infectious diseases depends. This discovery opens a wide vista in the field of the comprehension and treatment of infectious diseases.

II. The Assimilative and Formative Activity of the Cell.

The gases, the fluids, and the solid substances, which are introduced into the protoplasm as food, and through respiration, compose the very varying raw materials which are elaborated in the chemical workshop of the cell, and which are converted into an exceedingly large number of substances. Amongst these the most important for both plants and animals are: carbo-hydrates, fats, proteids, and their numerous compounds.

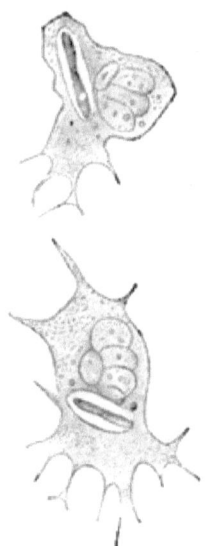

FIG. 61.—A leucocyte of a Frog, enclosing a Bacterium, which is undergoing digestion. The Bacterium is stained with vesuvine. The two figures represent two stages of one and the same cell. (After Metchnikoff, Fig. 54.)

Similarly the ways in which they are utilised in the vital processes of the cell vary very considerably. They serve partly to replace the substances, which, during the vital process, become decomposed in the cell, such as the substance which is oxidised during respiration, and which thus furnishes the vital energy necessary for the activity of the cell. They are also utilised for that growth and increase of the protoplasm which is absolutely indispensable for the function of reproduction. Farther, some of the substances formed in the chemical laboratory are stored up for future use in the cell-body in some form or other, thus constituting reserve material. Finally they may be set aside to fulfil some function inside or outside the cell.

Thus arise the different materials which, especially in the animal kingdom, are very numerous, and upon which the differentiation of tissues depends: glandular secretions, which are passed to the exterior, membranes, and intercellular substances of very varying chemical composition, and muscle and nerve fibres, which, in consequence of their peculiar organisation, are endowed in a special manner with contractility and the power of conducting stimuli. In the last case the chemical activity of the cell assumes a character which Max Schultze has designated as its formative activity. The protoplasm makes use of the raw material which is brought to it, and prepares from it often very wonderfully constructed substances, which answer special purposes. In this activity the cell appears, to a certain extent, like a builder, or, as Haeckel (V. 4b) has it, like a modeller or sculptor.

This formative activity of the cell, or, as it is better expressed, the power of the protoplasmic body to create different structures, is of extreme importance; for it is solely due to this power that there is so great a diversity of elementary particles, in consequence of which the animal body is able to attain to so high a degree of perfection. The division of labour, which is so successful amongst cells, is based solely upon this foundation, and by its means the capacity for work of the cell community is rendered much greater.

Hence this subject of the assimilation of material must be examined from two points of view; the first is a chemical one, in so far as it treats of the formation of innumerable substances by means of the protoplasm, whilst the second is more morphological, in so far as the various substances present in the protoplasm may be seen to differ from it, to occupy a definite position, to have a fixed form and structure, and to obey special laws of development.

One of the most important tasks for the biological chemist of the future is to render accessible to morphological investigation the various substances distributed throughout the cell body by means of differential staining mixtures.

1. **Chemistry of Assimilation.** The chemical processes of the cell, which are at present shrouded in mystery, can only be treated here in so far as they are connected with fundamental problems, such as the synthesis of carbo-hydrates, fats, and proteids out of more simple elementary substances.

The chemical processes in the animal kingdom appear to differ considerably from those occurring in the vegetable kingdom. Only that protoplasm present in plant cells, which contains chlorophyll, is able to make high molecular ternary compounds out of carbon dioxide and water; the protoplasm which does not contain chlorophyll, and which is present in animals and certain colourless portions of plants, is only able to undertake further synthesis with this original material, and thus to produce quaternary compounds.

It is as yet impossible to say what chemical processes occur in the green protoplasm, when, under the influence of the sun's vital energy, carbon dioxide and water are taken up, and oxygen is given off. The first product of assimilation, which can be definitely made out, is starch, or perhaps, as a preliminary stage, sugar. It is almost inconceivable that either of these could be formed by a direct synthesis of carbon and water; apparently a number of intermediate substances are formed during the course of a complicated process. "Indeed, it is not impossible," as Sachs (IV. 32a) remarks, "that certain closely-connected constituents of the green plasma themselves participate in the process; that, for example, the molecules of the green protoplasm become split up, and that certain atoms are given up and others substituted for them. The theory has a certain degree of probability from the observation that in many, though not all cases, the mass of chlorophyll substance gradually decreases, and finally quite disappears, whilst the starch granules which it contains become larger and larger."

The carbo-hydrates (starch) which, by means of the chlorophyll function, have accumulated in the body of the plant, form the material which is converted in the protoplasm into the vegetable oils. The ternary non-nitrogenous, organic compounds supply further the basis for the synthesis of quaternary albuminous substances, and thus assist in the completion and increase of the protoplasm. However, for these processes, nitrates and sulphates are necessary, and these are obtained by the plants from the earth by means of their roots.

That proteid substances can be formed by the living cell out of such material has been experimentally proved by Pasteur. He cultivated low Schizomycetes, such as *Mycoderma aceti*, Yeast, etc., in artificially prepared nutrient solutions. Thus he showed that *Mycoderma aceti* can multiply actively in the dark, if only a few cells are placed in a nutrient solution, composed of a salt of ammonia, phosphoric acid, potash, magnesia, water, and alcohol or acetic acid of suitable strength. Hence the fungi cells, if they have multiplied to a considerable extent, must have formed proteid materials by means of the decomposition of these substances, in addition to cellulose and fats.

Thus plants, which by means of their chlorophyll produce carbo-hydrates, and convert these again into fats and albuminous substances, supply to the animal organism the ternary and quaternary substances which are necessary for its nutriment, and which it is unable to elaborate, as the plants do, from such simple substances. In this manner the vegetable and animal kingdoms constitute a life cycle, in which they assume opposite positions and complement each other. This antithesis may be formulated as follows:—

In the green plant cell the organic substance is formed synthetically from carbon dioxide and water, whilst the vital force which is obtained from the sunlight becomes potential; on the other hand, the animal cell uses as nutriment the ternary and quaternary compounds formed in the vegetable kingdom, for the most part oxidising them. By this means it reconverts the potential energy stored up in the complex compounds into vital energy whilst performing work and evolving heat. The plant, whilst its chlorophyll is exercising its function, absorbs carbon dioxide, and gives off oxygen; the animal breathes in oxygen, and breathes out carbon dioxide. In the chemical processes of the plant reduction and synthesis predominate, whilst in those of the animal oxidation, combustion and analysis are most important.

However, from this one example of antithesis occurring in the economy of nature between the animal and vegetable kingdoms, it must not be concluded that plant and animal cells are quite opposed in all their ordinary vital phenomena; for this is not true. Close investigation shows that there is universal unity in the fundamental processes of the whole organic world. The above-mentioned difference is only due to the fact that the plant cell has developed a special faculty which is lacking in animal cells, namely, the power of decomposing carbon dioxide by means of its chlorophyll. With the exception of this one function, exercised by chlorophyll, many of the metabolic processes which are essential for the maintenance of life are performed in the protoplasm in a perfectly similar manner in both plant and animal cells.

In both the protoplasm must breathe, take up oxygen, evolve heat, and give up carbon dioxide if the vital processes are to be carried on. In both plants and animals the decomposition and reconstruction of protoplasm follow one another, and complicated processes of correlated chemical analysis and synthesis occur.

This similarity can be more easily understood when it is remembered that a large proportion of plant cells, namely all those which do not contain chlorophyll, are in a position similar to that occupied by animal cells; these also, since they cannot assimilate directly, must obtain from the green cells, the material necessary for the maintenance of their life, for their growth, and for their reproduction. Thus the same antithesis, which is present in the economy of nature between plants and animals, also exists in the plant itself between its colourless and its chlorophyll-containing cells.

Claude Bernard has shortly and in a striking way expressed the relationship in the following words :

"If, in the language of a mechanician, the vital phenomena, namely the construction and destruction of organic substance, may be compared to the rise and fall of a weight, then we may say that the rise and fall are accomplished in all cells both plant and animal, but with this difference, that the animal element finds its weight already raised up to a certain level (*niveau*), and that hence it has to be raised less than it subsequently falls. The reverse occurs in the green plant cells. In a word, 'Des deux versants, celui de la descente est prépondérant chez l'animal; celui de la montée, chez le végétal'" (Claude Bernard, IV. 1a, vol. ii. p. 514).

Now, having placed the subject of the chlorophyll function in its true position, we will proceed to examine the important uniformity which exists in the chemistry of metabolism between plant and animal cells.

We must first lay stress upon the fact that a large number of the materials made use of in progressive and retrogressive metamorphosis are common to both plants and animals.

Further, the means by which certain important processes in plant and animal cells are carried out appear to be similar. Carbo-hydrates, fats and albuminous substances are not adapted in every condition for direct use in the laboratory of the cell and for conversion into other chemical compounds. It is necessary to prepare them by transforming them into a soluble and easily diffusible form. This occurs, for instance, when starch and glycogen are converted into grape sugar, dextrose and levulose; when fat is split up into glycerine and fatty acids, or when proteids are peptonised.

Sachs (IV. 32a) describes the above-mentioned modifications of carbo-hydrates, fats and proteids as their active condition, in distinction to their passive condition, when they either remain accumulated in the cell as fixed reserve materials—starch, oil, fat, albumen crystals—or are taken up as nourishment by animals. It is only when they are in the active condition that the plastic materials in both plant and animal bodies can accomplish their migrations, by means of which they reach the places where they are either to be temporarily stored up or immediately used.

For instance, the starch, which is accumulated in seeds or in portions of plants which are underground, such as tubers, was not assimilated at these spots. It originated in the assimilating green cells, from which it was transported, often through long distances, by means of intermediate cells to the tubers or seeds. Now, since starch grains cannot pass through the cell-membrane, this migration can only occur when the substances are in a soluble form (sugar); when they reach the place where they are to be stored up, they are re-converted into the insoluble form (starch). If now the germ develops, either in the tuber or in the seed, the passive reserve materials assume the active form and make their way to the place where they are needed, namely, to the cells of the developing germ. Similarly the carbo-hydrates, fats and proteids which enter the body in the form of food, must be rendered soluble, so that they may be able to reach the place where they will be used, and the fats which are stored up in fatty tissues must be altered before they can be used in any part of the body.

In plant and animal cells this important transformation of carbo-hydrates, fats and proteids from a passive into an active condition is efficiently accomplished by means of very peculiar chemical substances called ferments. These are allied to the albumens, and indeed are derived from them; they are present in very minute quantities in the cell, but nevertheless produce powerful chemical effects, and induce chemical processes without being essentially altered themselves. This process of fermentation is very characteristic of the chemistry of the cell. There are special ferments for carbo-hydrates, others for proteids, and others for fats.

Whenever starch is rendered soluble in plants, the process is effected by means of a ferment, diastase, which can easily be obtained from germinating seeds. Its efficacy is so great, that one part by weight of diastase is sufficient to convert in a short time 2,000 parts of starch into sugar. Another ferment, invertin, which acts upon carbo-hydrates, is present in some fission fungi and moulds; it splits cane sugar up into dextrose and levulose.

The salivary ferment in the animal, ptyalin, which converts starch into dextrin and maltose, corresponds to the diastase in the plant. Similarly the non-diffusible glycogen, which in consequence of its properties has been called animal starch, must, if it is to be utilised further, be converted by means of a sugar-forming ferment, wherever it occurs, into sugar (liver, muscles).

Albuminous bodies are peptonised before they can be absorbed. In the animal body this takes place chiefly by means of a ferment, pepsine, which is secreted by the cells of the gastric glands. A small quantity of pepsine is able either in the stomach or in a test-tube to dissolve a considerable amount of coagulated albumen in the presence of free hydrochloric acid, thus converting it into such a form that it is able to diffuse through membranes.

Peptonising ferments have been also demonstrated in plant cells. For example, one has been extracted in the form of a digestive juice from those organs of carnivorous plants which are adapted for the capture of insects, such as the glandular hairs of the leaves of the Drosera; in this manner the small dead animals are partially dissolved and absorbed by the plant cells. A ferment resembling pepsine has also been demonstrated in germinating plants, where it serves to peptonise the proteid bodies which are stored up as reserve material in the seed. The peptonising ferment from the milky juice of the *Carica papaya* and of other species of Carica is well known on account of its energetic action. Finally, a similar ferment has been discovered in the body of the Myxomycetes by Krukenberg.

In the animal body fats are split up into glycerine and fatty acids. This result is effected mainly by the pancreatic juice. Claude Bernard endeavoured to trace this back to a fat decomposing ferment secreted by the pancreas. Further, it is supposed that during the germination of fat-containing plant seeds the oils are split up into glycerine and fatty acids by means of ferments (Schützenberger).

Thus even from these few data it may be seen that, although at present so little is known about the subject, there appears to exist a far-reaching uniformity throughout the whole organic kingdom as regards the elaboration of material in the cell.

One of the points which is least understood concerning the metabolism of the cell is the part played by the protoplasm.

This is especially true of all the processes which are described above as belonging to the formative activity of the cell. What relationship does the protoplasm bear to its organised products, such as the cell membrane, the intercellular substance, etc.?

Two quite opposite views have been suggested upon this subject. According to the one, the organised substances are formed by the transformation of the protoplasm itself, that is to say, through the chemical rearrangement or splitting up of the protoplasmic molecules; according to the other, on the contrary, they are supposed to be formed of plastic materials, carbo-hydrates, fats, peptonised proteids, etc., which are taken up during metabolism by the protoplasm, conveyed to the place where they are required, and there brought into a suitable condition for secretion.

This difference may be best explained by an example, such as the formation of the cellulose membrane of the plant cell.

According to a hypothesis which has been strongly supported by Strasburger (V. 31–33) amongst others, the microsome containing protoplasm becomes directly transformed into cellulose lamellæ; that is to say, cellulose, as a firm organised substance, is formed directly out of the protoplasm.

Another theory is, that some non-nitrogenous plastic substance, such as glucose, dextrin, or some other soluble carbo-hydrate, forms the materials from which the cell membrane is constructed. These materials are conveyed by the protoplasm to the place where they are required, and are here converted into an insoluble modification, cellulose. Since this cellulose acquires a fixed structure from the beginning, the protoplasm must, in a manner at present unknown to us, assist in its construction; this process is described by the expression "formative activity."

According to the first hypothesis, the cellulose membrane may be described shortly as a metabolic product of the protoplasm, and, according to the second, as a separation product of it.

The question of the formation of chitinous skin, of the ground substance of cartilage and bone, of calcareous and gelatinous substances, may also be regarded from the same two points of view; in fact, all conceptions of the metabolism of the cell present the same difficulty.

Claude Bernard (IV. 1a) described this relationship in the following words: "From a physiological standpoint it may be conceived that in the organism only one synthesis occurs, that of protoplasm, which grows and develops itself at the expense of the substances which it absorbs. Then, from the splitting up of this most complex of all organised bodies, all the complicated ternary and quaternary compounds must arise, the formation of these being ordinarily ascribed to a direct synthesis. Hence Sachs was obliged to allow that it was possible, although he considered it improbable, that in the assimilation of starch decomposition and restitution occur in the molecules of the green protoplasm."

These remarks show how difficult the whole subject is in so far as it concerns the chemical processes in question.

If it is allowable to draw conclusions from analogous cases, I must certainly decide in favour of the second hypothesis, according to which the protoplasm participates more indirectly than in the first in the formation of the greater number of intercellular substances. For in the cases where organisms construct a silicious or calcareous membrane the nature of the substance itself distinctly shows that it could not proceed directly as a firm organised substance out of protoplasm. This latter in such a case, in consequence of its chemical composition, can only play the part of an intermediary, by selecting the substances from its environment, absorbing them, accumulating them at the places where they are required, and depositing them in a distinct form as firm compounds, which are invariably joined to an organic substratum.

Such a conception appears to me to be nearer the truth in the case of the formation of the cellulose membrane also, if the facility with which various carbo-hydrates become transformed into one another is taken into account, as well as the complicated process, which would be necessary if protoplasm were to be converted into cellulose. And even those intercellular substances which are chemically more nearly related to protoplasm, such as chondrin, gluten, etc., may be governed by the same laws of construction. For, apart from the organised proteid substances, protoplasm and nuclear substance, there are always present in each cell a large number of unorganised proteids; these serve as formative material, and occur in a condition of solution in the cell sap of plant cells, in the nuclear sap, and in the blood and lymph of animals. Instead of the protoplasm itself being directly seized upon and used up in the formation of nitrogenous intercellular substances, it is possible that the unorganised proteid materials may be utilised by the formative activity of the cell, in the same way as has been suggested above, that other substances are used for the formation of the cellulose membrane.

In what way the protoplasm executes its above-mentioned function of adoption is quite beyond our comprehension at this present time, when the majority of the bio-chemical processes escape our observation. This function of the protoplasm, however, may consist in this, that certain particles of its substance may unite, through molecular addition, with particles of other substances present in the nutrient solutions, and thus become transformed into an organic product. Thus soluble silicious compounds may unite with molecules of organic substance to form a silicious skeleton; thus particles of cellulose may be formed through the influence of particles of protoplasmic substance from soluble carbo-hydrates, forming with them a compound (probably permanent, but possibly only temporary), and becoming organised to form a cell-membrane. This conception is quite in accordance with the fact that in many objects freshly-formed layers of cellulose are found to pass imperceptibly into the neighbouring protoplasm.

2. **The Morphology of Metabolism. The formative activity of the Cell.** The substances which are formed during the metabolism of the cell may be included under the head of morphology, in so far as they can be optically distinguished from the protoplasm. They may be differentiated out in a formed or unformed condition, either in the interior of the protoplasm, or upon its surface; according to their position they are distinguished as internal or external plasmic products. However, as is so often the case in biological classifications, a sharp line of distinction cannot be drawn between the two groups.

a. **Internal Plasmic Products.** Substances dissolved in water may separate out as larger or smaller drops in the protoplasm, and thus cause cavities or vacuoles. These play a most important part, especially in the morphology of plants. As has already been described in detail on p. 31, a plant cell (Fig. 62) is able by secreting sap to increase its size in a short time more than a hundred-fold. It is by means of the simultaneous action of a large number of such cells that in spring-time certain organs of plants are able to grow to such a considerable size. The solid substance contained by a plant very rich in water may be as little as 5 per cent., or even only 2 per cent.

The cell sap, however, is not pure water, but a very complex, nutrient solution containing vegetable acids and their salts, nitrates and phosphates, sugar, and small quantities of dissolved proteids, etc. Thus between the protoplasm and the sap material is interchanged to a considerable extent, substances for use being extracted from the one, which in return receives other substances in exchange. Since the sap represents a concentrated solution of osmotic substances, it exerts a powerful attraction upon water, and also an internal pressure, which is often considerable, upon the envelope surrounding it, thus producing a tense condition, which was described on p. 141 as turgor.

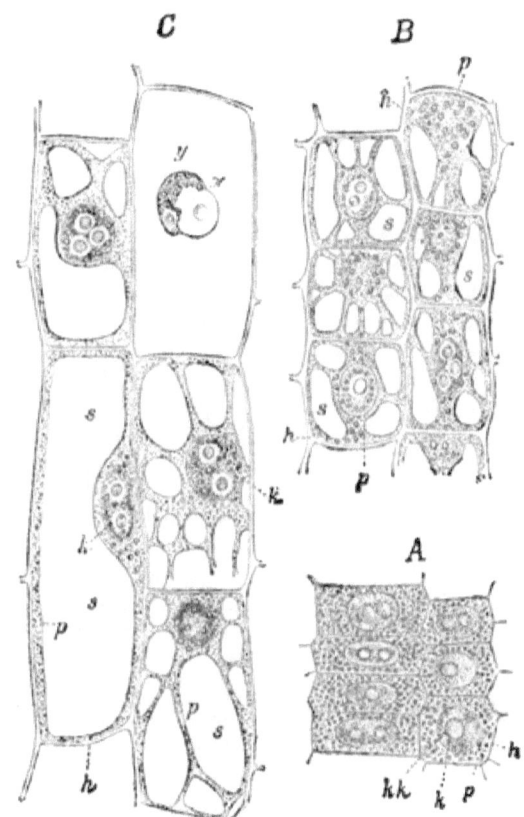

FIG. 62.—Parenchyma cells from the cortical layer of the root of *Fritillaria imperialis* (longitudinal sections, × 550; after Sachs II. 33, Fig. 75): *A* very young cells, as yet without cell-sap, from close to the apex of the root; *B* cells of the same description, about 2 mm. above the apex of the root; the cell-sap (*o*) forms in the protoplasm (*p*) separate drops between which are partition walls of protoplasm; *C* cells of the same description, about 7–8 mm. above the apex; the two lower cells on the right hand side are seen in a front view; the large cell on the left hand side is seen in optical section; the upper right hand cell is opened by the section; the nucleus (*ry*) has a peculiar appearance, in consequence of its being distended, owing to the absorption of water; *k* nucleus; *kk* nucleolus; *h* membrane.

Many botanists, especially de Vries (V. 35) and Went, consider the vacuoles to be special cell organs, which are not of accidental formation in the cell-body, but which can only be produced by division. Even in the youngest plant-cells, according to their opinion, minute vacuoles are present, which multiply continually by fission, and which are distributed amongst the daughter cells when cell division occurs. Here all the vacuoles of the whole plant would originate from those of the meristem. This theory however is disputed by other investigators. Just as the protoplasm is bounded externally by a peripheral layer, the vacuoles, in de Vries' opinion, possess a special wall (the tonoplast), which regulates the secretion and accumulation of the dissolved substances present in the cell sap.

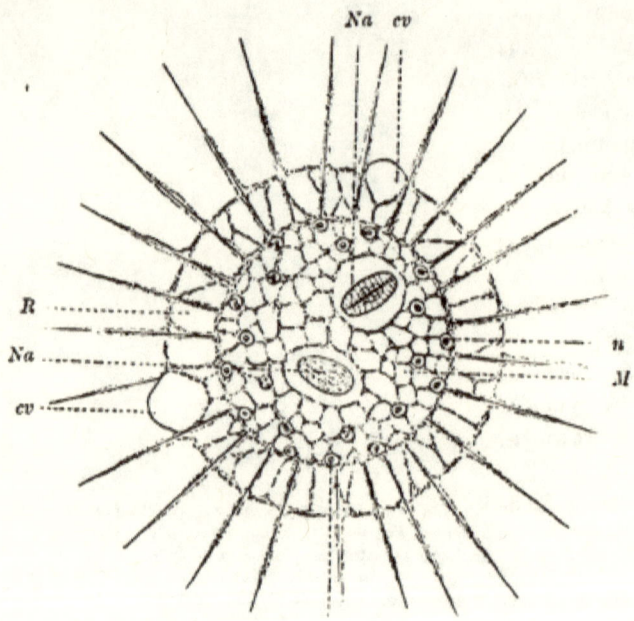

FIG. 63.—*Actinosphærium Eichhorni* (after R. Hertwig, *Zoologie*, Fig. 117): *M* medullary substance with nuclei (*n*); *R* peripheral substance with contractile vacuoles (*cv*); *Na* nutrient material.

The formation of vacuoles also occurs to a considerable extent in the lower organisms. In *Actinosphærium*, for example, the protoplasmic body has quite a foamy appearance, in consequence of the large number of great and small vacuoles present in it. A few vacuoles, the number of which is constant, acquire a specially contractile peripheral layer; they are then described as contractile vacuoles or reservoirs (p. 85). This occurs with especial frequency in Ciliata.

Finally, it occasionally, although rarely, happens that the sap collects into special vacuoles; this may occur in various kinds of animal cells, and especially in structures which have a supporting function in the body. In the tentacles of many Cœlenterates, in certain appendages of Annelids, and also in the *chorda dorsalis* of Vertebrates, there are comparatively large vesicular cells, which are separated from the exterior by a thick membrane, and which contain hardly anything but cell sap, only a very minute quantity of protoplasm being present. This is spread out in a very thin layer over the membrane, extending threads here and there across the sap space; the nucleus is generally embedded in a somewhat denser collection of protoplasm, either in the peripheral layer, or in the network. Here also, as in plants, the firm cell-wall is tensely distended in consequence of the osmotic action of the substances in the sap. Although no experimental investigations have yet been made concerning the turgescence of the organs in question, yet it can only be explained in this manner: that the notochord functions in the body of a Vertebrate as a supporting organ. The very numerous small turgescent notochord cells being built up into one organ, and also shut off from the exterior by means of a firm elastic sheath, their individual tensions are summed up, and through the internal pressure of the sheath the structure is kept rigid.

The absorption and secretion of sap occur in nuclear substance, just as in protoplasm. The sap serves the same purpose in both cases, namely to offer a large surface to the active substances, and to put them into direct communication with the nutrient fluid.

Although the formation of sap vacuoles occurs but rarely in animal cells, various substances, such as fat, glycogen, mucin, albuminates, etc., frequently separate out from the protoplasm.

The fat is seen to occur at first as small drops in the protoplasmic body, resembling the drops of cell sap in young plant cells. Just like such vacuoles, the droplets increase in size, and run together, producing, finally, one single large drop, which fills the whole internal space of the cell, and which is surrounded by a delicate cell-membrane, and by a thin layer of protoplasm, which contains the nucleus.

Glycogen collects in separate particles in the liver cells; these drops, when a solution of iodine in iodide of potassium is added to them, acquire a mahogany-brown coloration, by means of which they can be easily seen.

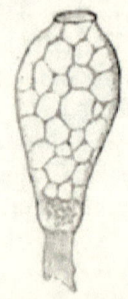

FIG. 64.—Goblet-cell from the bladder epithelium of *Squatina vulgaris*, hardened in Müller's fluid. (After Last, Plate I., Fig 9.)

Mucigenous substances often fill up the interior of the cells, by which they are secreted (Fig. 64) in such quantities that the cells swell up into vesicles, or assume the form of goblets. The greater part of the protoplasm is collected at the base of the cell, where the nucleus also is situated, whilst the remainder surrounds the mucigenous substance with a thin envelope, and extends into it a few threads which unite together to form a net. The mucigenous substances can be clearly distinguished from protoplasm when the cell is stained with one of several aniline dyes.

The internal plasmic products very frequently acquire greater solidity in egg-cells, which are loaded in the most various ways with reserve materials. These are grouped according to their form as yolk-globules (Fig. 65), yolk granules, and yolk lamellæ, and from a chemical point of view chiefly consist of a mixture of albuminates and fats. The more numerous, small, and closely packed these yolk-elements are, the more the plasmic body assumes a foamy or net-like appearance.

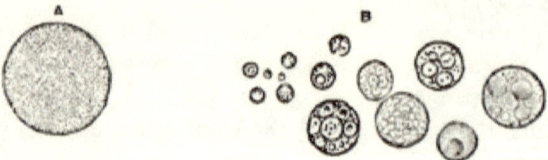

FIG. 65.—Yolk elements out of a Hen's egg (after Balfour): *A* yellow yolk spheres; *B* white yolk spheres.

Many plasmic products are crystalline in character, such as the guanin crystals, to which the glistening silvery appearance in the skin and peritoneum of fishes is due, or as the pigment granules in the pigment cells.

Plasmic products, similar to those in animal cells, occur also in plant cells; however, in this case they are generally present in a few special organs, which are utilised either for the storing up of reserve material, or, as with seeds, for purposes of reproduction.

Under such circumstances the cells are filled with drops of oil (oily seeds), with granules of various albuminous substances (vitellin, gluten, aleuron), with crystalloids of proteinaceous substance, or with starch granules, about which more will be said later.

The above-mentioned internal plasmic products being only temporarily accumulated during metabolism before being utilised, vary considerably in composition, but there are others which attain a higher degree of organisation, and which participate permanently in the functions of the cell. To such belong the internal skeletal structures of the protoplasm, the various substances in plant cells, described under the common name of trophoplasts, the cnidoblasts of Cœlenterata, and, finally, the sheaths of the muscle and nerve fibres, etc.

Internal skeletons are found in the bodies of a large number of Protozoa, but especially in great variety and beauty in Radiolarians. They consist sometimes of regularly arranged spicules, sometimes of a fine, open trellis-work, and sometimes of a combination of the two kinds of structures (Fig. 66). In some families of Radiolarians they are composed of an organic substance which

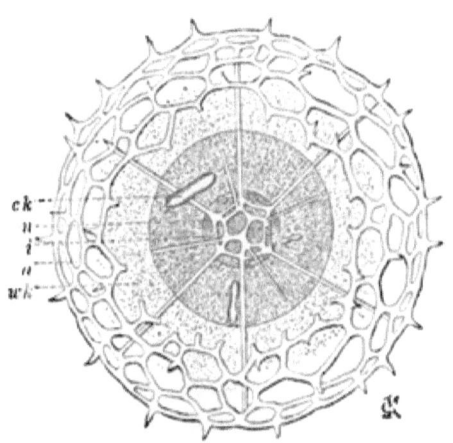

FIG. 66.—*Haliomma erinaceus* (from R. Hertwig, Zool., Fig. 82): *a* external, *i* internal trellis work; *ck* central capsule; *wk* soft extra capsular body; *n* internal vesicle (nucleus).

is soluble in acids and alkalies, but in most cases, on the contrary, they consist of silicious material which is united to an organic substratum, just as, in the bones of Vertebrates, the phosphates are united with the ossein. In each species the skeleton has a constant and characteristic structure, and follows certain fixed laws during the process of its development (Richard Hertwig, 5, 40).

Under the name trophoplasts, the highly organised differentiated products of vegetable protoplasm are included; these occur as constantly as the nucleus, and possess great functional independence. They are of great importance in the nutrition of plants, for the whole process of assimilation and the formation of starch takes place in them (Meyer V. 9-11).

Trophoplasts are small bodies, which are generally either globular or oval in shape; they are composed of a substance very similar to and yet distinct from protoplasm. They are easily destroyed, whilst the preparation is being made, by either water or reagents, and are most successfully fixed by means of tincture of iodine, or concentrated picric acid. They acquire a steely blue coloration in nigrosin, and thus stand out clearly from the protoplasmic body. They often occur in great numbers in the cell, and may actively change their form. According to the investigations of Schmitz (V. 29), Schimper (V. 27, 28), and Meyer (V. 9-11), trophoplasts are not direct new formations in the protoplasm, but on the contrary reproduce themselves, like nuclei, from time to time by division. According to this conception, all the trophoplasts in the generations of cells which spring from the original vegetable egg cell are derived from those trophoplasts which were originally present.

Various kinds of trophoplasts may occur, fulfilling various functions; these are distinguished as starch-forming corpuscles, as chlorophyll corpuscles, and as pigment-granules (amylo- or leucoplasts, chloroplasts, chromoplasts).

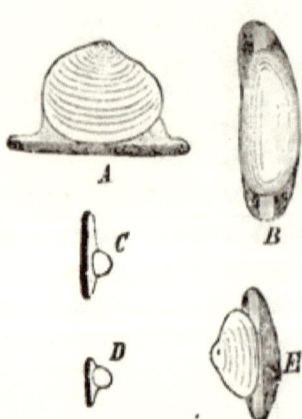

FIG. 67.—*Phajus grandifolius*, amyloplasts from the tuber (after Strasburger, *Botanisches Prakticum*, Fig. 30): *A*, *C*, *D*, and *E* are seen from the side, *B* from above, *E* is coloured green. (×540.)

Most starch-forming corpuscles (amyloplasts) (Fig. 67) occur in the non-assimilating cells of young plant organs, and in all underground portions, as also in stems and petioles. In the pseudo-tubers of *Phajus grandifolius*, which are especially suitable for investigation, they form, when viewed on the flat, ellipsoidal finely granular discs, whilst when viewed from the side they look like small rodlets; these when treated with picro-nigrosin stain a steely blue colour, and so stand out clearly from the surrounding protoplasm. On one of the flat sides of the disc, a starch granule is situated. When this is small, it is completely covered with a thin coating of the substance of the amyloplast; when it is somewhat larger, only the side turned to the amyloplast is so coated. Further, a concentric stratification may occur; under these conditions the hilum, which is surrounded by the concentric layers, is situated near the surface, which is turned away from the amyloplast. Hence the layers on this surface are very thin, becoming gradually thicker and thicker as they approach the starch-forming corpuscle, which is only natural, since they grow out of it, and are formed by it. Frequently a rod-shaped crystal of albumen may be seen embedded in the substance of the amyloplast, on the surface which is turned away from the starch granule.

Now since starch, as has been already mentioned, can only be produced synthetically in the green portions of plants, these white amyloplasts cannot be regarded as its true places of origin. It is much more likely to be true that they have obtained the starch, in a soluble form, probably as sugar (Sachs), from those places where assimilation occurs, so that their only function is to reconvert this soluble substance into a solid, organised body.

The chlorophyll granules (Fig. 68) must be closely connected with the starch-forming corpuscles, since the latter may be converted directly into them—this occurs when chlorophyll under the influence of sunlight develops in them. In such a case the amyloplasts turn green, increase in size, and part with their starch granules, which become dissolved. In addition, chlorophyll granules are formed from the colourless trophoplasts, which are developed at the growing points in the form of undifferentiated corpuscles; finally they multiply by division in the following manner

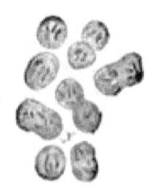

FIG. 68.— *Chlorophyll granules from the leaf of Funaria hygrometrica*, both in a resting condition and undergoing division. (× 540: after Strasburger, *Pract. Bot.*, Fig. 17.)

(Fig. 68): to start with, their substance increases in size, and they elongate themselves; they next become biscuit-shaped, and finally divide into two equal portions.

The chlorophyll granules consist of two substances: a ground substance, which reacts like albumen, and a green colouring matter (chlorophyll), which saturates the stroma. This may be extracted by means of alcohol, when it is seen to be distinctly fluorescent, appearing green with transmitted, and bluish red with reflected, light.

Several small starch granules are generally enclosed in the chlorophyll corpuscles, being formed in them through assimilation. They are most easily seen, if, when the chlorophyll has been extracted by means of alcohol, tincture of iodine is added to the preparation.

As has been proved by Stahl's investigations, the chlorophyll granules, quite apart from the changes of position brought about by the streaming movements of the protoplasm (*vide* p. 104), are able to change their shape under the stimulating influence of the sun's rays, to a surprising extent. Whilst in diffused daylight they assume the shape of polygonal discs with their broad sides directed towards the source of light, in direct sunlight they contract up into little round balls or ellipsoidal bodies. By this means they effect a change which is necessary for the performance of the chlorophyll function, by "offering to direct sunlight a small surface, and to diffused daylight a larger one, for the absorption of the rays of light. In this, they offer us an insight into the high degree of the differentiation that they have attained which we could never have arrived at simply by the study of their chemical activity" (de Vries V. 46). As regards their mode of multiplication by division, their active motility, their functions in the processes of assimilation, etc., they appear, like nuclei, to be very highly specialised plasmic products.

Finally another variety of trophoplasts, the colour-granules, must be mentioned: the red and orange red coloration of many flowers is caused by their presence. They consist of a protoplasmic substratum which may assume very various forms, occurring sometimes in the shape of a spindle and sometimes of a sickle, a triangle or a trapezium. In this substratum crystals of colouring matter are deposited. In this case also colourless trophoplasts may, in suitable objects, be seen to develop gradually into colour granules. Further Weiss has observed spontaneous movements and changes of form in these granules also.

We will conclude this review of the various kinds of trophoplasts by describing in more detail the structure of the starch grains, which have acquired considerable theoretical importance in consequence of Nägeli's (V. 17, 20) researches, and the conclusions which have been deduced from them.

The starch grains (Fig. 69) in a plant cell may vary considerably as to size. Sometimes they are so small that even with the strongest powers of the microscope they only appear as minute points, whilst at others they may be as large as 2 mm. in circumference. Their reaction towards iodine solution is characteristic; they become either dark or light blue according to the strength of the solution. In warm water they swell up considerably, and if further heated turn into a paste.

Their shape also varies, being sometimes oval, sometimes round, and sometimes irregular. When strongly magnified they are seen to be distinctly stratified, and in an optical section bright broad bands are seen to alternate with more narrow dark ones. Nägeli explains this appearance by the supposition that the starch grain is composed of lamellæ of starch substance, which are alternately rich and poor in water. Strasburger (V. 31), on the other hand, is of opinion, that "the darker lines represent the specially marked adhesion surfaces of consecutive lamellæ, which," he considers, "are more or less identical with each other in composition."

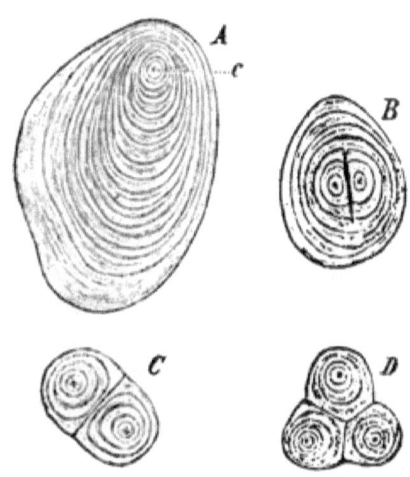

FIG. 69.—Starch grains from a *Potato* tuber (after Strasburger, *Pract. Bot.*, Fig. 3): *A* simple grain; *B* semi-compound grain; *C* and *D* compound grains; *c* the hilum. (×540.)

The lamellæ (Fig. 69) are arranged round a hilum, which is either situated in the centre of the whole grain (*B*, *C*) or, as is more frequently the case, is eccentric in position (*A*). Further it is not rare to find starch grains, which consist of two (*B*, *C*) or three (*D*) systems of lamellæ, united together; these are termed compound grains, in contradistinction to others which contain one single hilum. When the hilum is in the centre, the strata of starch surrounding it are fairly uniform in thickness. On the other hand when its position is eccentric, only the inner layers surround it completely, whilst the peripheral layers are of greatest thickness on that side which is turned away from the hilum, and grow thinner and thinner as they approach it, becoming finally so narrow, that they either fuse with neighbouring lamellæ, or end freely.

In each starch grain the amount of water contained is greatest at the centre, and diminishes as the surface is approached. The hilum is richest in water, whilst the superficial layer, bordering on the protoplasm, is most dense in composition. To this cause we can trace the fissures which occur in the hilum of the starch grain as it dries, and which extend outward from it towards the periphery (Nägeli V. 17).

As has been already mentioned, the starch grains of plants do not, as a rule, arise directly in the protoplasm, but in certain special differentiation products of it, the starch-forming corpuscles (amyloplasts, and chlorophyll bodies). According to the investigations of Schimpfer (V. 27), the special variety of stratification which occurs in the grain depends upon whether it is situated in the interior or upon the surface of one of these corpuscles. In the first case, the starch lamellæ arrange themselves evenly around the hilum since they receive equal accretions on every side from the starch-forming corpuscle. In the second case, that portion of the grain, which adjoins the free surface of the amyloplast, is under less favourable conditions for growth, for the surface of the grain, which is directed towards the centre of the starch-forming corpuscle, acquires the most substance, and in consequence the layers are thicker at this point, and grow gradually thinner as they approach the opposite side.

Hence the hilum, about which the layers are arranged, becomes pushed further and further beyond the surface of the amyloplast, assuming a more and more eccentric position in the stratification.

That the starch grains grow by the deposition of new layers upon the surface, that is by apposition, may be deduced from a statement of Schimpfer's. He observed, that around the corroded centres of starch grains whose surfaces had been dissolved away new layers had been deposited.

Strasburger is of opinion that starch grains may be occasionally produced in the protoplasm itself, without the intervention of special starch-forming corpuscles. He found them in the cells of the medullary rays of *Coniferæ*, during their early stages of development, as minute granules, embedded in the strands of the plasmic network. As they grew larger they were to be plainly seen situated in the plasmic cavities. These cavities have highly refracting walls, upon which microsomes are situated.

One of the most remarkable of the internal plasmic products is the nematocyst (Fig. 70), which functions in *Cœlenterata* as a weapon of attack, in the cnidoblasts, which are distributed throughout the ectoderm. It consists of an oval capsule (*a* and *b*), which is formed of a glistening substance, and which has an opening in that end which is directed towards the external surface. The internal surface is lined with a delicate lamella which, at the edge of the opening, merges with the sheath of the capsule; the structure of this sheath is frequently very complicated (*cf.* Fig. 70 *a*, *b*). In the figure, this sheath consists of a very delicate filament and of a broad, conical, proximal portion, which is situated in the interior of the capsule, and is provided with shorter and longer barbs. The filament stretches from the end of the conical portion, and is wound spirally round and round it several times; the free, internal cavity is filled with an irritating secretion; the protoplasm, which borders on the nematocyst, is differentiated to form a contractile envelope, which also has an opening to the exterior (Schneider V. 45).

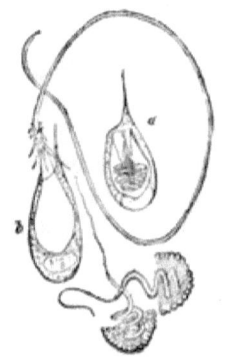

Fig. 70.—Thread cells of a Cnidarian (from Lang; Hertwig, Zool., Fig. 161): *a* cell with cnidocil, and the thread coiled up in the capsule; *b* thread evaginated from the capsule, and armed at its base with barbs; *c* prehensile cell of a *Ctenophore*.

Near the opening of the capsule a rigid, glistening, hair-like process, the cnidocil, stretches out from the free surface of the cell. If this is touched by any foreign body, it communicates the stimulus to the protoplasm. In consequence, the cnidoblast, enclosing the nematocyst, contracts suddenly and forcibly, thereby compressing it, and forcing out the thread which is in the interior, so that it is turned inside out, like the finger of a glove (Fig. 70 *b*). At first the conical proximal portion is protruded with the barbs extended outwards, next comes the delicate, rolled-up thread. The irritating secretion is apparently poured out through an opening in the capsule.

Some light is thrown upon the formation of this extraordinary apparatus by the history of its development. First of all, an oval secretion cavity is formed in the cnidoblast; this cavity is separated from the protoplasm by a delicate membrane, then a delicate protoplasmic process grows into the secretion cavity from the free end of the cell; it gradually assumes the position and form of the internal thread apparatus, separating upon its surface the delicate enclosing membrane. Finally, the shining, tough, external wall of the capsule, with its opening, becomes differentiated, and around it the contractile sheath develops.

b. **External Plasmic Products.** The external plasmic products may be divided into three groups,—cell membranes, cuticular formations, and intercellular substances.

Cell membranes are structures which separate out, and envelop the whole surface of the cell-body. In the vegetable kingdom they are very important, and easily seen, whilst in the animal kingdom they are frequently absent, or are so slightly developed that they can hardly be made out even with the strongest powers of the microscope.

In plants, the cell membrane is composed of *cellulose*, a carbohydrate very nearly allied to starch. The presence of this substance may generally be easily demonstrated by a very characteristic reaction. If a section of a plant tissue, or a single plant cell, is saturated first with a dilute solution of iodine in potassic iodide, and then (after the excess of the iodine solution has been removed) the preparation is immersed in sulphuric acid (2 parts acid to 1 part water), the cell membranes assume a lighter or darker blue coloration. Another reaction for cellulose is seen when chlorzinc-iodine solution is used (Schulze's solution).

The membranes of plant cells often become thick and firm, and then they show, in section, a distinctly marked striation, being composed, like starch grains, of alternate bands of high and low

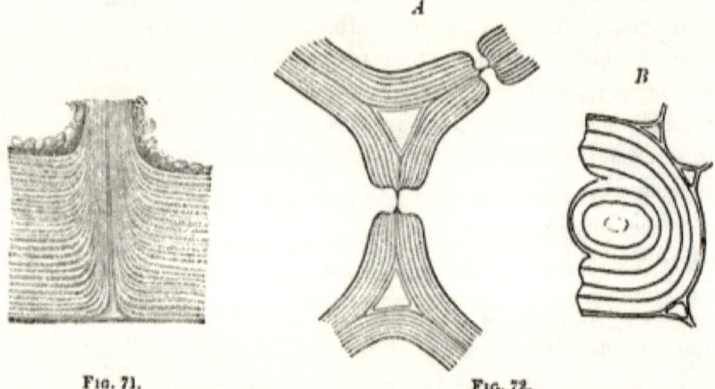

FIG. 71. FIG. 72.

FIG. 71.—Transverse section through the thallus of *Caulerpa prolifera* at the place where a branch is inserted. (After Strasburger, Pl. I., Fig. 1.)

FIG. 72.—*A* Portion of a fairly old pith cell, with six layers from *Clematis vitalba* (after Strasburger, Pl. I., Fig. 13); *B* a similar cell after it has been swollen up by sulphuric acid. (After Strasburger, Pl. I., Fig. 14.)

refractive power (Figs. 71, 72 A and B). However, when the surface is examined, a still more delicate structure can frequently be seen. The cell membrane is faintly striated, looking as though it were composed of a large number of parallel layers; these are crossed by others running in an opposite direction. They run either longitudinally and transversely—that is to say, like rings round the cell—or are arranged diagonally to the longitudinal axis of the cell. Nägeli and Strasburger hold different opinions concerning the relation of this delicate striation towards the separate cellulose lamellæ.

Nägeli (V. 19) considers that both systems of striation are present in each lamella; further that, as in starch grains, the lamellæ, as well as the intersecting bands, consist of substances alternately rich and poor in water, and hence are alternately dark and light in appearance. In consequence, a lamella is, as it were, divided into squares or rhomboids, like a parquetted floor. "These may assume one of three appearances; they may consist of substances of greater, of less, or of medium density, according as to whether they occur at the point of intersection of two denser, of two less dense bands, or of one dense and one less dense band." Hence Nägeli is of opinion that the whole cell membrane "is divided in three directions into lamellæ, which consist of substances alternately rich and poor in water, and which intersect in a manner similar to that seen in the intersecting laminæ of a crystal. The laminæ in one direction compose the layers, those in the others the two striated systems. These latter may intersect at almost any angle; they both meet the lamellæ of the layers, apparently, in most cases at right angles."

On the other hand, in opposition to Nägeli, Strasburger (V. 31–33) and other botanists, whose statements are not to be disputed, consider that *intersecting striæ never belong to the same lamella*; they think it much more likely that if one lamella is striated in a longitudinal direction, the next one is striated transversely, and so on alternately. Strasburger does not believe that the difference, either in the lamellæ or the striæ, is due to the varying amount of water which they contain. The lamellæ and the striæ in them are separated from one another by their surfaces of contact, which, in consequence of being seen at different angles (cross section and surface view), appear as darker lines. Thus the arrangement is similar, in the main, to that seen in the cornea, which consists of laminæ formed of bundles of white fibres which cross one another at right angles in alternate laminæ.

Not infrequently cellulose membranes show delicate sculpturings, especially upon the inner surface. Thus thickenings may originate in the interior; these may run into each other to form a spiral, or may be arranged in large numbers transversely to the long axis of the cell, or finally, may be united together in an irregular fashion to form a network. On the other hand, the thickenings may be absent at various places, where neighbouring cells touch, and thus pits or perforations are produced (Fig. 72 *A*), by means of which neighbouring cells can interchange nutrient substances with greater ease.

Moreover, as regards its composition, the cell-wall can alter its character in various ways soon after its original formation; this may be produced by the deposition of various substances upon it, or by its transformation into wood or cork.

Lime salts or siliceous substances are not infrequently deposited in the cellulose, thus producing greater solidity and hardness of the walls. When portions of such plants are burnt, the cellulose is destroyed and a more or less perfect skeleton of lime or silica remains in the place of the framework of the cell. Lime is deposited in *Corallineæ*, in *Characeæ*, and in *Cucurbitaceæ*; and silica in *Diatomaceæ*, *Equisitaceæ*, Grasses, etc.

Similarly the cell-wall obtains very great strength through the formation of wood. Here the cellulose becomes mingled with another substance, woody substance (lignin and vanillin), this may be dissolved away by means of potassic hydrate, or with a mixture of nitric acid and chlorate of potash, after which a framework, which gives the reaction, of cellulose remains.

In the formation of cork the cellulose becomes united in larger or smaller quantities with corky substance or suberin. In this case, also, the physical properties of the cell-wall are altered, it being no longer permeable to water. Thus cork cells are formed on the surface of many parts of plants in order to prevent evaporation.

Whilst it is evident, that in the deposition of lime and silica, the particles of these substances must be conveyed by the protoplasm to the place where they are required, and where they are deposited between the particles of cellulose, whereupon molecular combinations are again called into play, two explanations may be given concerning the formation of wood and cork. Either the wood and cork substances are constructed in a soluble form, by means of the protoplasm, and, like the lime and silica particles, are deposited as an insoluble modification in the cellulose membrane, or both substances originate on the spot, through a chemical transformation of the cellulose. This is another problem which must be decided by means of physiological chemistry rather than through morphological investigations (*vide* p. 153).

The question as to how the cell membrane grows is a very important problem, and has led to much discussion; it is very difficult to come to any decision on the subject. Two methods of growth may be distinguished, a superficial and an interstitial method. The delicate cellulose coating, which at first is scarcely measureable, may by degrees attain a very considerable thickness, growing by the addition of numerous laminæ, the number of which varies with the thickness. It is most probable that layer after layer is deposited by the protoplasm of the outer layer which was at first differentiated off. This method of growth is termed " growth by apposition," in contradistinction to " growth by intussusception," which, according to Nägeli, is the way in which the cell-wall grows, that is to say, by deposition of particles in the interstices between the particles already present.

The apposition theory is supported by the following three observations: (1) Before the ridge-like thickenings are formed upon the inner surface of a cell-wall, the protoplasm is seen to collect together at those places, where thickening of the wall is about to occur, in masses, which exhibit active streaming movements. (2) When, in consequence of plasmolysis, the protoplasmic body has receded from the cell-wall, a new cellulose membrane is seen to appear on its naked surface (Klebs IV. 14). If the plasmolysing agent be removed, and the cell-body be made to increase in size by the absorption of water, so that its new cellulose membrane comes into close contact with the original cell-wall, they unite with one another. (3) When a plant cell divides, it may often be plainly seen that each daughter cell surrounds itself with a new wall of its own, so that the two newly-formed walls of the daughter-cells are enclosed by the old wall of the mother-cell.

It is more difficult to explain the growth in superficial area of the cell-wall. This may be effected by two different processes, working either singly or in unison. The membrane may become stretched, like an elastic ball which is inflated with air; or it may grow by intussusception, that is to say, by the deposition of new cellulose particles between the old ones.

That such a stretching of the cellulose membrane does actually occur is proved by several phenomena. The turgescence already mentioned causes distension. When a cell is plasmolysed it at first contracts somewhat as a whole, in consequence of the loss of water, before the outer layer of the protoplasm becomes separated from the cell-wall. This indicates that it was subjected to internal pressure. It may be observed in many *Algæ*, that the cellulose lamellæ, which are first formed, are eventually ruptured by the stretching, and discarded (*Rivularia, Glœocapsa, Schizochlamys gelatinosa*, etc.). Each distension and contraction must be connected with a change of position of the most minute particles, which become located either on the surface or in the deeper layers.

Thus the way in which a membrane increases in size when stretched offers many points of resemblance to growth by intussusception. The difference consists in this, that in the first case particles of cellulose already present are deposited in the surface, whilst in the second case particles in process of formation are so deposited.

However, I do not wish to totally disregard growth through intussusception, as Strasburger formerly did (V. 31). On the contrary, I consider it to form, in addition to apposition, a second important factor in the formation of the cell-wall, although it is certainly not the only factor, as is dogmatically stated in Nägeli's theory.

Many phenomena in cell-growth may be most easily explained by means of intussusception, as has been done by Nägeli, whilst the apposition theory presents numerous difficulties.

It does not often occur that the cell-wall becomes ruptured by stretching, and yet the increase in size which occurs in nearly all cells from their initial formation until their full growth, is quite out of proportion to the elasticity of the cell-wall, which, as it is composed of cellulose, cannot be assumed to be very great. Many plant cells grow until they are a hundred or even two hundred times as long as they were originally (*Chara*).

The fact that many cells are very irregular in form would be very difficult to explain if the cell membrane were considered to increase superficially solely by stretching, like an indiarubber bladder. For example, *Caulerpa, Acetabularia*, etc., are apparently differentiated, like multicellular plants, into root-like, stem-like, and leaf-like structures, although each plant consists of only a single cell-cavity. The growth of each of these parts proceeds according to a law of its own. Many plant cells grow only at one point: either at the apex or near the base, or they develop lateral outgrowths and branches. Others undergo during growth complicated changes of direction, as in the internodes of the *Characeæ*.

Finally, Nägeli states, as a point in favour of the theory of growth by intussusception, that many membranes increase considerably both superficially and in thickness after they have become separated from the protoplasmic body, in consequence of the formation of special membranes around the daughter-cells; "*Glœocapsa* and *Glœocystis* appear first as simple cells with a thick gelatinous cell-wall. The cell divides into two, whereupon each develops for itself a similar enclosing cell-wall, and in this manner the enveloping process proceeds." The outermost gelatinous cell-wall must in consequence become larger and larger. According to Nägeli's computation, their volume during successive developmental stages may increase from 830 cubic micromillimetres to 2,412, to 5,615, and finally to 10,209 cubic micromillimetres.

In another species the gelatinous cell-wall was seen to increase from 10 to 60 micromillimetres, that is to say, it became six times as thick. "In *Apiocystis* the pear-shaped colonies, which consist of cells embedded in a very soft gelatinous matrix, are surrounded by a thicker membrane. In this case, moreover, the membrane increases with age, not only in circumference but also in thickness; for whilst in smaller colonies it is barely 3 micromillimetres thick, in larger ones it is 45 micromillimetres thick; in the former it is 27,000 square micromillimetres in area, and in the latter 1,500,000 square micromillimetres. Thus the thickness of the sheath increases at a ratio of 1 to 15, the superficial area of 1 to 56, and the cubic contents of 1 to 833. That apposition should take place upon the inner surface of this sheath is out of the question, for its smooth internal surface never comes into contact with the small spherical cells, or only does so in a few isolated spots."

In all these cases I am obliged to agree with Nägeli, who considers that we have to make too many improbable assumptions, if we attempt to explain the superficial growth of the cell membrane solely by the deposition of new layers, whereas the above-mentioned "*phenomena (variations in form and direction, uneven growth of various parts, torsions) may be explained in the simplest and easiest fashion by intussusception. Everything depends upon this,* that the new particles become deposited in definite positions, in definite quantities, and in definite directions, between those already present."

Moreover, the process of intussusception is not to be disregarded in those cases where calcium and silicon salts are deposited in the cell-wall, for this mostly occurs at a later period, the salts being frequently only found in the superficial layers. It could only be proved that it is impossible for particles of cellulose to be deposited in a similar manner, if it could be shown that cellulose is actually only produced by the direct metamorphosis of layers of protoplasm. However, up till now this is anything but proved; and, moreover, it seems that the study of plant anatomy, by means of microscopic observation alone, is insufficient to establish this theory, and that in addition a very much improved and advanced knowledge of micro-chemistry must be reached, as in the case mentioned on pp. 153, 154. Consideration of the statements made there shows especially, that under certain conditions in the formation of cellulose there is not the marked difference that is frequently considered to exist between growth by apposition and growth by intussusception.

Cuticular structures are the skin-like formations with which a cell covers its external surface—not all over, however, but only on one side. In the animal kingdom, those cells which are situated on the surface of the body, or which cover the internal surface of the alimentary canal, are frequently provided with a *cuticle*, which protects the underlying protoplasm from the hurtful influences of the surrounding media. The cuticle usually consists of thin lamellæ, intersected by fine parallel pores, into which delicate processes stretch from the underlying protoplasm. As cuticular formations of a peculiar kind, which exhibit at the same time a very marked structure, the outer portions of the rods and cones in the retina may be cited.

Cuticular membrane-like formations, consisting of cells united

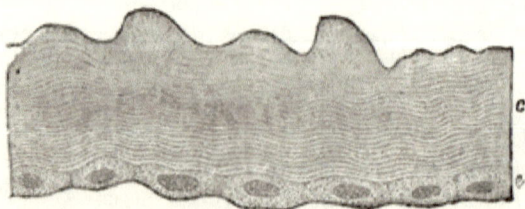

FIG. 73.—Epithelium with cuticle of a Saw-fly (*Cimbex coronatus*) (from R. Hertwig; Fig. 21*f*): *c* cuticle; *e* epithelium.

together, form by their coalescence extensive structures (Fig. 73), which, especially in Worms and Arthropods, serve as a protection to the whole surface of the body. This skin consists chiefly of chitin, a substance which is only soluble in boiling sulphuric acid. In its minute structure it very closely resembles cellulose membranes, especially in its stratification, which indicates that growth has taken place by the deposition of new lamellæ upon the inner surface of those already formed.

Occasionally the old chitinous sheaths are ruptured and discarded after they have developed beneath them a younger, more delicate skin to take their place; this process is termed sloughing. Calcium salts may be deposited, by means of intussusception, in the chitinous skin in order to strengthen it.

Finally, intercellular substances are formed, when numerous cells secrete from their entire surfaces solid substances, which, however, do not remain isolated as in cell membranes, but which coalesce to form a coherent mass, it being impossible to recognise from which cells the various portions of it originated (Fig. 74). Thus, in tissues with intercellular substance, the individual cells cannot be separated from one another, as they can be in plant tissue. In the continuous ground-substance, which may consist of very different chemical substances (mucin, chondrin, glutin, ossein, elastin, tunicin, chitin, etc.), and which further may be either homogeneous or fibrous, small spaces are present, which contain the protoplasmic bodies. Now, since the area of intercellular substance in the neighbourhood of the cell space is controlled to a considerable extent by the protoplasmic bodies it contains, it

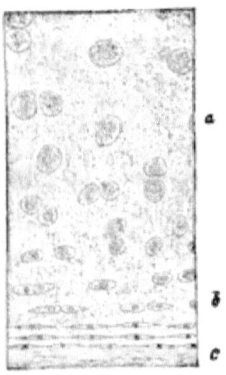

FIG. 74.—Cartilage (after Gegenbaur): c superficial layer; b intermediate layer passing into a, typical cartilage.

has been called by Virchow (I. 33) a cell territory. Such a cell territory, however, is of necessity not marked off from neighbouring ones.

Amongst the cell products, which may be classed as external or internal according to their position, the muscle and nerve fibres must be mentioned. Being composed of protein substance, they come next after protoplasm in the consideration of the substances of which tissues are composed; they must be classed with the above-mentioned structures, since they are quite distinct from protoplasm, and may be described as peculiar formations which perform a definite function in the life of the cell. Their more delicate structure will be discussed in another volume dealing with the tissues.

Literature V.

1. BAUMANN. *Ueber den von O. Löw und Th. Bokorny erbrachten Nachweis von der chemischen Ursache des Lebens. Pflügers Archiv. Bd. XXIX.* 1882.
2. BUNGE. *Physiological and Pathological Chemistry, trans. by Wooldridge.*
3. ENGELMANN. *Neue Methode zur Untersuchung der Sauerstoffausscheidung pflanzlicher und thierischer Organismen. Botan. Zeitung.* 1881.
4. HAECKEL. *Die Radiolarien.* 1862.
   HAECKEL. *Générale Morphologie.*
5. HESS. *Untersuchungen zur Phagocytenlehre. Virchows Archiv. Bd.* 109.
6. LANGHANS. *Beobachtungen über Resorption der Extravasate und Pigmentbildung in denselben. Virchows Archiv. Bd.* 49. 1870.
7. LÖW U. BOKORNY. *Die chemische Ursache des Lebens. München.* 1881.
8. MARCHAND. *Ueber die Bildungsweise der Riesenzellen um Fremdkörper. Virchows Archiv. Bd.* 93. 1883.
9. ARTHUR MEYER. *Ueber die Structur der Stärkekörner. Botan. Zeitung.* 1881.
10. ARTHUR MEYER. *Ueber Krystalloide der Trophoplasten und über die Chromoplasten der Angiospermen. Botan. Zeitung.* 1883.
11. ARTHUR MEYER. *Das Chlorophyllkorn in chemischer, morphologischer und biologischer Beziehung. Leipzig.* 1883.
12. METCHNIKOFF. *Untersuchung über die intracellulare Verdauung bei wirbellosen Thieren. Arbeiten der zoologischen Institute in Wien. Bd. V. Heft* 2.
13. METCHNIKOFF. *Ueber die Beziehung der Phagocyten zer Milzbrand-bacillen. Archiv. für patholog. Anatomie u. Physiologie. Bd.* 96 u. 97. 1884.
14. METCHNIKOFF. *Ueber den Kampf der zellen gegen Erysipelkokken. Ein Beiträg zur Phagocytenlehre. Archiv. für patholog. Anatomie u. Physiologie. Bd.* 107.
15. METCHNIKOFF. *Ueber den Phagocytenkampf bei Rückfalltyphus. Virchows Archiv. Bd.* 109.
    METCHNIKOFF. *Lectures on Inflammation, trans. by Starling.* 1893.
16. NÄGELI. (1) *Primordialschlauch.* (2) *Diosmose der Pflanzenzelle. Pflanzenphysiologische Untersuchungen.* 1855.
17. NÄGELI. *Die Stärkekörner. Pflanzenphysiologische Untersuchungen. Heft* 2. 1858.
18. NÄGELI. *Theorie der Gährung.* 1879.
19. NÄGELI. *Ueber der inneren Bau der vegetabilischen Zellenmembran. Sitzungsber. der bairischen Akademie. Bd. I. u. II.* 1864.

20. Nägeli. *Das Wachsthum der Stärkekörner durch Intussusception.* Botan. Zeitung. 1881.
21. Nägeli. *Ernährung der niederen Pilze durch Kohlenstoff- u. Stickstoffverbindungen. Untersuch. über niedere Pilze aus dem pflanzenphysiolog. Institut in München.* 1882.
22. *Pathological Society's Transactions. Discussion on Phagocytosis and Immunity.* Vol. XLIII. 1892.
22a. W. Pfeffer. *Ueber intramoleculare Athmung. Untersuchungen aus dem botan. Institut zu Tübingen.* Bd. I.
22b. W. Pfeffer. *Ueber Aufnahme von Anilinfarben in lebende Zellen. Untersuchungen aus dem botan. Institut zu Tübingen.* Bd. II.
23. W. Pfeffer. *Pflanzenphysiologie.* 1881.
24. W. Pfeffer. (1) *Ueber Aufnahme und Ausgabe ungelöster Körper.* (2) *Zur Kenntniss der Plasmahaut und der Vacuolen nebst Bemerkungen über den Aggregatzustand des Protoplasmas und über osmotische Vorgänge. Abhandl. der Mathemat. physik. Classe d. kgl. sächs. Gesellsch. d. Wissenschaft.* Bd. XVI. 1890.
25. Pflüger. *Ueber die Physiolog. Verbrennung in den lebendigen Organismen. Archiv. f. Physiologie.* Bd. X. 1875.
26. Pflüger. *Ueber Wärme und Oxydation der lebendigen Materie.* Pflügers Archiv. Bd. XVIII. 1878.
27. W. Schimper. *Untersuchungen über das Wachsthum der Stärkekörner.* Botan. Zeitung. 1883.
28. W. Schimper. *Ueber die Entwickelung der Chlorophyllkörner und Farbkörner.* Botan. Zeitung. 1883.
29. Fr. Schmitz. *Die Chromatophoren der Algen. Vergleichende Untersuch. über Bau und Entwickelung der Chlorophyllkörper und der analogen Farbstoffkörper der Algen.* Bonn. 1882.
30. Schützenberger. *Die Gährungserscheinungen.* 1876.
31. Strasburger. *Ueber den Bau und das Wachsthum der Zellhäute.* Jena. 1882.
32. Strasburger. *Ueber das Wachsthum vegetabilischer Zellhäute. Histologische Beiträge.* Heft 2. 1889.
33. Strasburger. *Practical Botany,* trans. by Hillhouse.
34. A. Weiss. *Ueber spontane Bewegungen und Formänderungen von Farbstoffkörpern Sitzungsber. d. kgl. Akademie d. Wissensch. zer Wien.* Bd. XC. 1884.
35. Hugo de Vries. *Plasmolytische Studien über die Wand der Vacuolen.* Pringsh. Jahrb. f. wissensch. Botanik. Bd. 16. 1885.
36. Hugo de Vries. *Untersuch. über die mechanischen Ursachen der Zellstreckung.* 1877.
37. Went. *Die Vermehrung der normalen Vacuolen durch Theilung.* Jahrb. f. wissensch. Botanik. Bd. 19. 1888.
38. Jul. Wortmann. *Ueber die Beziehungen der intramolecularen u. normalen Athmung der Pflanzen. Arbeiten des botanischen. Instituts zu Würzburg.* Bd. II. 1879.
39. Wiesner. *Die Elementarstructur u. das Wachsthum der lebenden Substanz.* 1892.

40. Richard Hertwig. *Die Radiolarien.*
41. Ehrlich. *Ueber die Methylenblaureaction der lebenden Nervensubstanz. Biologisches Centralblatt. Bd. VI. 1887.*
42. R. Heidenhain. *Physiologie der Absonderungsvorgänge. Handbuch der Physiologie. Bd. V.*
43. Max Schulze. *Ein reizbarer Objecttisch u. seine Verwendung bei Untersuchungen des Blutes. Archiv. f. mikrosk. Anatomie. Bd. I.*
44. Oscar Schulze. *Die vitale Methylenblaureaction der Zellgranula. Anat. Anzeiger, 1887, p. 684.*
45. Camillo Schneider. *Histologie von Hydra fusca mit besonderer Berücksichtigung des Nervensystems der Hydropolypen. Archiv. f. mikrosk. Anatomie. Bd. XXXV.*
46. Hugo de Vries. *Intracellulare Pangenesis. Jena. 1889.*

CHAPTER VI.

THE VITAL PHENOMENA OF THE CELL.

I. Reproduction of the Cell by Division.—One attribute of the cell, which is of the greatest importance, since the maintenance of life depends upon it, is its power of producing new forms similar to itself, and by this means maintaining its species. It is becoming daily more and more clearly evident, as the result of innumerable observations, that new elementary organisms can only arise through the division of the mother-cell into two or more daughter-cells (*Omnis cellula e cellula*). This fundamental law, which is of paramount importance in the study of biology, has only been established after much laborious work along the most diverse lines, and after many blunders.

1. History of Cell Formation. Schleiden and Schwann (I. 28, 31), in developing their theories, asked themselves the natural question, "How do cells originate?" Their answer, based upon observations both faulty and insufficient, was incorrect. They held that the cells, which they were fond of comparing to crystals, formed themselves, like crystals, in a mother-liquor. Schleiden named the fluid inside the plant cell *Cytoblastem*. He considered it to be a germinal substance, a kind of mother-liquor. In this the young cells were supposed to originate a solid granule, the nucleolus of the nucleus developing first, around which a layer of substance was precipitated; this, they considered, became transformed into the nuclear membrane, whilst fluid penetrated between it and the granule. The nucleus thus formed constituted the central point in the formation of the cell, in consequence of which it was termed the *Cytoblast*. The process of cell development was then supposed to be similar to the one described above when the nucleus was formed round the nucleolus. The cytoblast surrounded itself with a membrane which was composed of substances precipitated from the cell-sap. This membrane was at first closely in contact with the nucleus, but later on was pushed away by the in-pressing fluid.

Schwann (I. 31), whilst adopting Schleiden's theory, fell into a second, and still greater error. He considered that the young cells developed, not only within the mother-cell (as propounded by Schleiden), but also outside of it, in an organic substance, which is frequently present in animal tissues as intercellular substance, and which he called also Cytoblastem. Thus Schwann taught that cells were formed spontaneously both inside and outside of the mother-cell, which would be a genuine case of spontaneous generation from formless germ substance.

These were indeed grave fundamental errors, from which, however, the botanists were the first to extricate themselves. In the year 1846 a general law was formulated in consequence of the observations of Mohl (VI. 47), Unger, and above all, Nägeli (VI. 48). This law states, that new plant cells only spring from those already present, and further that this occurs in such a manner, that the mother-cell becomes broken up by dividing into two or more daughter-cells. This was first observed by Mohl.

It was much more difficult to disprove the theory, that the cells of animal tissues arise from cytoblasts, and this was especially the case in the domain of pathological anatomy, for it was thought that the formation of tumours and pus could be traced back to cytoblasts. At last, after many mistakes, and thanks to the labours of many investigators, amongst whom Kölliker (VI. 45, 46), Reichert (VI. 58, 59), and Remak (VI. 60, 61) must be mentioned, more light was thrown upon the subject of the genesis of cells in the animal kingdom also, until finally the cytoblastic theory was absolutely disproved by Virchow, who originated the formula, "*Omnis cellula e cellula.*" No spontaneous generation of cells occurs either in plants or animals. The many millions of cells of which, for instance, the body of a vertebrate animal is composed, have been produced by the repeated division of *one* cell, the ovum, in which the life of every animal commences.

The older histologists were unable to discover what part the nucleus played in cell-division. For many decades two opposing theories were held, of which now one and now the other obtained temporarily the greater number of supporters. According to the one theory, which was held by most botanists (Reichert VI. 58; Auerbach VI. 2a, etc.), the nucleus at each division was supposed to break up and become diffused throughout the protoplasm, in order to be formed anew in each daughter-cell. According to the other (C. E. v. Baer; Joh. Müller; Remak VI. 60; Leydig;

Gegenbaur; Haeckel V. 4b; van Beneden, etc.), the nucleus was supposed to take an active part in the process of cell-division, and, at the commencement of it, to become elongated and constricted at a point, corresponding with the plane of division which is seen later, and to divide into halves, which separate from one another and move apart. The cell body itself was supposed to become constricted, and to divide into two parts, in each of which one of the two daughter-nuclei formed the attraction centre.

Each of these theories, so diametrically opposed, contains a grain of truth, although neither describes the real process, which remained hidden from the earlier histologists, chiefly on account of the methods of investigation used by them. It is only during the last two decades, that our knowledge of the life of the cell has been materially advanced by the discoveries made by Schneider (VI. 66), Fol (VI. 18, 19), Auerbach (VI. 2a), Bütschli (VI. 81), Strasburger (VI. 71, 73), O. and R. Hertwig (VI. 30–38). Flemming (VI. 13–17), van Beneden (VI. 4a, 4b), Rabl (VI. 53), and Boveri (VI. 6, 7). These discoveries have revealed to us the extremely interesting formations and metamorphoses, which are seen in the nucleus during cell-division. These investigations, to which I shall have occasion to refer frequently in this section, have all pointed to the same conclusion, that the nucleus is a permanent and most important organ of the cell, and that it evidently plays a distinct rôle in the cell life during division. Just as the cell is never spontaneously generated, but is produced directly by the division of another cell, so the nucleus is never freshly created, but is derived from the constituent particles of another nucleus. The formula, "*omnis cellula e cellula*," might be extended by adding "*omnis nucleus e nucleo*" (Flemming VI. 12).

After this historical introduction, we will consider more in detail, first, the changes which take place in the nucleus during division, and next, the various methods of cell multiplication.

II. **Nuclear Division.**—The nucleus plays an important and most interesting part in each process of cell-division. Three methods of nuclear reproduction have been observed: indirect, or nuclear segmentation, direct (Flemming), or nuclear fission, and endogenous nuclear formation.

1. **Nuclear Segmentation.** Mitosis (Flemming). Karyokinesis (Schleicher). The phenomena which occur during this process are very complicated; nevertheless they conform to certain laws which are wonderfully constant in both plants and animals.

The main feature of the process consists in this, that the various chemical substances (*vide* p. 40), which are present in the resting nucleus, undergo a definite change of position, and the nuclear membrane being dissolved, enter into closer union with the protoplasmic substance. During this process the constant arrangement of the nuclein becomes especially apparent; and, indeed, the changes, which occur in this substance, have been most carefully and successfully observed, whereas we are still very much in the dark concerning what takes place in the remaining nuclear substances.

The whole mass of nuclein in the nucleus becomes transformed during division into fine thread-like segments, the number of which remains constant for each species of animal. These segments are generally curved, and vary in form and size according to the individual species of plant or animal; they may appear as loops, hooks, or rodlets, or if they are very small, as granules. Waldeyer (VI. 76) proposed the common name of *chromosomes* for all these various forms of nuclein segments. As a rule I shall employ the more convenient name of *nuclear segments*, which applies equally to them all, whilst, at the same time, the expression indicates the most important part of the process of indirect division, which consists chiefly in this, that the nuclein breaks up into segments. Similarly the term *nuclear segmentation* appears to me to be preferable to the longer and less significant expression of indirect nuclear division, or the terms mitosis and karyokinesis, which are incomprehensible to the uninitiated.

During the course of division each nuclear segment divides longitudinally into two daughter segments, which for a time lie parallel to one another, and are closely connected. Next, these daughter segments separate into two groups, dividing themselves equally between the two daughter-cells, where they form the foundation of the vesicular daughter nuclei.

The following phenomena are also characteristic of the process of nuclear segmentation: (1) the appearance of the two so-called *pole corpuscles* (centrosomes), which function as central points, around which all the cell constituents arrange themselves; (2) the formation of the so-called *nuclear spindle*; and (3) the development of the *protoplasmic radiation* figures around the centrosomes.

As regards the two centrosomes, they make their appearance in the vesicular nucleus at an early stage, before the membrane has been dissolved, being situated in that portion of the protoplasm which is directly in contact with the membrane. At this period they are close to one another, and are in the form of two extremely small spherules. They are composed of a substance which is only stained with difficulty, and which is, perhaps, derived from the substance of the nucleolus. These spherules are the pole or central corpuscles (corpuscules, poles, centrosomes), which have been already described. Gradually they separate from one another, describing a semicircle round the upper surface of the nucleus, until they take up their position at opposite ends of the nuclear diameter.

The nuclear spindle develops itself between the centrosomes. It consists of a large number of very delicate fibrils, which are parallel to one another, and which are probably derived from the linin framework of the resting nucleus. These fibrils diverge somewhat at their centres, and converge at their ends towards the centrosomes, in consequence of which the bundle assumes more or less the shape of a spindle. At first, when the centrosomes are just commencing to separate, the spindle is so small, that it can only be made out with difficulty, as a band connecting them together. However, as the centrosomes separate from one another, the spindle increases in size, and becomes more clearly defined.

The protoplasm also commences to arrange itself around the poles of this nuclear figure as though attracted by them. Thus an appearance, similar to that seen at the ends of a magnet, which has been dipped in iron filings, is produced. The protoplasm forms itself into a large number of delicate fibrils, which group themselves radially around the centrosome as a middle point or centre of attraction. At first they are short and confined to the immediate neighbourhood of the attraction centre. However, during the course of the process of division they increase in length, until finally they extend throughout the whole length of the cell. This arrangement of the protoplasm around the pole is variously described as the plasmic radiation, radiated figure, star, sun, etc., in consequence of its resemblance to the rays of light, attraction spheres, etc.

These are briefly the various elements out of which the nuclear division figures are built up. The centrosomes, the spindle, and the two plasmic radiations have been grouped together by Flemming under the name of the *achromatin portion* of the dividing nuclear figure, in contradistinction to the various appearances which are produced by the re-arrangement of the nuclein, and which constitute the *chromatin portion* of the figure.

All the individual constituent portions of the division-figure as a whole vary according to fixed laws, by grouping their elements in various ways during the course of the process of division.

For the sake of convenience it is well to distinguish four different phases, which succeed each other in regular sequence.

During the first stage the resting nucleus undergoes changes preparatory to division, resulting in the formation of the nuclear segments and the nuclear centrosomes, whilst at the same time the spindle commences to develop. During the second stage the nuclear segments, after the nuclear membrane has become dissolved, arrange themselves into a regular figure, midway between the two poles, at the equator of the spindle. During the third the daughter-segments, into which during one of the former stages the mother-segments have divided by longitudinal fission, separate into two groups, which travel in opposite directions from the equator until they reach the neighbourhood of the centrosomes. During the fourth stage reconstruction takes place, vesicular resting daughter nuclei being formed out of the two groups of daughter-segments, whilst the cell body divides into two daughter-cells. In the next few sections a more minute description will be given of the process of cell division as it occurs in some individual cases, and finally a special section will be devoted to the discussion in detail of certain disputed points.

The most convenient, and at the same time the commonest, subjects for examination in the animal kingdom are the tissue cells of young larvæ of *Salamandra maculata*, of *Triton*, the spermatozoa of mature animals, the segmentation spheres of small transparent eggs, especially of Nematodes (*Ascaris megalocephala*), and of Echinoderms (*Toxopneustes lividus*). Amongst plants the protoplasm of the endosperm of the embryo sac, especially of *Fritillaria imperialis*, and the developing pollen cells of *Liliaceæ*, are especially to be recommended.

a. **Cell division, as it occurs in the *Salamandra maculata*, as an example of the division of the sperm-mother-cell.**

First Stage. Preparation of the Nucleus for Division.

In the *Salamandra maculata* certain preliminary changes occur in the resting nucleus some time before division actually commences. The nuclein granules, which are distributed all over the linin framework (Fig. 75 A), collect together at certain places and arrange themselves into delicate spiral threads, which are covered

 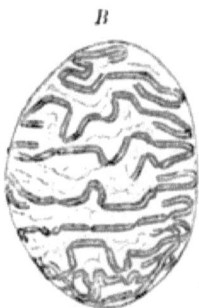

FIG. 75.—*A* Resting nucleus of a sperm-mother-cell of *Salamandra maculata* (after Flemming, Pl. 23, Fig. 1; from Hatschek). *B* Nucleus of a sperm-mother-cell of *Salamandra maculata*. Coil stage. The nuclear threads are already commencing to split longitudinally (diagrammatic, after Flemming, Pl. 26, Fig. 1; from Hatschek).

with small indentations and swellings. From these, innumerable most delicate fibrils branch off at right angles; these fibrils, which consist of strands of the linin framework, only become visible as the nuclein withdraws itself from their surface. Later on the nuclein threads become still more clearly defined, and, as the indentations and swellings disappear, develop a perfectly smooth surface (Fig. 75 *B*). Now since they surround the nuclear space on every side, they produce an appearance described by Flemming as the coil figure (*spirem*, *skein*). The coil is much more dense in the epithelial cells of *Salamandra* than in sperm cells, whilst at the same time the threads are much finer and longer (Fig. 76).

It is as yet undecided, whether at the outset the coil consists of a single long thread or of several such threads. I agree with Rabl (VI. 53) that the latter is more probable.

A striking difference is now seen in the way the various nuclear

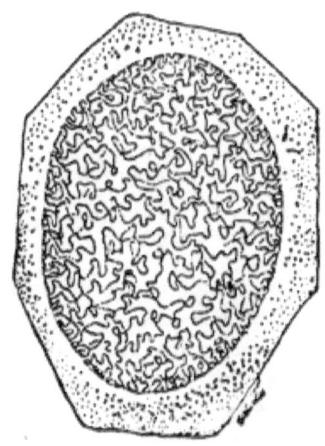

FIG. 76.—Nucleus of an epithelial cell at the commencement of division; from a Salamander larva. Fine coil formation. The remains of two nucleoli are still present. (After Flemming.)

constituents absorb staining solutions, compared to that observed in former stages. The more distinctly and sharply defined the threads grow, the more strongly stained do they become, and the more energetically do they retain the colouring matter, whereas the network of the resting nucleus exhibits these properties to a much less degree. This may be especially well demonstrated if Graham's method of staining be employed, for whilst the resting nuclei are completely decolourised, those that are preparing to divide, or are actually undergoing the process, are so strongly stained that they cannot fail to attract the attention of the observer.

During the first stage of coil formation the nucleoli are still present; however, they gradually diminish in size, until after a short time no trace of them can be seen. Up till now it has not been determined with certainty what is formed from them.

Whilst the coil is developing, careful observation reveals a small spot on the surface of the nucleus. This becomes more and more distinctly defined as the process progresses: it has been designated by Rabl *the polar area* (Fig. 77). The opposite surface of the nucleus is the anti-polar area. The nuclein threads become gradually more and more distinct, and arrange themselves so as to point towards these two areas.

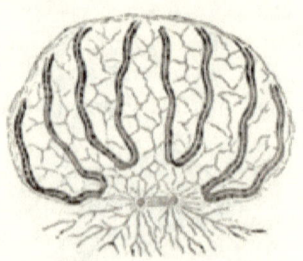

Fig. 77.—Diagrammatic representation of a nucleus with a polar area, in which the two centrosomes and the spindle are developing. (After Flemming, Pl. 39, Fig. 37.)

Starting from the anti-polar region they collect in the neighbourhood of the polar area. "Here they bend round upon themselves in a loop-like fashion, and then return, by means of several small, irregular indented loops, to the neighbourhood of their starting point." Later on the threads become shorter and correspondingly thicker; they are less twisted, and cling less closely together, so that the whole skein looks much looser. In the meantime their arrangement in loops gradually grows more and more distinct. In favourable cases it has been ascertained that there are twenty-four such loops or nuclear segments; this number is constant for the tissue cells and sperm-mother-cells of *Salamandra* and *Triton*.

Meanwhile the two centrosomes and the spindle—most important portions of the nuclear figure—have developed in the polar area. However, on account of the difficulty in staining them, and their minute size and extreme delicacy, these appearances are not easily made out at this stage; further, they may be more or less concealed by granules, which collect in the protoplasm in their neighbourhood. According to Flemming and Hermann, two centrosomes may be made out in successful preparations. These are situated very close together, and have probably been formed by the division of an originally single centrosome. Between them the connecting fibrils, which later on develop into the spindle, can be seen.

Second Stage of Division.

The second stage may be said to date from the time when the nuclear membrane grows indistinct and dissolves. The nuclear sap then distributes itself evenly throughout the cell body, whilst the nuclear segments come to lie freely in the middle of the protoplasm (Fig. 78). The two centrosomes, which are now further apart from one another, are situated near them. The spindle increases proportionately in size and distinctness, and is seen to consist of a number of most delicate fibrils, stretching continuously from one centrosome to the other, as is clearly shown in Hermann's preparation represented in Fig. 78. The centrosomes of the nuclear figure commence

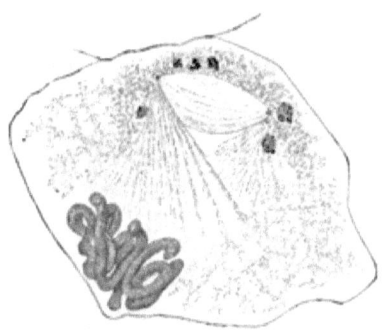

FIG. 78.—Nucleus of a sperm-mother-cell of *Salamandra maculata* preparatory to division. The spindle is situated between the two centrosomes. (After Hermann (VI. 29), Pl. 31, Fig. 7.)

at this stage to exercise an influence upon the surrounding protoplasm. Around each centrosome as centre, innumerable protoplasmic fibrils group themselves radially, stretching out principally towards that region where the nuclear segments are situated, and appearing to adhere to their surface. From now on, the spindle commences to increase rapidly in size until it has attained the dimensions seen in Fig. 79.

Meanwhile the chromatin figure becomes markedly altered (Fig. 79). The nuclear segments have grown considerably shorter and thicker, and are grouped around the spindle in the form of a complete ring, the arrangement being that described by Flemming as the mother-star. The loop-like shape of the segments is now most clearly defined. They are invariably so arranged that the angle of the loop is directed towards the axis of the spindle, whilst its arms point towards the surface of the cell. All of the twenty-four loops lie pretty accurately in the same plane, which, since it bisects the spindle at right angles, is called the equatorial plane; it is identical with the plane of division which develops later. When seen from either of the poles the chromatin figure has "the shape of a star whose rays are formed of the arms of the V-shaped loops, and whose centre is traversed by the bundle of achromatin fibrils which compose the nuclear spindle." This point of view is the most convenient one for counting the nuclear segments, and for determining their number to be twenty-four.

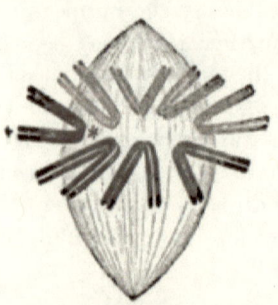

FIG. 79.—Diagrammatic representation of the segmentation of the nucleus (after Flemming). Stage in which the nuclear segments are arranged in the equator of the spindle.

Another most important process occurs during the second stage. If the nuclear segment of a well-preserved preparation be examined with a high power of the microscope, it will be seen that each mother segment is cleft longitudinally, and is thus split up into two parallel daughter segments, which lie close together. Now since no sign of this longitudinal division could be seen in the original nuclear network, it follows that it must have occurred after karyokinesis had commenced. Generally the longitudinal cleft may be first seen when the nuclear threads have arranged themselves in the form of a coil (Fig. 75 *B*), but it is always completed during the second stage (mother-star), when it is most clearly defined. This was first observed by Flemming (VI. 12, 13), in *Salamandra*; and his statements have been corroborated by v. Beneden (VI. 4a), Heuser (VI. 39), Guignard (VI. 23), Rabl (VI. 53), and many others, who made observations upon the same and other objects. This longitudinal splitting appears to occur invariably in indirect nuclear division, and is of the greatest importance for the comprehension of the process, as will be shown later on, when the subject is discussed theoretically.

Third Stage of Division.

The third stage is characterised by the division of the single group of mother-segments in the equatorial plane into two groups of daughter-segments, which retreat in opposite directions from one another, until they are situated in the neighbourhood of the two poles of the nuclear figure (Fig. 80 A, B, C). The two

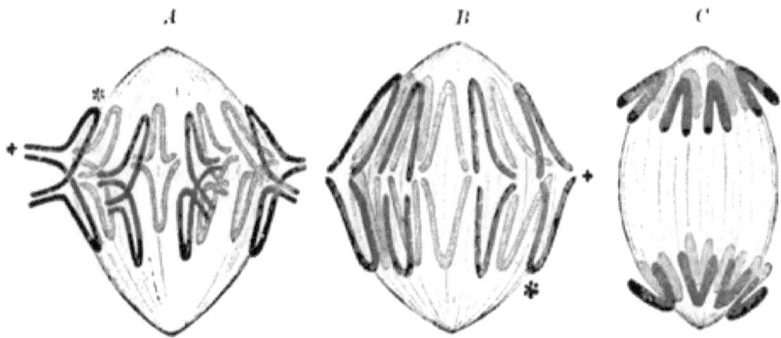

Fig. 80.—Diagrammatic representation of nuclear segmentation (after Flemming). The daughter-segments are retreating in two groups towards the poles. (From Hatschek.)

daughter-stars are formed, as Flemming expresses it, from the mother-star. The details of the process, which can only be observed with difficulty, are as follows:—

The daughter-segments, which have been produced by the splitting of a mother-segment, separate from one another at the angle of the loop, which is directed towards the spindle, and commence to retreat towards the poles, whilst for a time the ends of the arms of the loop remain undivided. Finally these also split up. From out of the 24 original loops two groups, each containing 24 daughter-loops, have developed; these move towards the centrosomes, until they come quite close to them, when they stop, for they never actually reach the poles themselves. Between these two groups fine "connecting fibrils" stretch; these are probably derived from the spindle fibrils.

Each loop, or daughter-segment, has "its angle directed towards the pole, whilst its free ends are turned either obliquely, or perpendicularly, to the equatorial plane." As might be expected, to start with, they are much thinner than the mother-segments; however, they soon begin to shorten and to become proportionately thicker. When the daughter-star is first formed, the segments lie somewhat far apart, but they soon begin to draw more closely together, so that it becomes very difficult to count them and to trace their further development; in fact, it can only be accomplished in exceptional cases.

Fourth Stage of Division.

During this stage each group of daughter-segments becomes gradually re-transformed into a vesicular resting nucleus (Fig. 81).

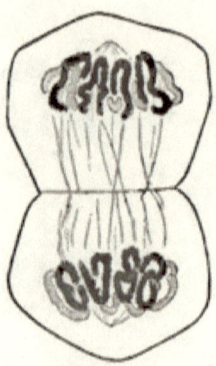

Fig. 81.— Diagrammatic representation of nuclear segmentation (after Flemming). The resting nucleus has commenced to build itself up out of the daughter-segments. (From Hatschek.)

The threads draw still more closely together, become more bent and thicker; their surfaces grow rough and jagged, and small processes become developed externally upon them, whilst a delicate nuclear membrane develops around the whole group. The radiated appearance around the centrosomes gradually grows less and less distinct, until it soon quite disappears. Finally, also, the centrosomes and the spindle fibrils can no longer be distinguished. It has not yet been decided what they develop into. In fact, their origin and their disappearance are equally shrouded in mystery. Near to the place where the centrosome was situated a depression may be seen in the newly forming daughter nucleus. Rabl considers it to be the above-described polar area of the nucleus which is seen preparatory to division, and is of opinion that the centrosome has ensconced itself within it, being enclosed in the protoplasm of the cell-body. The nucleus gradually swells up more and more through the absorption of nuclear sap, and becomes globular in form, whilst the framework of the resting nucleus, with its irregularly distributed nuclein granules of various sizes, is reconstructed. Further, one or more nucleoli have made their appearance in the framework during the process of reconstruction, but as yet no one has succeeded in discovering their origin.

When, at the commencement of the fourth stage, the two daughter-stars are separated as far as possible from one another, and have taken the preliminary steps towards becoming transformed into the resting daughter nuclei, the cell-body itself begins to divide. The radiations at the centrosomes have now attained their greatest size. At this period a small furrow becomes visible on the surface of the cell-body, corresponding to a plane, which passes perpendicularly through the centre of the nuclear axis, uniting the two centrosomes; this has already been referred to as the plane of division. "The furrow commences on one side, and gradually extends itself round the equator; however, it remains somewhat deeper on the side where it commenced than on the opposite one" (Flemming). This ring-like constriction gradually cuts more and more deeply into the cell body, until finally it divides it completely into two nearly equal parts, each of which contains a daughter nucleus, undergoing the process of reconstruction. As soon as division is complete, the polar radiations commence to fade away.

The above-mentioned connecting fibrils between the daughter nuclei may be distinguished, in many objects, until division is completed. They are then severed in their centres by the cutting through of the cell-body. Sometimes a number of spherical swellings, which become intensely stained, may be seen at this time to develop at the centres of the spindle fibrils; these Flemming (VI. 13″) has named separation bodies, and he considers that they probably represent the equatorial plates of plants, which are much better developed.

b. **Division of the egg-cells of *Ascaris megalocephala* and *Toxopneustes lividus*.** The nuclei of the eggs of *Ascaris* are remarkable for the size and distinctness of their centrosomes, and for the small number of their nuclear segments, of which in one species only four, and in another only two, are present. Another very important phenomenon, the multiplication of the centrosomes by division, may be especially clearly seen in this object. It is best to commence our investigations at that point when the egg has just developed the furrow, and when the four nuclear loops on either side of the plane of division have transformed themselves into a vesicular nucleus of irregular outline (Fig. 82). The side of the nucleus, which is directed towards the pole, has several ragged processes, the nuclein being spread out upon its loose network. The centrosome may still be distinguished in the neighbourhood of

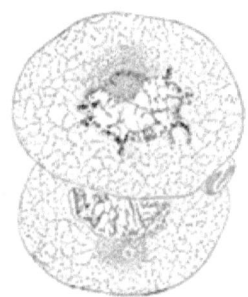

Fig. 82.—Egg of *Ascaris megalocephala* undergoing the process of double division. Nuclei are resting; the centrosomes as yet undivided. (After Boveri, Pl. IV., Fig. 74.)

what was formerly the pole of the division figure; it is enclosed in granular protoplasm, which contracts with the yolk substance of the egg, and has been named by van Beneden the attraction sphere, and by Boveri the archoplasm.

Before the nucleus has quite returned to the resting condition, and even sometimes before the first division is completed, it commences to make preparations to divide a second time; these start with changes in the centrosome (Fig. 84), which extends itself

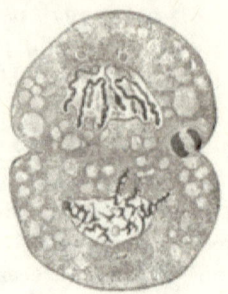
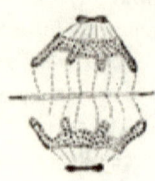

Fig. 83. Fig. 84.

Fig. 83.—Dividing egg of *Ascaris megalocephala*. The nuclei are preparing to divide; the centrosomes are divided. (After Boveri, Pl. IV., Figs. 75, 76.)

Fig. 84.—Two daughter-nuclei with lobulated processes commencing to reconstruct themselves. The centrosomes are multiplying by self-division. (After van Beneden and Neyt, Pl. VI., Fig. 13.)

longitudinally parallel to the first division plane, becomes biscuit-shaped, and divides itself by a constriction into two daughter centrosomes, which for a time are enclosed by one common granular sphere; these phenomena were discovered by van Beneden (VI. 4b) and Boveri (VI. 6, 1888). Next, the two centrosomes separate somewhat from one another (Fig. 83), in consequence of which their common radiation sphere becomes converted into two spheres.

This division of the centrosomes gives the signal, as it were, for the occurrence of the following changes in the nucleus, although the latter is not yet completely at rest (Fig. 83). The nuclein withdraws itself out of the framework, and collects in four long loops, the surfaces of which are at first uneven, but later on become smooth. The four loops are turned in the same direction as the daughter-segments after the first division, so that Boveri (IV. 6) agrees with the opinion expressed by Rabl (VI. 53), that they are derived directly from the substance of the segments, and that even when the nucleus is resting they have an independent individuality. The angles of the loop are turned towards the original pole (the polar area in the *Salamandra*), whilst the ends of the loop, which are knob-like and swollen, are directed towards the region of the anti-pole.

The second stage of division now commences. The centrosomes, with their spheres, separate and travel for some distance, until their common axis lies either somewhat obliquely or parallel to the first division plane. The nuclear membrane dissolves. The four segments arrange themselves in the equator between the two centrosomes in the manner described above, whilst a distinct radiation develops around the centrosomes in the protoplasm, so that the appearance, seen from the pole, resembles that depicted in Fig. 85 *A*. The four segments then split longitudinally

Fig. 85.—*A* Four mother-segments seen from the pole of the nuclear figure (after van Beneden and Neyt, Pl. VI., Fig. 16). *B* Longitudinal splitting of the four mother-segments into eight daughter-segments (after van Beneden and Neyt, Pl. VI., Fig. 17).

—that is to say, the third stage commences (Fig. 85 *B*). The daughter segments thus formed separate from one another, and travel towards opposite poles. E. van Beneden (VI. 4b) and Boveri (VI. 6) consider that the spindle fibrils play an active part in this process. In their opinion, the spindle in *Ascaris* is composed of two independent portions, each of which consists of a large number of protoplasmic fibrils. These converge towards the centrosome and attach their ends to it, whilst the opposite ends diverge, approach the nuclear loops, and fasten themselves at various points to the daughter-segments, which are turned towards them. These threads by gradually contracting, and thus becoming shortened, cause, in van Beneden's and Boveri's opinion, the separation of the four daughter-segments, which are thus gradually drawn towards the centrosomes.

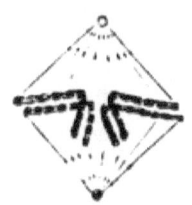

Fig. 86.—The construction of the spindle out of two half-spindles, the fibrils of which have attached themselves to the daughter-segments. After van Beneden and Neyt, Pl. VI., Fig. 8.)

During the fourth stage the cell-body divides, and the daughter-nucleus becomes built up again. This, according to van Beneden, takes place in the following manner (Fig. 87): the four chromatin

FIG. 87.—*A* A group of four daughter-segments seen from the pole, the swellings at the ends, forming the loops, are especially well marked (after van Beneden and Neyt, Pl. VI., Fig. 19). *B* Reconstruction of the nucleus from the four daughter-segments, diagrammatic (from van Beneden and Neyt, Pl. VI., Fig. 20). *C* Resting condition of the nucleus, seen from the pole (from van Beneden and Neyt, Pl. VI., Fig. 21).

loops (*A*) absorb fluid, which becomes nuclear sap, out of the protoplasm; they become saturated with it, as a sponge with water, and thus swell up into thick vesicular bodies (*B*). The nuclein divides up into granules, which are connected together by delicate threads, which are situated chiefly upon the surfaces of these vesicles. The inner surfaces of these latter come close together and fuse. Thus a vesicular nucleus, irregular in shape, and saturated with nuclear sap, is formed; it is separated from the protoplasm by a membrane, and contains a delicate framework, upon which the chromatin substance is distributed.

The eggs of *Ascaris* afford us special advantages for the study of centrosomes and nuclear segments, but the small eggs of *Echinoderms* (Hertwig VI. 30a; Fol VI. 19a) are also of great use, particularly for observing radiation phenomena in the protoplasm of the living cell. More will be said about this later on.

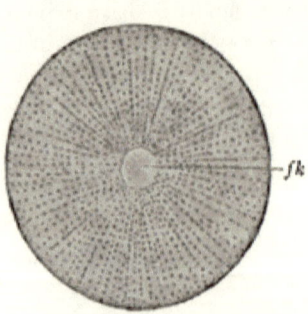

FIG. 88.—Egg of a Sea-urchin just after fertilisation has been completed (from O. Hertwig, *Embryology*, Fig. 20). Egg nucleus and sperm nucleus are fused to form the cleavage nucleus (*fk*) which occupies the centre of a protoplasmic radiation.

In the egg-cell of a living *Echinoderm*, a few minutes after fertilisation (Fig. 88), the small globular cleavage-nucleus is seen to be situated in the centre of the yolk; it looks like a clear vesicle, and is surrounded by rays of protoplasm, like a sun with rays of light. This radiation is so distinct in this object during life, as the large number of small granules, which are situated in the yolk, are arranged in rows, passively following the arrangement of the protoplasm. After a short time this radiated appearance, which is the result of the processes which occur during fertilisation, begins to fade, and to become metamorphosed into two radiated systems, which are found at opposite points of the nucleus. These are small at first, but become momentarily larger and more distinct, until finally they extend all over the whole yolk-sphere, dividing it up into two radiated masses, each arranged around its own attractive centre (Fig. 89).

A small homogeneous spot can be distinguished in the middle of each radiation from the very beginning; this spot adheres closely to the nuclear surface, and is free from granules. It contains the centrosome, which, however, cannot be distinguished at all in the living object.

As the radiations become more distinct and more spread out, the collections of homogeneous non-granular protoplasm in the neighbourhood of the centrosomes become larger, whilst at the same time they gradually retreat farther and farther apart, carry-

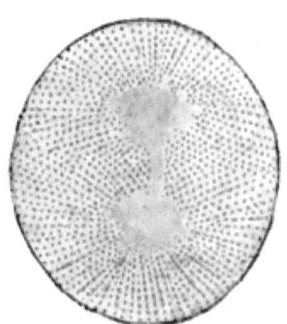

FIG. 89.—Egg of a Sea-urchin preparing to divide; taken from the living object (from O. Hertwig, Embryologie, Fig. 27). The nucleus is invisible, the dumbbell figure having taken its place.

ing the poles with them. At this period the nucleus loses its vesicular properties, and assumes the spindle structure which has been described in other objects, but which, on account of its minuteness, cannot be distinguished here during life. In consequence, the very characteristic dumb-bell appearance, depicted in Fig. 89, develops in the granular yolk. The two collections of homogeneous protoplasm, enclosing the poles of the division figure, form the heads of the dumb-bell; the non-granular connecting portion indicates the place where, during the preceding stages, the now invisible nucleus was situated. This has been replaced by the spindle, the ends of which extend right up to the centrosomes. The granular yolk mass is arranged in two radial systems around this homogeneous dumb-bell figure. These systems have been named *amphiaster*, or double star, by Fol.

The egg, which at the outset was perfectly round, now commences to extend itself longitudinally in the direction of the axis of the dumb-bell, and quickly enters the last stage of division (Fig. 90 A). A ring-like furrow corresponding to a plane, which

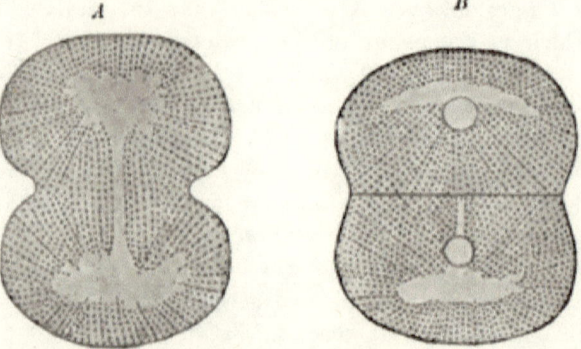

Fig. 90.—Egg of a Sea-urchin when division is just taking place (from O. Hertwig, Embryology, Fig. 29). *A* A circular furrow cuts into the yolk and divides it in a plane which is perpendicular to the centre of the nuclear axis and to the long axis of the dumb-bell. *B* Egg of a Sea-urchin after division has taken place. In each of the division products a vesicular daughter nucleus has been formed. The radial arrangement of the protoplasm is commencing to become indistinct. Both figures are drawn from the living object.

might be carried through the dumb-bell at right angles to its longitudinal axis, develops upon the surface of the egg. This rapidly penetrates more and more deeply into the egg-substance, quickly dividing it into two equal portions, each of which contains half of the spindle with a group of daughter segments, that is to say half of the dumb-bell, and a radial system of protoplasm.

When the division in two is nearly completed, the two portions of the egg are in contact at a small portion only of their surfaces, at the middle of the handle of the dumb-bell. When, however, cleavage is quite finished, the whole of their division surfaces come closely into contact with one another, so that they flatten each other into nearly hemispherical bodies (Fig. 90 *B*).

Meanwhile the nucleus has become visible in the living object. Somewhere near the place where the head and the handle of the dumb-bell merge, that is to say, at some little distance from the centrosome, a few small vacuoles make their appearance, being caused by the saturation of the daughter nuclear segments with nuclear sap. After a short time these fuse together to form a globular vesicle, the daughter nucleus (Fig. 90 *B*). The radiated arrangement of the protoplasm grows gradually less distinct, and makes way, if the cell prepares to divide a second time, for a new double radiation.

For examination with reagents, and especially for studying chromatin figures, the eggs of *Echinoderms* are not so suitable as those of *Ascaris*. The loop-like nuclear segments are especially small and numerous in them, so that even with the strongest powers they only look like small granules. Fig. 91 represents a spindle, which has been treated with reagents and staining solutions; it corresponds somewhat to Fig. 89, where the living egg is depicted, and may therefore be considered to complete it.

The process of segmentation may take a fairly long time in very large eggs, such as Frogs' eggs, where a considerable amount of

Fig. 91.

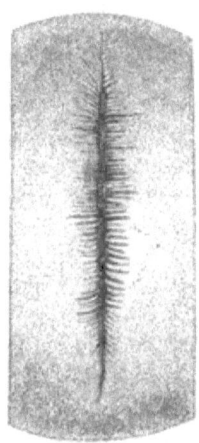
Fig. 92.

Fig. 91.—Nuclear figure of an egg of *Strongylocentrotus*, one hour and twenty minutes after fertilisation. Reagents have been used.

Fig. 92.—A portion of the upper hemisphere of an egg of *Rana temporaria* a quarter of an hour after the appearance of the first furrow, when the coronal radiation is most sharply and plainly defined. (After Max Schultze, Pl. I., Fig. 2.)

yolk has to be divided. Consequently a second process of division may commence before the first is completed. In Frogs' eggs an interesting appearance may be observed, which has been described under the name of the coronal furrow (VI. 68) (Fig. 92). This first furrow commences to appear on a small area of the black pigmented hemisphere of the egg, which is directed upwards; as it penetrates into the substance, it increases in length, and, during the course of half an hour, extends itself round the whole periphery of the globe, appearing last upon the bright surface, which is turned downwards. At this place it penetrates less deeply into the yolk. When it first appears, it is not smooth in appearance, but is seen—most distinctly at that period when it has extended itself around one third of the circumference of the egg—to be provided with a large number of small grooves, which open into it on both sides for the most part at right angles (60-100 on either side, Fig. 92). Thus a very pretty picture is produced, like a long deep valley in the mountains, with a large number of shorter, narrower valleys opening into it on either side. As the process of division progresses, and the main furrow deepens, the side furrows diminish in number, and finally quite disappear.

The appearance of this peculiar and clearly marked coronal furrow is a phenomenon which is connected with the contraction of the protoplasm during cleavage.

c. **Division of Plant Cells.** The protoplasmic coating of the wall of the embryo-sac of *Fritillaria imperialis* affords an instructive illustration of the great uniformity of the process of nuclear division as it occurs in plants and animals. This, as well as the embryo-sacs of other *Liliaceæ*, is particularly suitable for the study of nuclear figures, for the layer of protoplasm is extremely thin, and, if examined at the right time, is seen to contain a large number of nuclei at various stages of division (Strasburger VI. 71-73; Guignard VI. 23).

The large resting nucleus contains a linin framework with small meshes (Fig. 93 *A*), upon the surface of which a large number of small nuclein granules are pretty evenly distributed. In the majority of cases nucleoli are present. These vary in size, and lie between the meshes of the framework, to which they are attached. Strasburger is of opinion that, when the nucleus is preparing to divide, the whole framework becomes transformed into a few fairly thick threads, which are much twisted; he describes in them a diagonal striation (*c*) similar to that observed by Balbiani (II. 3) in the nuclei of *Chironomus* larvæ (Fig. 27). He accounts for this striation by the statement, that each thread is composed of numerous discs of nuclein arranged one after the other, and separated by their partition walls of linin.

In the course of time, as the process advances, the nuclear membrane dissolves, and the nucleoli break up into smaller granules and disappear, whilst the nuclein threads grow shorter and thicker, and produce twenty-four nuclear segments; a typical spindle composed of a large number of most delicate fibrils develops, in the centre of which the nuclear segments arrange themselves in a circle (Fig. 93 *D*). Guignard has lately demonstrated the presence of two centrosomes with their radiation spheres situated at either end of the spindle.

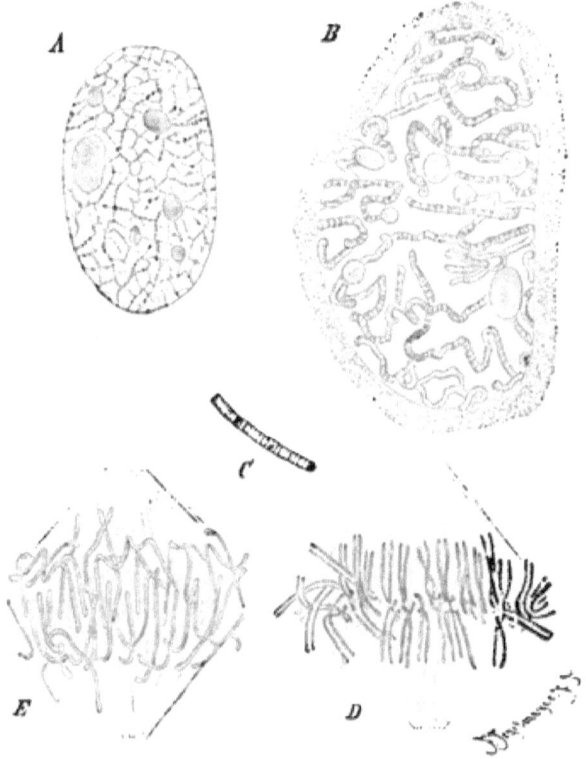

Fig. 93.—*Fritillaria imperialis.* A resting nucleus and other nuclei at various stages of division, taken from the free protoplasmic lining of the wall of the embryo-sac depicted in Fig. 128 (after Strasburger, *Practical Botany,* Fig. 191). *A* A resting nucleus; *B* a coil of thick threads, as yet unsegmented; *C* a portion of a nuclear thread, more highly magnified; *D* a nuclear spindle, with segments longitudinally split; *E* the separation and change of position of the daughter-segments. *A, B, D, E* × 800; *C* × 1100.

When the process of division has reached its highest point, the nuclear segments split longitudinally. The daughter segments then travel towards the two poles, twenty-four on each side (*E*), and thus form the foundation for the daughter nuclei, which develop in a manner similar to that described as occurring in *Salamandra maculata.* As soon as the daughter nuclei become vesicular, several nucleoli appear in them.

Up to this point the resemblance shown by the process to that seen in animal nuclear division has been complete; however, now, at the end of the whole process, a peculiar and interesting deviation is shown in the formation of the so-called cell plate. In order to study this phenomenon, it is better to watch the process of division as it occurs in pollen mother-cells, and in various other objects, rather than to study the embryo-sac of *Fritillaria*, which up till now has formed the basis of our description; for in the latter nuclear division is not immediately followed by cell division.

The following description refers to pollen mother-cells of *Fritillaria persica* (Fig. 94). After the daughter-segments have

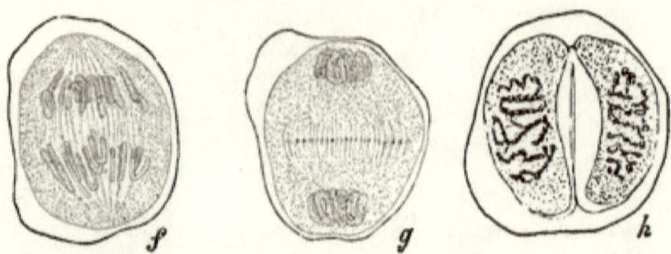

FIG. 94.—Three stages in the division of the pollen mother-cells of *Fritillaria persica* (after Strasburger, Fig. 114, Eng. Edition): *f* separation of the daughter segments; *g* formation of daughter coils and of the cell-plate; *h* position of the nuclear segments in the daughter nuclei and in the developed partition wall. (× 800.)

separated into two groups, delicate connecting fibrils are seen to be stretched between them; these, according to Strasburger (VI. 73), are derived from the central portions of the spindle fibrils (Fig. 94 *f*). After a time, in the middle of the connecting fibrils, small swellings, which look like glistening granules, are formed (Fig. 94 *g*). They are most regularly arranged, so that they are seen in optical section to lie close to one another in a row. Thus collectively they form a disc, composed of granules, and situated in the division plane between the two daughter-nuclei; this disc has been called the cell plate by Strasburger. Flemming (VI. 13″) considers, that these are represented in a rudimentary form in animal cells in the above-mentioned (p. 189) central granules, which are found in a few objects. The cell plate is of the greatest importance in plants, in connection with the formation of the cellulose partition wall, which is the final stage in the whole process of division (Fig. 94 *h*). "The cell plate," as described by Strasburger, "ultimately extends over the whole diameter of the cell, its elements fusing together to form a partition wall, which divides the mother-cell into two daughter-cells." A thin layer of cellulose may soon be distinguished. Meanwhile the connecting fibrils disappear, first around the daughter-nuclei, and then also in the neighbourhood of the cellulose partition wall.

The minute, definite particles, which collect as granules in the middle of the connecting fibrils, and form a cell plate, may be designated as cell-wall formers, in accordance with the above-mentioned conception, which will be entered into at more detail later on.

d. **Historical remarks and unsolved problems concerning nuclear segmentation.**—In the commencement of the year 1870, in consequence of the labours of Bütschli (VII. 6), Strasburger (VI. 71), Hertwig (VI. 30a), and Fol (VI. 19a), the changes experienced by the nucleus during division were described on the whole correctly, although somewhat vaguely. The fibrillous nuclear spindle, the collection of shining granules, which is stained with carmine, in its centre (Strasburger's nuclear plate), the subsequent division of the granules into two groups, or two daughter nuclear plates, and the development of the vesicular daughter nuclei from these latter, had all been discovered by then. Further, the radiation figures—stars, or amphiaster (Fol)—at the ends of the spindle were known, and Fol and myself had already described the presence of more strongly glistening granules, the centrosomes, in them; diagrams had been made of them, and their functioning as attraction centres had been pointed out. Further it had been satisfactorily established that during cell-division the nucleus did not become dissolved (karyolysis, Auerbach, VI. 2a), but became metamorphosed. Further, through my investigations on mature eggs, especially on those of *Asteracanthion* and *Nephelis*, and in consequence of the discovery of the internal phenomena which occur during fertilisation, I showed, at the same time, that the nucleus is not a new development in the egg, but that it is derived from definite portions of the germinal vesicle, which united themselves with the male pronucleus, derived from the head of the spermatozoon (the altered nucleus of the sperm cell), to form the division nucleus. As a result, the important proposition was formulated that all nuclei may be traced back in an unbroken line of descent from the nucleus of the egg-cell, just as all cells of the animal organism are derived from a fertilised egg-cell (*Omnis nucleus e nucleo*. Flemming VI.).

The theory of nuclear and cell division, which was founded in consequence of the above-mentioned investigations, has been proved subsequently to be right in the main, whilst at the same time it has formed a good foundation for many further discoveries, and has suggested a number of problems, which have not yet been definitely solved. These problems may be expressed in a single sentence: it was necessary, and to a certain extent is still necessary, to follow more closely in every detail the movements which, during nuclear division, and during the formation of the characteristic figures, take place in the individual micro-chemical particles of substance, which can be distinguished in the nucleus and in the division figures; that is to say, to trace the rearrangements which occur in the nuclein granules, the linin framework, the spindle fibrils, the centrosomes, and the nucleoli, etc. The discovery of suitable objects for examination, such as the nuclei of tissue cells of *Salamander* larvæ (Flemming), and the eggs of *Ascaris megalocephala* (van Beneden), as well as the use of the newer oil immersion and apochromatic lenses, and the improvement in the manipulation of reagents and staining solutions, have rendered progress in this direction possible.

The greatest advance has at present been made in the investigation of the figures produced by the changes of place of the nuclein, thanks in the main to the excellent experiments of Flemming (VI. 12-17), and the supplementary investigations of van Beneden (VI. 4), Rabl (VI. 53), Boveri (VI. 6), Strasburger (VI. 71-73), and Guignard (VI. 23).

Flemming, who has made his observations chiefly upon tissue cells of *Salamander* larvæ, distinguishes clearly between the achromatin and chromatin portions of the nuclear figure, that is to say, the unstainable spindle fibrils and plasmic radiations, and the stainable nuclear loops, or segments, which rest upon their surfaces. He was the first to make the important discovery that these latter split longitudinally. The explanation of these interesting phenomena was afforded by the discoveries of Henser, Guignard, van Beneden, and Rabl, who all observed independently, on different objects, that the halves of the divided segments (chromosomes) separate, and move towards the nuclear poles, forming the foundation for the daughter-nuclei.

The changes of position of those substances, which are connected with the development of the spindle and the centrosomes, and with the disappearance of the nucleoli, have been much less accurately investigated.

As concerns the spindle, very various opinions are held, both as

THE VITAL PHENOMENA OF THE CELL 201

to its construction and origin. Whilst the first observers considered that the spindle consisted of most delicate fibrils, which stretched continuously from pole to pole, van Beneden (VI. 46) and Boveri (VI. 6) are of opinion that these fibrils are broken at the equator, and that, in consequence, the spindle is composed of two separate and distinct half-spindles (Fig. 95). They contend that the half-spindles are attached directly with the ends of their fibrils to the nuclear segments, and in consequence are of mechanical use in nuclear division, in that they shorten or contract like muscle fibres after the segments have divided into daughter-segments, and thus draw the daughter-segments, which are attached to them, in opposite directions.

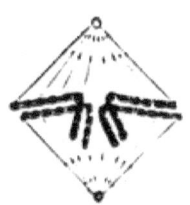

FIG. 95. — Construction of the spindle out of two half-spindles, the fibrils of which are attached to the daughter-segments. (From van Beneden and Neyt, Pl. VI., Fig. 8.)

On the other hand, Flemming (VI. 14) for the tissue cells of *Salamandra*, and Strasburger (VI. 72) for plants, still adhere to their old theory, that spindle fibrils, stretching uninterruptedly from pole to pole, do exist. The observations made by Hermann, which have been already mentioned, are especially convincing concerning the undivided condition of the spindle; they call to mind my description and representation of the formation of the spindle in the germinal vesicle of *Asteracanthion* (VI. 30a, Pl. VIII., Figs. 3, 4). In both cases a very small, undivided spindle may be observed between the poles, which are situated near to one another (Fig. 96), at that period when the nuclear segments are a good way off, and so cannot hide it at all; it is seen to grow gradually, as its fibrils increase in length, until it reaches its full size.

The explanation of this discrepancy, as has been suggested by Hermann, is that the structure described by van Beneden and Boveri as the half-spindle is something quite different from the spindle of the earlier

FIG. 96. — Nucleus of a sperm-mother-cell of *Salamandra maculata* preparing to divide. Position of the spindle between the two centrosomes. (After Hermann, Pl. XXXI., Fig. 7.)

observers. The half-spindles, described by van Beneden and Boveri, consist of a portion of the protoplasmic radiation figure proceeding from the poles, namely, all those fibrils which are situated in the equator around the nuclear segments. The true spindle lies in the centre of these protoplasmic fibrils and nuclear segments. Hermann, to distinguish it from van Beneden's spindle, has given it the name of *central spindle*. The prefix "central," however, appears to me to be quite superfluous; for one thing, it is better to decide to limit the name of spindle once for all to this portion of the nuclear figure, and to give, if necessary, some other name to the protoplasmic polar rays, which are connected with the nuclear segments, and which are described by van Beneden and Boveri as half-spindles; indeed, the name spindle is not suitable to them.

Another moot point is the derivation of the spindle fibrils. Many investigators are inclined to trace them back to that protoplasm, which forced its way in between the nuclein threads when the nuclear membrane was dissolved (Strasburger VI. 72; Hermann VI. 29, etc.). I have already advocated, and am still inclined to hold the view, that, with the exception of the polar radiations, which belong to the protoplasmic body of the cell, the various structural portions of the nuclear figure are derived from the various substances in the resting nucleus. I consider that the substance of the spindle and of the connecting fibrils is derived from the linin framework. This view is supported also by Flemming, and to some extent by the micro-chemical investigations of Zacharias. However, the most important facts in its favour appear to me to be the following :—

In many unicellular organisms the nuclei, during certain stages of division, remain separated from the protoplasm by a delicate membrane; this occurs in *Euglypha* (Schewiakoff VI. 65b), and in the nuclear divisions of *Ciliata* and *Actinosphæria* (Rich. Hertwig, VI. 82, 83). Under these conditions there can be no doubt but that the spindle threads have sprung from the achromatin portion of the nucleus itself. Similar cases are occasionally met with in the animal kingdom as well. In some molluscs (*Pterotrachea, Phyllirhoë*), as Fol (VI. 19a) and I myself (VI. 30a) have observed, the polar spindle, as long as the nuclear membrane remains, is situated in the interior of the germinal vesicle (Fig. 97 *A, B*), which, in this case, is of small size. The assumption that, under these circumstances, protoplasm has made its way into the nuclear space from the exterior, appears to me, at the least, forced. Further, in my opinion, it can no longer be doubted that the connecting

FIG. 97.—*A* A germinal vesicle, in which a spindle is developing, taken from a newly laid egg of *Phyllirhoë*. Acetic acid preparation (Hertwig, Pl. XI., Fig. 2). *B* Germinal vesicle from a freshly laid egg of *Phyllirhoë*, in which the spindle is seen in optical section. Acetic acid preparation (Hertwig, Pl. XI., Fig. 6).

threads, which, in the dividing sperm-mother-cells of *Ascaris*, extend between the separating nuclear segments, are derived from the linin framework. I was not able to observe a typical spindle development in this object.

Another point under discussion is the origin of the centrosomes. These were first described and depicted at the commencement of the year 1870, but they were only brought into prominence as a distinct component part of the nuclear division figure by van Beneden (VI. 4a), when he succeeded in differentiating them clearly from their environment by means of a staining solution of aniline dyes dissolved in 33 per cent. glycerine solution. Soon afterwards both van Beneden and Boveri made simultaneously and independently of each other (VI. 4b, 6) the important discovery, that centrosomes multiply by self-division; later on I was able to verify this statement for the sperm cells of *Ascaris* (VI. 34). Van Beneden came to the following conclusion as a result of his observations: that the centrosomes, like nuclei, are permanent organs of the cell, and must therefore always occur in the protoplasm as independent forms. This view was supported to a certain extent by the discoveries of Flemming (VI. 17), Solger (VI. 70), and Heidenhain (II. 16), who stated that in many kinds of cells, such as lymph corpuscles and pigment cells, a centrosome with a radiation sphere may be demonstrated in the protoplasm, even when the nucleus, which is frequently situated some little distance off, is completely at rest. (See p. 56, Figs. 34–36.)

Our knowledge of the centrosomes was as early as 1834 much advanced by the study of the processes of fertilisation. I expressed the opinion (VI. 85) that during fertilisation a centrosome was introduced into the egg with the spermatozoon, and that to all appearance it was really the so-called middle portion, or neck, which functions as the attraction centre in the protoplasmic radiation preceding the sperm nucleus. I compared this to "the small quantity of substance present at the end of the nuclear spindle (the polar substance and the centrosome), which, although only stained with difficulty, can yet be distinguished from the protoplasm," and hence I came to the conclusion that if the comparison is correct, the radiations of the protoplasm, which occur during fertilisation and cell-division, have a common cause in the presence of one and the same substance.

Richard Hertwig (VI. 84) repeatedly pointed out that the polar substance, the middle portion of the spermatozoon, and the substance of the true nucleoli are similar in composition. Boveri (VI. 7) was of opinion that the spermatozoon carried a pole corpuscle or centrosome with it into the egg. The question was definitely decided by Fol (VII. 14) and Gnignard (VI. 23b), whose important discoveries will be described later on. According to them the nucleus of the egg, as well as that of the spermatozoon, has a centrosome of its own. Whilst the nuclei coalesce, each centrosome splits up into two parts; half of the one then unites with one half of the other, and thus the two new centrosomes, which are situated at the ends of the division spindle, are formed.

In spite of this discovery, one problem still remains unsolved. Are the centrosomes to be regarded as permanent cell organs of the protoplasm, and if so, are they contained in it during rest, only coming into correlation with the nucleus during division; or are they to be regarded as special elementary portions of the nucleus, such as the nuclear segments, spindle threads, nucleoli, etc.? In the latter case they must be enclosed during rest in the nucleus itself, and only come into relation with the protoplasm during division.

The material for observation, which we have at present, does not suffice for the solution of this question. It is extremely difficult to follow the movements of the centrosomic substance during and after nuclear division as closely as we can observe those of the nuclear substance, for the centrosomes are so excessively small; and further, it is not always possible to be sure of rendering them visible under all circumstances by means of certain definite staining solutions. During division they are chiefly recognised by means of their radiation figures, but these are not seen during rest.

Several data seem to point to the conclusion that the centrosomes originate in the nucleus; firstly, with a few exceptions, nothing corresponding to a centrosome can be found in the protoplasm during rest; secondly, at the commencement of division, the centrosome is seen to be in immediate contact with the surface of the nuclear membrane (Fig. 98), and only later on to move further away from the nucleus into the protoplasm; thirdly, subsequent to this appearance of the centrosome, the nuclear membrane frequently collapses, just as if nuclear sap had exuded through a small aperture; and fourthly, in many objects the appearance of the centrosome is simultaneous with the disintegration of the nucleoli.

FIG. 98.—Nucleus of a sperm-mother-cell of *Ascaris megalocephala bivalens*. The nuclein substance is arranged in threads which are separated from one another in two groups. Appearance of the centrosomes. Breaking up of the nucleolus. (Pl. III., Fig. 7.)

I have frequently occupied myself with this question of the origin of the centrosomes, and have expended in vain a great deal of energy upon it. Latterly, during my experiments upon the construction of the eggs and spermatozoa of Nematodes, I have again gone into the subject, but have been unable to arrive at any definite conclusions. However, although at the present time the majority of investigators consider that they belong to the protoplasm, yet a certain amount of importance must be attached to the opposite view, namely, that they have a nuclear origin.

Finally, another point, which is as yet unexplained, is the fate of the nucleoli, which disappear at the commencement of nuclear division, and reappear in the daughter nuclei. What interchanges of substances can have occurred in this process? There are exceptional difficulties in the way of the solution of this question, since in many cases the nucleoli are composed of two chemically different substances (vide p. 51).

It appears probable to me that if we disregard the above-mentioned connection with the centrosomes, the nucleoli, during the preparation for division, become split up into small portions, and become distributed upon the nuclear segments.

In sperm-mother-cells of *Ascaris*, which have been hardened with Flemming's weak solution, the nuclein loses its power of becoming stained, whilst the nucleoli become stained dark red in

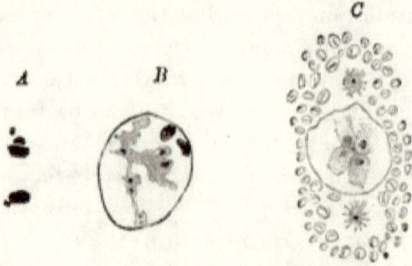

Fig. 99.—*A* Nucleoli, with granules, which are dissolving (Pl. III., Fig. 4). *B* Nucleus of a sperm-mother-cell of *Ascaris megalocephala bivalens* from the end of the growth zone. Preserved in Flemming's weak solution of chromo-osmic acid. Stained with acid fuchsine (Pl. III., Fig. 5). *C* Nucleus of a sperm-mother-cell of *Ascaris megalocephala bivalens* from the middle of the division zone. Preserved in Flemming's weak solution of chromo-osmic acid. Stained with acid fuchsine (Pl. III., Fig. 9).

acid fuchsine (Fig. 99 *A*, *B*). By this means I was able to observe that during the preparatory stages the nucleolus breaks up into several pieces, that small portions of these dissolve off, and that similar particles, stained a deep red, are deposited upon the nuclear threads. Later on, when the nuclear segments are fully formed, and the nucleolus has quite disappeared (Fig. 99 *C*), the centrosomes become visible upon the surface of the nucleus, and moreover, each nuclear segment is seen to enclose a dark red granule, which reacts towards staining solutions like the substance of the nucleolus.

Several interesting reactions with staining solutions seem to point to the fact that the nucleolar substance is taken up into the nuclear segments, although probably in an extremely finely divided state. As Wendt has discovered by his experiments on plants, the nuclein framework of the nucleus from the embryo sac of any one of several species of the *Liliaceæ* is stained blue green when treated with fuchsine iodine-green, whilst the nucleoli are coloured red. On the other hand, during the division stages, when the nucleoli are dissolved, the nuclear segments are stained violet. Further, later on, after the nucleoli have reappeared in the daughter nuclei, the nuclear threads are again stained bluish green. Wendt explains this varying reaction towards staining solutions by assuming that during division the nuclear segments absorb the nucleolar substance, and give it up again after division, so that the nucleoli may be found in the daughter nuclei.

Flemming (VI. 13, 1891) and Hermann, by means of double staining with safranin-hæmatoxylin, safranin-mauvine, safranin-gentian, etc , have obtained a similar alteration of staining reactions in animal cells, varying according to the condition of the nucleoli. "It appears to me important," says Flemming on this occasion, "that in those stages when nucleoli are still present, or have only just disappeared, or have just reappeared, the chromatin figure inclines towards a blue coloration, whereas in those cases where the nucleoli are quite disintegrated the figures are distinctly safranophil, just like the nucleoli."

2. **Direct Nuclear Division**. (Direct nuclear multiplication, fragmentation, amitosis, amitotic division.)

As a contrast to the complicated processes connected with segmentation, nuclear division may take place apparently in a very simple manner. This is called fragmentation, or direct nuclear division, and is seen in a few kinds of cells. Under these circumstances spindle threads, nuclear segments, and protoplasmic radiations are not seen. The division of the nucleus appears rather to proceed in a manner resembling that described by the earlier histologists. It can be most easily observed in the lymph corpuscles, both when alive, and when fixed by means of reagents.

There are various ways in which good preparations may be made: a drop of lymph may be drawn up from the dorsal lymph sac of a Frog into a fine capillary tube, and then placed upon a slide and covered with a cover-glass, the edges of which should be smeared with paraffin, in order to protect the preparation from evaporation. Or a small glass chamber may be prepared according to Ziegler's method, by fastening together by their four corners, or by two of their sides, two extra thin cover-glasses, so that there is a capillary space between them. The glass chamber is then placed for one or more days in the dorsal lymph sac of a Frog, during which time a large number of lymph cells make their way between the two cover-glasses, where they undergo changes. The third method, recommended by Arnold, is to place a thin pervious disc of elder pith in the lymph sac. After a few hours numbers of leucocytes have attached themselves to its surface, and are thus available for observation. Later on, thin layers of fibrin, produced by coagulation, are deposited upon the disc of elder pith; these may be removed, and, with the cell elements which are attached to them, may be easily examined.

Ranvier (VI. 54) observed all the phenomena of division take place in a lymph cell during the course of three hours, the preparation being kept at a temperature varying from 16° to 18°. Arnold (VI. 1) and others have verified his statements, and have amplified them in various ways. The vesicular nucleus can change its form actively, and can cover itself with excrescences and protuberances. Under such circumstances constrictions frequently occur, after which the nuclei break up into two, three, or more pieces (Fig. 100 A, B). The nuclear fragments move apart from one another, not infrequently remaining joined together for a considerable time by delicate connecting threads. Cell division often closely follows nuclear division, as is seen in Figs. 100 A, B.

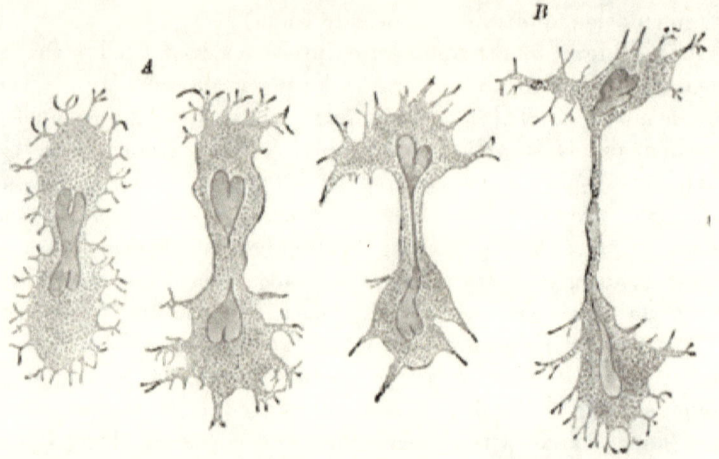

Fig. 100.—A A migratory cell from a disc of elder pith which has lain for ten days in the lymph sac of a Frog. When first observed the nucleus was somewhat constricted in its middle, whilst its ends were bilobed. After five minutes the nuclear division was completed (after Arnold, Pl. XII., Fig. 1). B Migratory cell during division. Fig. A developed into Fig. B during the course of thirty minutes (after Arnold, Pl. XII., Fig. 3).

The protoplasmic body also becomes constricted between the nuclear fragments, which move apart, but are still joined by a fine thread. The two nuclear fragments move in opposite directions by means of a large number of amœboid processes. In consequence, the connecting bridge between them is sometimes drawn out to a long fine thread, after the daughter-nuclei have separated from one another.

"No law can be laid down as to the time when the various stages of division follow one another during fragmentation; very frequently nuclei and cells linger in one or other stage" (Arnold).

It is in consequence of this delay in completing the process of cell division after the nucleus has divided that cells containing several nuclei are found. Sometimes, during inflammatory processes, such cells become so large that they are called giant cells (Fig. 101); the small nuclei vary considerably both as to form and arrangement. Sometimes they are globular vesicles, sometimes oval, sausage-shaped, or lobulated bodies; they may occur singly and evenly distributed throughout the protoplasm, or they may be arranged in chains and circles; finally, isolated small nuclei are occasionally found arranged one after another in rows. As time goes on, small cells may become detached from the giant cells, as has been observed by Arnold. This may occur in one of two ways. "Sometimes the giant cell protrudes knob-like processes containing nuclei, which, after having been withdrawn and again protruded several times, sooner or later become separated; sometimes they become detached without any or only very slight movement on the part of the cell."

FIG. 101.—A large multinucleated cell, with nucleated cells becoming constricted off peripherally. (After Arnold, Pl. XIV., Fig. 13.)

Cell division, accompanied by the phenomenon of direct nuclear division, has been observed in epithelial cells, as well as in lymph corpuscles; this occurs with especial frequency in Arthropods. They have been described by Johnson (VI. 41) and Blochmann (VI. 86) in the embryonic cells of the Scorpion; by Platner (VI. 52) in the cells of the Malpighian tubes, and by other investigators in other objects.

A peculiar method of nuclear constriction has been described by Göppert (VI. 22), Flemming (VI. 16), von Kostanecki (VI 46), and others. The most suitable object for observing it appears to be the lymphoid tissue on the surface of the liver of Amphibians. According to Göppert, the nucleus of a lymph cell develops a funnel-shaped invagination, which grows deeper and deeper until it reaches the opposite surface of the nuclear membrane, where it opens to the exterior by a minute aperture (Fig. 102 A, B). Thus a ring-shaped nucleus, perforated by a narrow canal, is formed. This ring becomes first constricted, and then cut asunder at a certain point, whilst at the same time it transforms itself into a semicircle, which becomes divided by superficial constrictions into several portions (Fig. 102 C). As the disintegration progresses, it may be broken up into a larger number of smaller

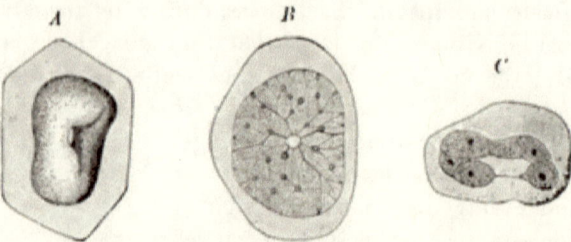

FIG. 102.—*A* Side view of a perforated nucleus from the lymphatic peripheral layer of the liver of *Triton alpestris*. The nucleus is flattened in the direction of the perforation (after Göppert, Pl. XX., Fig. 4). *B* Perforated nucleus with distinct radial arrangement of the nuclein framework (after Göppert, Pl. XX., Fig. 4). *C* Ring-shaped nucleus of a lymph cell divided into several portions by constrictions (after Göppert, Pl. XX., Fig. 10).

nuclei, which are sometimes connected for a long time by delicate connecting bridges. Similar "perforated nuclei" have been observed in other objects by Flemming (VI. 16); for instance, in the epithelium of the Frog's urinary bladder. However, in this case, division of the cell body does not appear to occur.

Direct nuclear division occurs also occasionally in the vegetable kingdom. Certain objects, like the long internodal cells of the Characeæ, or older cells of more highly organised plants, are most suitable for observing it; thus Strasburger (II. 41) observed in the older internodes of *Tradescantia* more or less irregular nuclei which are divided into portions of varying size and shape. "If the indentation is one-sided, the cell nuclei appear kidney-shaped; but if they are indented all round, they look biscuit-shaped, or irregularly lobulated. In many cases the fragments have quite separated from one another, either still remaining in contact, or lying at a greater or less distance from one another. These nuclear fragments may number as many as eight to ten in one cell." In Characeæ the nuclei may temporarily assume the appearance of a string of pearls in consequence of several constrictions having occurred. This appearance passes away when the fragmentation is completed.

However, even if constrictions of the nucleus are observed, it cannot be immediately taken for granted that direct division is commencing, unless this method of multiplication has been already observed in all its stages in the object in question. Thus in ova and in sperm-mother-cells, mulberry-shaped or irregularly lobulated nuclei are frequently seen, and yet fragmentation does not appear to occur in these cases, so that the lobulation must not

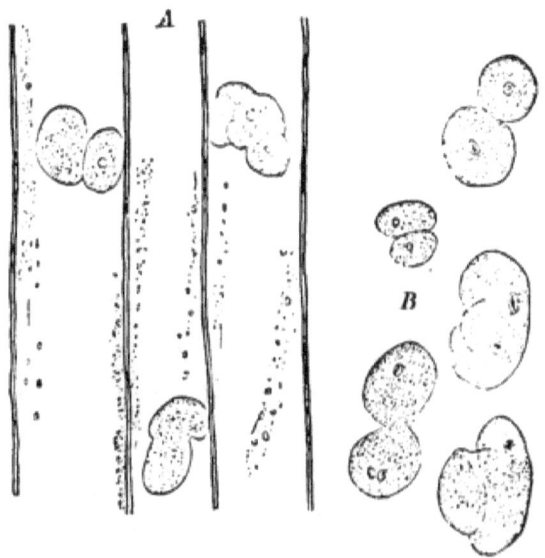

FIG. 103.—*Tradescantia virginica.* Cell nuclei of older internodes undergoing direct division (after Strasburger, Fig. 193): *A* from life; *B* after treatment with acetic-acid-methyl green.

be considered to be the commencement of direct division. It is apparently connected with metabolic processes in the nucleus (cf. what is said upon the subject in Chapter VIII.).

Nuclear multiplication by direct division occurs also amongst Protista; it is seen with especial frequency in the group of Acinetæ, of which the *Podophrya gemmipara* (Fig. 104), described on p. 229, is an instructive example.

3. **Endogenous Nuclear Multiplication, or the Formation of Multiple Nuclei.**

A third, very different

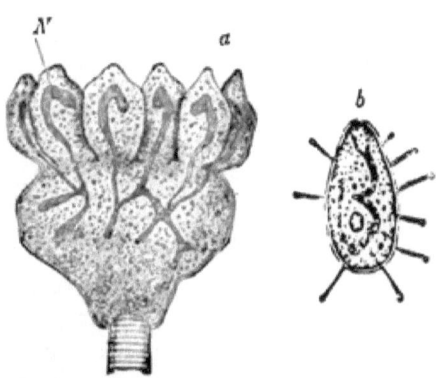

FIG. 104.—Cell-budding. *Podophrya gemmipara* with buds (R. Hertwig, Zoology, Fig. 21): *a* buds which are becoming detached and developing into zoospores *b*; *N* nucleus.

method of nuclear multiplication, to which I should like to attach the above name, has been observed by Richard Hertwig (VI. 36) amongst a group of Radiolarians, the *Thalassicollidæ*; these observations have been corroborated by Carl Brandt (VI. 8), who has followed them up in greater detail.

The *Thalassicollidæ*, which are the largest in size of all the Radiolarians, the diameter of their central capsule being nearly as long as that of the Frog's egg, possess during the greater part of their lives one single highly differentiated giant nucleus, the so-called internal vesicle; this is about $\frac{1}{2}$ mm. in diameter, and possesses a thick porous nuclear membrane. It is very similar to the multinucleated germinal vesicle of a Fish or of an Amphibian. A large number of variously shaped nuclein bodies, generally compressed together into a heap in the centre, are present in its interior (Fig. 105). Amongst these, a bright central corpuscle (centrosome), surrounded by a radiation sphere, may very frequently be seen. This was observed and depicted by R. Hertwig, and has recently been more closely investigated by Brandt. The latter observer was able to follow how, at the time of reproduction, the centrosome, which appears to me to correspond with the body of that name in plant and animal cells, betakes itself to the surface of the internal vesicle, drawing the radiation sphere after it. Here, after passing through the nuclear membrane, it enters into the surrounding protoplasm of the central capsule; however, as yet nothing has been reported as to its further fate.

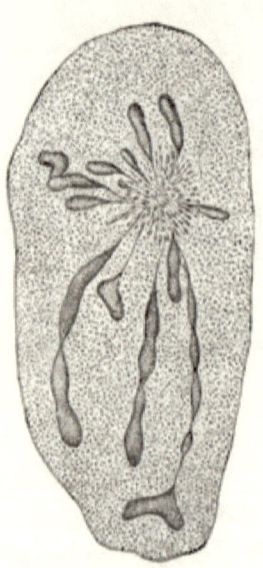

Fig. 105.—A small portion of a section through a great vesicular nucleus, the so-called internal vesicle of *Thalassicolla nucleata* with funicular internal bodies (nuclear bodies) which radiate from a common point. (R. Hertwig, Pl. V., Fig. 7.)

About this time a large number of small nuclei make their appearance outside of the internal vesicle, being situated in the protoplasm of the central capsule, which originally was quite free from nuclei; these function as centres around which nucleated zoospores develop, whose number finally may amount to some hundreds of thousands. Meanwhile, the internal vesicle begins to shrink up and loses its nuclei, which pass into the protoplasm outside. Finally it is quite dissolved. Brandt has observed that this nuclear multiplication varies according to whether isospores or anisospores are formed.

From the whole process R. Hertwig and Brandt draw the following conclusion, which is certainly correct: that the nuclei which function in the formation of zoospores, and which occur in the central capsule, at first but sparsely, but which gradually increase in number, are derived from the substance of the internal vesicle (nuclear corpuscles). "This explanation," remarks R. Hertwig, "leads me to adopt a theory of nuclear multiplication which differs fundamentally from the generally accepted one, and which is not supported by any observations which up till now have been made in animal or vegetable histology. For if we try to explain this process histologically, we must conclude not only that nuclei can multiply by division or budding, but that they may be produced by the nuclear substance of a nucleus multiplying itself by division, the portions thus produced making their way into the protoplasm to which they belong, and there developing into independent nuclei. Hence such a cell containing many nucleoli may be regarded as potentially multinuclear, just as a multinucleated cell may be regarded as potentially multicellular; and thus the gradual transition between individual cells, and the groups of cells which are derived from them by division, is by these intermediate stages rendered easier than it would otherwise be."

The extraordinary phenomena of nuclear multiplication, observed by Fol (VI. 20), Sabatier, Davidoff (VI. 87), and others, in rather young immature eggs of Ascidians, and which have been shown to be connected with the development of follicle cells, may be mentioned here. Compare also the similar processes observed by Schäfer (VI. 65a) in young mammals.

III. Various Methods of Cell Multiplication.

1. General Laws.

In addition to the process mentioned in the last section under the names of nuclear segmentation, direct nuclear division, and endogenous nuclear formation, cell multiplication may assume very various appearances according to the way in which the protoplasmic body behaves during division. Before classifying the various kinds of cell multiplication, it is necessary to mention certain general relationships which exist between the nucleus and the protoplasm, and to which I have drawn attention in my paper upon the influence exerted by gravitation upon cell division (VI. 31).

In the resting cell the nucleus may occupy various positions; it may also change its place, as, for instance, in plant cells, where it may be carried along by the protoplasmic stream. However, under certain conditions, of which only those connected with cell division will be entered into here, whilst others will be mentioned later on in Chapter VIII., the nucleus occupies a definite constant position in relation to the protoplasmic body.

Certain interactions take place between the protoplasm and the nucleus during division, similar to those which (to use a familiar illustration) exist between iron filings and a magnet suspended loosely over them. The magnetic influence polarises the iron filings, causing them to group themselves radially about the poles. On the other hand, the whole mass of the polarised particles of iron has a directing influence upon the position of the magnet. These metastatic reactions between protoplasm and nucleus receive their evident expression in the appearance of the pole centres and the radiation figures, which have been already described. The result of the reaction is that the nucleus always endeavours to occupy the centre of the reaction sphere.

No objects are more suitable for demonstrating this than animal ova, which may vary considerably as regards size, shape, and internal organisation.

In most small ova, in which protoplasm and yolk substance are more or less evenly distributed, the nucleus, before fertilisation (Fig. 106 A), does not occupy any definite position. On the other hand, when, after fertilisation, it commences to be active and to divide (Fig. 106 B), it places itself exactly in the geometrical median point, that is to say, if the egg is spherical in the centre, or if it is oval (Fig. 110) in the point of intersection of the two longitudinal axes. The nucleus surrounded by a radiation sphere may be seen to travel through the protoplasm to this point.

Variations from the normal are seen when the protoplasm and yolk granules, of which the latter, as a rule, have the greater specific gravity, are unevenly distributed in the egg cavity. Very frequently the eggs undergo a polar differentiation, which is partly produced directly by gravity, the various substances being separated out according to the weights, and partly by other processes such as are brought about by the fertilisation and the maturation of the ova.

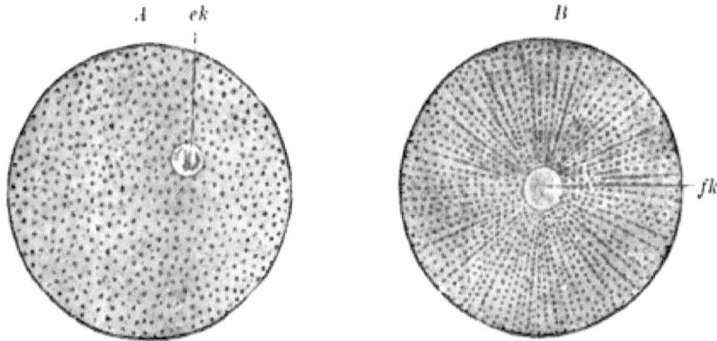

Fig. 106.—*A* Mature Egg of an *Echinoderm*, containing in its yolk a very small nucleus (*ek*) (O. Hertwig, *Embryol.*, Fig. 14). *B* Egg of a Sea-urchin, immediately after the close of fertilisation. Female pro-nucleus and male pro-nucleus have united to form the cleavage nucleus (*fk*), which occupies the centre of a protoplasmic radiation.

Polar differentiation consists in this, that the lighter protoplasm collects at one pole, and the heavier yolk substance at the other. They may be more or less sharply separated from one another. For instance, sections through the eggs of Amphibians do not show any striking separation, the only thing being, that in the one half the yolk plates are smaller, and are separated from each other by

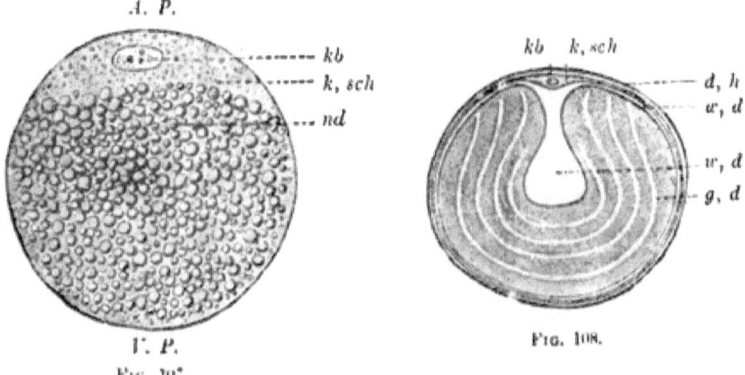

Fig. 107.—Diagram of an Egg with the nutritive yolk in a polar position (O. Hertwig, *Embryol.*, Fig. 3). The formative yolk constitutes at the animal pole (*A, P*) a germ disc (*k, sch*), in which the germinal vesicle (*kb*) is enclosed. The nutritive yolk (*nd*) fills the rest of the egg up to the vegetative pole (*V, P*).

Fig. 108.—Egg-cell (yolk) of the Hen, taken from the ovary (O. Hertwig, *Embryol.*, Fig. 6 *A*): *k, sch* germinal disc; *kb* germinal vesicle; *g, d* yellow yolk; *w, d* white yolk; *d, h* vitelline membrane.

a larger amount of protoplasm than in the other half, where they are larger and more closely packed together.

In other cases a small portion of protoplasm, more or less free from yolk, has separated itself from the yolk-containing portion of the egg, and, as in birds and reptiles (Fig. 108 *k, sch*), has assumed the form of a disc.

The two poles in an egg are distinguished from one another by the names *animal* and *vegetative*; at the former most of the protoplasm collects, and at the latter most of the yolk substance; hence the former has a smaller specific gravity than the latter. In consequence, eggs in which polar differentiation has occurred must always endeavour to attain a certain position of equilibrium. Thus, whilst in small cells, in which the substance is equally divided, the centre of gravity coincides with the centre of the sphere, the result being that the eggs can readily take up different positions, in eggs, on the other hand, in which polar differentiation has taken place, the centre of gravity has become eccentric, having approached the vegetative pole to a greater or less degree. Hence the egg so arranges itself in space that the animal pole is directed upwards, and the vegetative downwards. A line joining the two poles, the egg-axis, must, if the egg is allowed to move freely, assume a perpendicular position.

Frogs' eggs and Hens' eggs furnish us with useful examples of this. In the Frog's egg (Fig. 115) the unequal portions can be clearly distinguished externally, since the animal part is pigmented and of a deep black colour, whereas the vegetative is whitish yellow in appearance. If such an egg is placed in water after fertilisation has occurred, in a few seconds it takes up a position of equilibrium, the dark side being always turned upwards, and the specifically heavier light side downwards.

Similarly, in whatever way a Hen's egg (Fig. 108) may be turned about, the germinal disc (*k, sch*) will be seen to occupy the highest point in the yolk sphere, for the latter rotates in its albuminous sheath with every movement, keeping its vegetative pole always directed downwards.

Polar differentiation occurs both in oval and spherical eggs. The egg of the worm *Fabricia* (Fig. 109) may serve as an example. Here, at the one end more protoplasm is seen, at the other more yolk substance.

In eggs with polar differentiation it is useless to look for the cleavage nucleus in the place where it is seen in eggs poor in yolk.

However, this is only an apparent exception to the law already mentioned, for reflection shows that the nucleus, in seeking to occupy the centre of its sphere of action, only affords an example which confirms the law. Interactions take place between the nucleus and the protoplasm, not between it and the yolk-substance, for the latter during all the processes of division behaves like an inert mass. Thus the unequal distribution of the protoplasm must, in consequence of the above law, affect the position of the nucleus, forcing it to make its way to those places where the protoplasm is chiefly collected, that is to say, away from the centre of gravity. The nearer the latter approaches the vegetative pole, the nearer the cleavage nucleus approaches the animal pole.

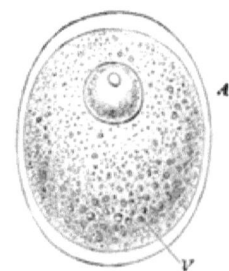

Fig. 109.—Egg from *Fabricia* (after Haeckel): *A* animal portion; *V* vegetative portion.

Actual examination shows the truth of this statement. In the Frog's egg (Fig. 115), the cleavage nucleus is somewhat above the equatorial plane of the sphere in the animal half, whilst in eggs, where the protoplasm is more sharply differentiated as a germinal disc from the yolk (Fig. 108), the cleavage nucleus has risen quite close to the animal pole, and has taken up a position inside the germinal disc itself (Reptiles, Birds, Fishes, etc.). Similarly in the egg of *Fabricia* (Fig. 109), the cleavage nucleus has been pushed towards that portion of the oval body which is rich in protoplasm.

Further, the reaction between protoplasm and nucleus, affecting the position of the latter, becomes more marked from the moment when the poles develop. Thus the second general law may be stated here, that the two poles of the division figure come to lie in the direction of the greatest mass of protoplasm, somewhat in the same way as the poles of a magnet are influenced as to their position by the iron filings in their neighbourhood.

According to the second law, in a spherical egg, for instance, in which protoplasm and yolk are evenly distributed, the axis of the centrally laid nuclear spindle may coincide with the direction of any radius whatever; whereas, on the contrary, in an oval protoplasmic body it can only coincide with the longest diameter. In a circular protoplasmic disc the spindle axis is parallel to the surface in any of the diameters, but in an oval disc it is parallel only to the longest diameter.

The phenomena observed during cell division, and especially during the formation of the furrows, are almost without exception in accordance with these laws. Two facts, however, are especially confirmatory of the truth of the second law; one was discovered by Auerbach, through his experiments on the eggs of *Ascaris nigrovenosa* and *Strongylus auricularis* (VI. 2), and the other by Pflüger.

The eggs of both the Nematodes investigated by Auerbach are oval in shape (Fig. 110), so that two poles can be distinguished in

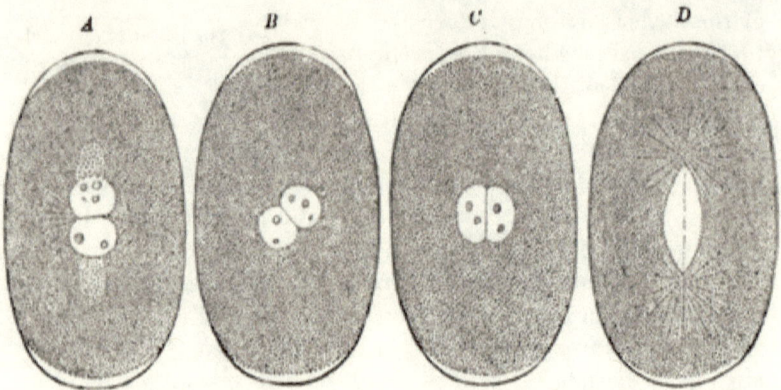

Fig. 110.—Eggs of *Ascaris nigrovenosa*, in four different stages of fertilisation. (After Auerbach, Pl. IV., Figs. 8-11.)

them, and these two poles play different rôles during fertilisation. At the one at which the germinal substance of the egg is situated, the pole cells are formed, and the female pro-nucleus develops, whilst at the other pole, which faces the mouth of the uterus, the spermatozoon enters, and fructification occurs; further, the male pro-nucleus makes its appearance here (*vide* Chap. VII.).

Whilst gradually increasing in size, both pro-nuclei approach each other, travelling in a straight line, which coincides with the axis of the egg; finally, after having grown into two vesicles of considerable size, they meet in the centre of the axis; they then come into such close contact that their contingent surfaces become flattened (Fig. 110 *A*).

As a rule, during the conjugation of the sexual nuclei, the axis of the spindle, which develops out of them, and at the ends of which the centrosomes are situated, lies somewhere in the plane of the contingent surfaces, that is to say, in the so-called conjugation plane. If this were to occur here, the spindle axis, contrary to the above-mentioned law, would cut the longitudinal axis at right angles, the centrosomes would be placed in the neighbourhood of the least amount of protoplasm, and finally, the first division plane would have to divide the egg longitudinally.

A proceeding so contrary to law does not occur here, for the protoplasm and nucleus, whilst reacting on each other, subsequently regulate their finally assumed positions, which are in accordance with the conditions present. The original position of the conjugating pair of nuclei, which is brought about by the process of fertilisation, and which is quite unsuitable for the purposes of division, becomes changed, whilst the two poles become more clearly defined. The nuclear pair commence to turn themselves through a right angle (Fig. 110 *B*), until the conjugation plane coincides with the longitudinal axis of the egg (Fig. 110 *C*).

" Sometimes they rotate in the same direction as the hands of a watch, sometimes in the opposite direction " (Auerbach).

In consequence of this interesting phenomenon of rotation, the two poles of the division figure come to be in the neighbourhood of the largest accumulation of protoplasm, in accordance with the law, whilst the smallest amount is situated near the division plane, which develops later (Fig. 110 *D*).

A second instance of the truth of this law is afforded by the experiments of Pflüger (VI. 49, 50) upon Frogs' eggs. He carefully compressed a freshly-fertilised egg between two vertical parallel glass plates, thus giving to it pretty nearly the form of " a much-flattened ellipsoid, of which the longest axis is horizontal, the one of medium length vertical, and the shortest again horizontal and perpendicular to the longest." In nearly every case the first division plane was vertical to the surface of the compressed plate, and at the same time perpendicular. Hence the nuclear spindle must again in this case, in accordance with the above-mentioned law, have placed itself in the direction of the longest diameter of the ellipsoid.

From this law, that the position of the nuclear axis in division is determined by the differentiation and form of the surrounding protoplasmic body, so that the poles place themselves in the direction of the greatest collection of protoplasm, we can deduce a third law, which Sachs (VI. 64) arrived at from a study of plant anatomy, and has described as the law of rectangular intersection of the dividing surfaces in bipartition. For, having once learnt the causes which determine the position of the spindle axes, we can know beforehand how the division plates must lie, in order to intersect the spindle axes at right angles.

As a general rule, unless the mother-cell is exceptionally long in any one direction, it happens that in each division that axis of the daughter-cell, which lies in the same direction as the chief axis of the mother-cell did, has become the shortest. Hence the axis of the second division spindle would never in such a case place itself in the direction of the preceding division spindle, but rather at right angles to it, according to the form of the protoplasmic body. In consequence, the second division plane must intersect the first at right angles.

Generally, the consecutive division surfaces of a mother-cell (which becomes split up into 2, 4, 8, and more daughter-cells by successive bipartitions) lie in the three directions of space, and so are more or less perpendicular to each other.

This is often very plainly to be seen in plant tissues, because here firm cell-walls, corresponding to the division planes of the cells, rapidly develop, and thus, so to speak, fix the places to a certain degree permanently. But in animal cells, which in the absence of a firm membrane frequently change their form during the processes of division, this is not the case; in addition the position of the cells to one another may change. "Fractures and displacements" of the original portions into which the mother-cell splits up occur, examples of which are afforded us by the study of the furrowing of any egg cell. This is entered into more fully on p. 224.

In botany, these three directions of space are designated as tangential or periclinal, transverse or anticlinal, and radial (Figs. 111, 112). Periclinal or tangential walls are parallel to the surface of the stem. Anticlinal or transverse walls intersect the periclinal walls, and at the same time the axis of growth of the stem at right angles. Finally radial walls, whilst being also at right angles to the periclinal ones, lie in the same plane as the axis of growth of the stem.

In order to render this clear by an example, we will select a somewhat difficult object, namely, the growing-point of a shoot. Sachs demonstrates the truth of his law with reference to this object in the following sentences which are taken from his lectures on plant physiology (II. 33):—

"Suitably prepared longitudinal and transverse sections of the growing-points of roots and shoots show characteristic cell-wall networks and cell arrangements, which agree with the type, even in the most various plants. This depends essentially upon the fact that the embryonic substance of the growing-point, as it increases in volume on every side and at all parts, becomes divided up into compartments or chambers by cell-walls, which intersect one another at right angles. The longitudinal section of a growing-point always shows a system of periclinal walls, intersected by anticlinal walls, which in their turn represent the right-angled trajectories of the former. If only the growing-points of flat structures be considered, then there will be only two systems of cell-walls present; if, however, the growing-point is hemispherical or conical, or of some other similar shape, that is to say not flat, but forming a solid mass, a third system of cell-walls must be taken into account; namely, the longitudinal walls, which stretch out in a radial direction from the longitudinal axis of the growing-point."

"It will facilitate a clear comprehension of the subject, if before proceeding farther we examine a diagram, which has been constructed arbitrarily, although according to fixed laws, and

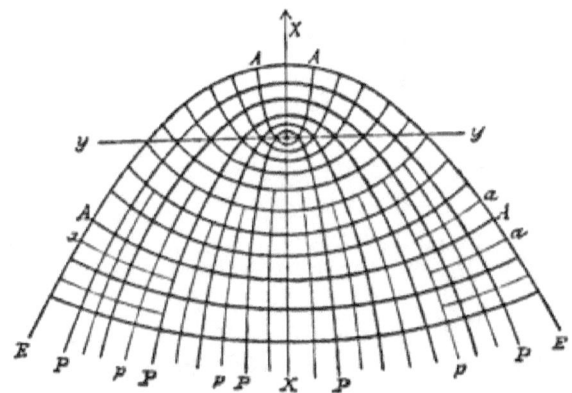

FIG. 111.—Diagram of the cell arrangement at a growing-point. (After Sachs, Fig. 281.)

for this purpose it will be well to consider as a starting-point a median longitudinal section through the growing-point (Fig. 111). Confining our attention, therefore, to our figure, of which the outline $E\ E$ represents the longitudinal section through a conical growing-point—which resembles fairly closely those met with in nature—it will be seen that it has the form of a parabola and that the space occupied by the embryonic substance is partitioned out, so that anticlinal and periclinal walls intersect at right angles. This being granted, the network of cells in Fig. 111 may be constructed according to a well-known geometrical law. Let $x\,x$ represent the axis, and $y\,y$ the direction of the parameter, then all the periclines, denoted by $P\,p$, form a group of confocal parabolas. Similarly, all the anticlines, $A\,a$, form another group of confocal parabolas, whose focus and axis coincide with those of the preceding group, but which run in the opposite direction. Two such systems cut one another everywhere at right angles.

"Let us now observe whether a median longitudinal section made through a dome-shaped, and approximately parabolic growing-point, does not present an arrangement of cells which corresponds in all essentials with our geometric diagram. We see at once, if we examine such a section, made from the growing-point of a Larch for example (Fig. 112), that the internal structure is identical, if

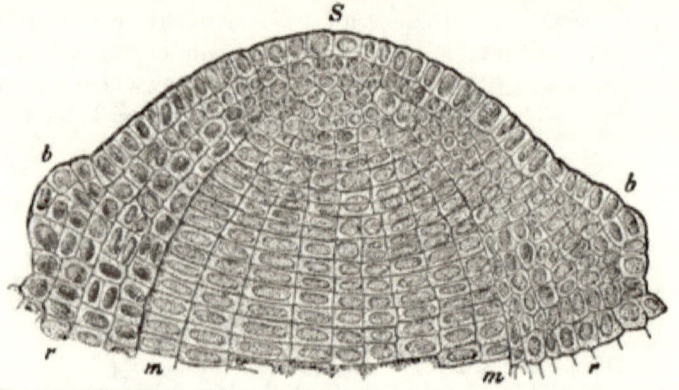

FIG. 112.—Longitudinal section through the growing-point of a winter bud of *Abies pectinata* (× about 200) (after Sachs, Fig. 285): S apex of growing point; $b\,b$ youngest leaves; r cortex; m pith.

we disregard the two protuberances, $b\,b$, which interfere somewhat with the symmetry of the figure. These are young leaf-rudiments, budding off from the growing-point. We recognise at once the two systems of anticlines and periclines, which it can scarcely be doubted cut each other at right angles, as in the diagram; that is to say, the anticlines are the right-angled trajectories of the periclines. As in the diagram, further, only a few periclines under the apex S run round the common focus of all the parabolas; the others, which come from below, only reach the neighbourhood of the focus; that is to say, the corresponding cell divisions only occur after the periclines below the centre of curvature have become sufficiently far apart from one another for it to be necessary for new periclines to intercalate themselves between them; and the same is true of the anticlines. It is easy to see in the diagram (Fig. 111), that the curvatures of the construction lines are especially sharp around the common focus of all the anticlines and periclines."

"Hundreds of median longitudinal sections, through the growing-points of roots and shoots, have been made by various observers, before the fundamental principle was at all understood, and all of these correspond with the construction which I have given, and thus prove its accuracy."

Finally, in order to understand certain variations from normal cell division, a fourth law must be mentioned, which has been formulated by Balfour (VI. 3) in the following words: "The rapidity with which a cell divides is proportional to the concentration of the protoplasm it contains. Cells rich in protoplasm divide more quickly than those which are poor in protoplasm and rich in yolk." This law is explained by the fact that, in the process of division, it is the protoplasm alone which is active, the yolk substance stored up in it being passive, and, so to speak, carried along by the active protoplasm. The greater the amount of yolk present, the more work is there for the protoplasm in division; indeed, in many cases there may be more to do than the protoplasm can accomplish. This occurs frequently in eggs, in which polar differentiation has occurred, the greater part of the protoplasm being concentrated at the animal pole. Then division is confined to this portion of the cell, the vegetative part being no longer broken up into cells. Thus an incomplete or partial division has resulted instead of a complete one. Both extremes are united in nature by intermediate forms.

2. Review of the Various Modes of Cell Division.

The following classification, upon which I have based my detailed accounts, may be made of the various methods of cell division.

I. COMPLETE OR HOLOBLASTIC SEGMENTATION.

        a. EQUAL.
        b. UNEQUAL.
        c. CELL-BUDDING.

II. Partial or Meroblastic Segmentation.
III. So-called Free Cell-Formation.
IV. Division with Reduction.

The most instructive examples of the various methods of cell division are afforded, for the most part, by animal egg-cells; for here the divisions follow so quickly one upon another, that the normal conditions may be clearly observed.

Ia. Equal Segmentation.

In equal division the egg, if, as is generally the case, it is spherical, is first split up into two hemispheres. According to the law explained above, in the division which follows, the nuclear spindle must place itself parallel to the base of the hemisphere, so that the latter is divided into two quadrants. Further, the spindle axis must coincide with the longitudinal axis of each quadrant, so that in each case a division into two octants is produced. In consequence, during the second and third stages of the cleavage process, the relative positions occupied by the second and third division planes towards one another, and towards the first division plane, are strictly according to law; that is to say, the second cleavage plane cuts the first at right angles, and halves it, whilst the third is perpendicular to the two first, and passes through the centre of the axis in which they intersect. If now the ends of this axis are considered as the poles of the egg, the two first division planes may be regarded as meridional, and the third as equatorial.

In many cases, after the second cleavage, the four portions may be seen to separate somewhat from one another, the result of which is that the furrows produced by the second division no longer intersect in one point, but meet the first formed meridional furrow at a little distance from the pole (Fig. 113). Thus a transverse line, the cleavage line, which varies in length, is produced. I have found this especially well marked (VI. 30b) in the eggs of *Sagitta* (Fig. 113).

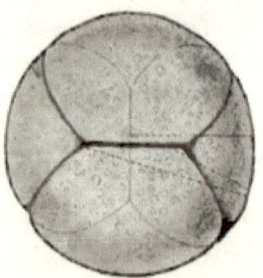

Fig. 113.—A four segmented egg of *Sagitta* seen from the animal pole. (× 100; Hertwig, Pl. V., Fig. 5.)

A short time after the termination of the second division of the egg of *Sagitta*, the four cells so arrange themselves (Fig. 113) that only two of them touch each other. At the animal pole they meet in a short transverse furrow, the animal cleavage line. The pointed ends of the two remaining cells, which do not come in contact with the pole, meet this line at its extremities. A similar arrangement is seen at the vegetative pole: here the two cells, which did not touch the animal pole, meet along a vegetative cleavage line, which is always in such a position that if both lines were projected upon a common plane they would intersect at right angles. Here the four cells, which are obtained by quartering the original cell, are not of the shape of ordinary quarters of a sphere. Each has a blunt and a pointed end, the latter being directed towards the pole of the egg. Each pair of cells formed from a hemisphere are so arranged that similar ends point in opposite directions.

A corresponding arrangement of the first four cleavage cells has been described by von Rabl in the eggs of *Planorbis*, and by von Rauber (VI. 56) in Frogs' eggs. The latter has entered into more details than the former.

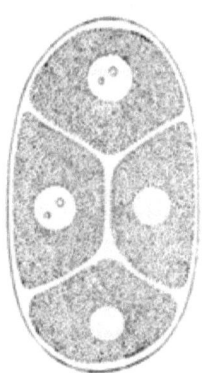

Fig. 114.—An egg of *Ascaris nigrovenosa* with four segments. (After Auerbach, Pl. IV., Fig. 19.)

Similarly in oval eggs, in which, according to our law, the first division plane is transverse to the longitudinal axis, distinct separations of the cells from each other occur during the second cleavage, which is vertical to the first. In consequence, well-marked cleavage lines appear, as is seen in Fig. 114 in the egg of *Ascaris nigrovenosa*.

1b. UNEQUAL SEGMENTATION.

Unequal division comes naturally after equal. It is most generally caused by the unequal distribution of the protoplasm and yolk substance in the cell. The Frog's egg, in which polar differentiation has occurred, will serve as an example of this. There, as has already been stated, the nucleus is situated in the upper or animal half of the sphere (p. 217). Now when division is about to occur, the axis can no longer lie in any one of the radii of the egg, for, in consequence of the unequal division of the protoplasm in the egg space, it is influenced by that part of the egg, which is pigmented and rich in protoplasm; this portion rests like a skull-cap upon the more transparent deutoplasm-con-

Q

taining portion, and, on account of its smaller specific gravity, floats upwards, and is spread out horizontally (Fig. 115 A). The

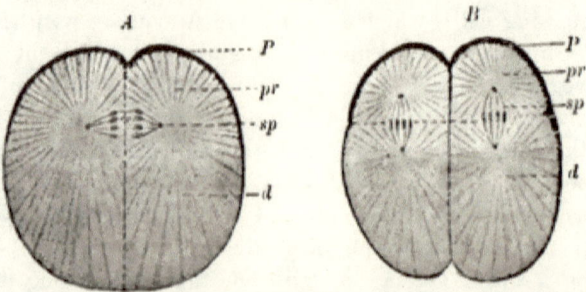

FIG. 115.—Diagram of the division of the Frog's egg (O. Hertwig, Embryology, Fig. 31): A first division stage; B third division stage; the four portions of the second stage of division are beginning to be divided by an equatorial furrow into eight portions; P pigmented surface of the egg at the animal pole; pr that part of the egg which is richer in protoplasm; d that part of the egg which is richer in deutoplasm; sp nuclear spindle.

nuclear spindle, however, lies horizontally, in a horizontal disc of protoplasm; hence the division plane must develop vertically. At first a small furrow appears at the animal pole, since this latter is especially influenced by the nuclear spindle which has approached it, and further because it contains more protoplasm, in which the movements occurring during division commence. The furrow slowly deepens, cutting downwards towards the vegetative pole.

The two hemispheres produced by this first division consist of an upper portion, rich in protoplasm, and of a lower portion, poor in protoplasm. By this means, in the first place the position of the nucleus, and in the second place its axis, are absolutely determined before it commences to divide a second time. The nucleus is to be looked for, according to the above-mentioned law, in that quadrant which is richest in protoplasm. The axis of the spindle must here lie parallel to the longitudinal axis of the quadrant, that is to say, it must lie horizontally. Hence the second division plane, like the first, is perpendicular, cutting the latter at right angles.

At the end of the second cleavage the amphibian egg consists of four quadrants which are separated from one another by vertical division planes, and which possess two unequal poles, the upper one being lighter and richer in protoplasm, and the lower one heavier and richer in yolk substance. In an egg where equal cleavage occurs, we saw that at the stage of the third division the axes of the nuclear spindles arrange themselves so as to be parallel to the longitudinal axis of the quadrants. The same thing occurs here in a somewhat modified form (Fig. 115 B). On account of the greater amount of protoplasm present in the upper half of each quadrant, the spindle is unable to lie in the centre, as in an egg in which equal cleavage occurs, but must approach nearer to the animal pole of the egg. Further, it is exactly perpendicular, for, on account of the unequal weight of their halves, the quadrants of the amphibian egg are firmly fixed in the egg space. In consequence, the third division plane must now be horizontal (Fig. 116 A), and further, it must be placed above the

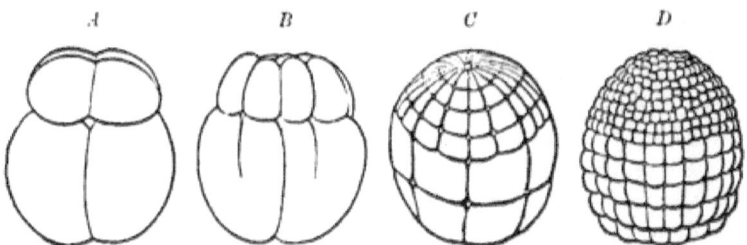

FIG. 116.—Stages in the cleavage of *Petromyzon*. (From Hatschek, Fig. 72; A, B after Shipley; C, D, after M. Schultze.)

equator of the sphere of the egg, being situated more or less towards the animal pole. The portions thus produced are very dissimilar both in size and constitution, and this is why this form of cleavage has been called unequal. The four upper portions are smaller, and poorer in yolk; the four lower much larger, and richer in yolk. They are called animal and vegetative cells according to whether they are directed towards the animal or vegetative pole.

As development proceeds (Fig. 116 B, C, D), the difference between the animal and the vegetative cells grows greater and greater, for the more protoplasm a cell contains, the more quickly and frequently does it divide, as has been already mentioned above.

Unequal cleavage can also occur in oval eggs. For instance, the egg of *Fabricia* (Fig. 117), as has been already mentioned (Fig. 109), in consequence of the collection of yolk around one pole, divides into one

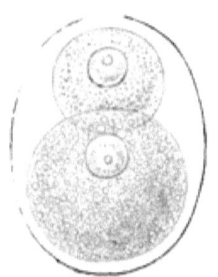

FIG. 117.—Egg of *Fabricia* divided into two cells. (After Haeckel.)

smaller cell, richer in protoplasm, and a larger one, richer in yolk; in these segmentation proceeds at different rates.

Ic. CELL-BUDDING.

When one of the portions produced by division is so much smaller than the other, that it appears as though it were only a small appendage to the original cell, scarcely causing any diminution of its substance, the process is called "cell-budding, or gemmative segmentation," the smaller portion being called the bud, and the larger the mother-cell. Two kinds of cell-budding are distinguished, according to whether one or more buds are formed.

In the animal kingdom this process of cell-budding occurs when the egg is mature, causing the development of the directive corpuscles, or polar bodies (polar cells). By this term we understand two or three small spherules, which are composed of protoplasm and nuclear substance, and hence are of the same value as small cells; they are frequently situated at the animal pole of the egg, within the vitelline membrane. The course of the process of cell-budding is as follows:—

Whilst the germinal vesicle is becoming broken up, a typical

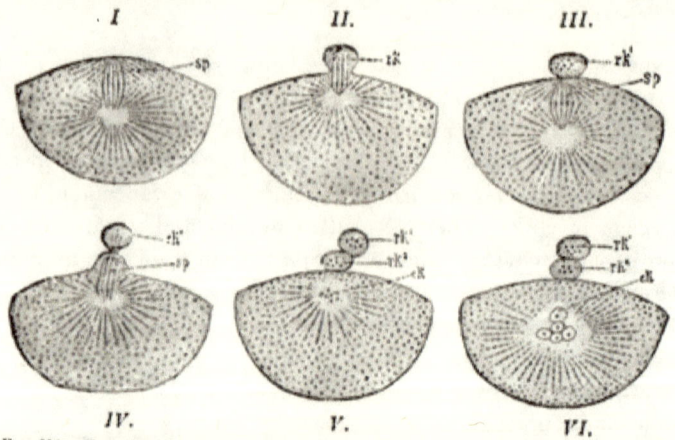

FIG. 118.—Formation of the polar cells in *Asterias glacialis* (O. Hertwig, Embryol., Fig. 13). In Fig. *I.* the polar spindle (sp) has advanced to the surface of the egg. In Fig. *II.* a small protuberance (rk^1) has been formed, which receives half of the spindle. In Fig. *III.* the protuberance is constricted off, forming a polar cell (rk^1). Out of the remaining half of the original spindle, a second complete spindle (sp) has developed. In Fig. *IV.* a second protuberance has bulged out below the first polar cell, which in Fig. *V.* becomes constricted off to form the second polar cell (rk^2). In Fig. *VI.* out of the remainder of the spindle the egg nucleus (rk) develops.

nuclear spindle, with a polar radiation at each end, develops out of its contents. This changes its position in the yolk (Fig. 118 I.), raising itself gradually towards the animal pole, until its end touches the surface. It then arranges itself with its longitudinal axis in the direction of a radius of the egg. Cell-budding soon commences at the place where one of the poles of the nuclear figure touches the surface; the yolk arches itself up to form a small knob, into which half of the spindle protrudes itself (Fig. 118 II.).

The protuberance then becomes constricted at its base, and, with half of the spindle, separates itself from the yolk, forming a very small cell (Fig. 118 III.). Then the whole process repeats itself (Fig. 118 IV.-VI.), the half of the spindle which has remained in the egg, without previously passing through a resting vesicular or nuclear condition, developing first into a complete spindle. This process, as far as it refers to the nuclear spindle, will be entered into at more detail on p. 237.

Cell-budding occurs frequently amongst certain species of unicellular organisms. I will select from amongst these a second example, which has been examined by Richard Hertwig (VI. 35), the *Podophrya gemmipara*, a marine Acineta, which attaches itself by means of a stalk at its posterior end to other objects. From eight to twelve cell-buds not infrequently develop at its free end, which is provided with prehensile tentacles and suction tubes; these cell-buds are grouped in a ring around the centre of the free surface. In this case, the nucleus divides in a peculiar fashion. As long as the *Podophrya* is young, and has not yet commenced to bud, the nucleus has, as in so many Ciliata, the form of a long horseshoe-shaped twisted band (Fig. 119 b). Later on, a large number of processes grow out in a vertical direction, towards the free surface of the body; their ends soon swell out into knobs, whilst the portion of the

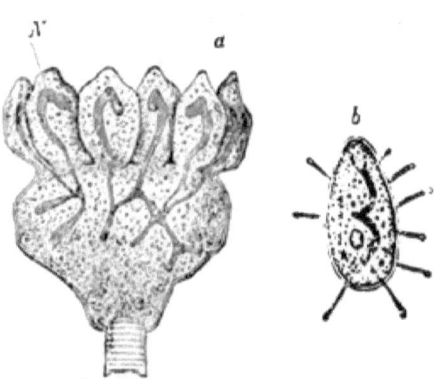

FIG. 119. — Cell-budding, *Podophrya gemmipara* with buds (O. Hertwig, Zoology, Fig. 21); a buds, which become detached and form zoospores b; N nucleus.

band connecting them with the main part of the nucleus generally becomes as fine as a hair. Small protuberances develop on the free surface whenever the knob-like nuclear ends touch it. Thus, as these ends grow, each is contained by a special protuberance or cell-bud of its own. The whole cell-bud then increases somewhat in size, and becomes constricted at its base from the mother-cell; the part of the nucleus, which it contains, takes the form of a horse-shoe, separating itself from the delicate connecting thread which united it to the mother-nucleus. The cell-buds are now mature, and after detaching themselves from the mother organism, move about for a time in the sea-water as zoospores.

II. Partial or Meroblastic Segmentation.

If we disregard the case of certain Protozoa (*Noctiluca*), partial segmentation occurs only in egg-cells. It may conveniently be considered after unequal division. It is found in all cases where the amount of yolk present is extremely great, and where the protoplasm is clearly separated from it, being collected together in a disc at the animal pole (Fig. 108). The nucleus, which is situated in the centre of this disc, must assume a horizontal position when it develops into a spindle. Hence the first division plane is in a vertical direction, and appears first at the animal pole in the centre of the disc (Figs. 120 *A*, 121 *A*), as in an egg, in which unequal cleav-

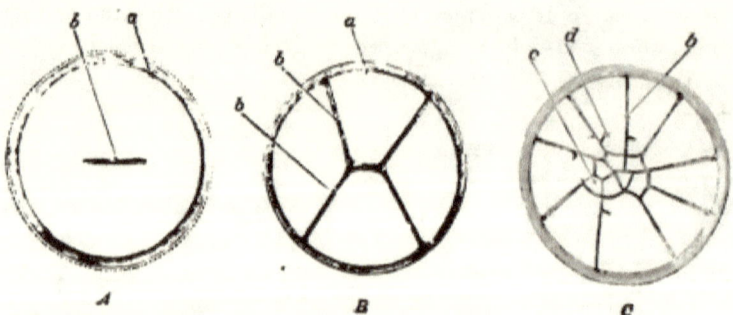

Fig. 120.—Surface view of the first cleavage stage of a Hen's egg (after Coste): *a* edge of germinal disc; *b* vertical furrow; *c* small central portion; *d* large peripheral portion.

age occurs (Fig. 92). Whilst, however, it gradually deepens and sinks in until it has cut its way through to the vegetative pole, the germinal disc is divided into two equal segments, which rest like two buds, with their broad bases upon the undivided yolk-mass, and are thus connected with one another. Soon afterwards a second vertical furrow makes its appearance, crossing the first at right angles, and terminating in a similar manner at the germinal disc, which is now split up into four segments (Figs. 120 *B*, 121 *B*). In this case also a cleavage line is formed.

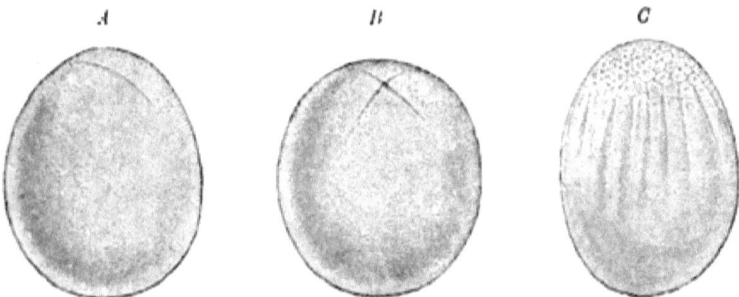

Fig. 121.—Discoidal cleavage of the egg of a *Cephalopod* (after Watase; from Hertwig, Fig. 99).

Each of the four segments is again halved by a radial furrow. The segments so produced correspond to sectors, whose pointed ends meet in the centre of the germinal disc, and whose broad ends are turned towards the periphery. The pointed ends are separated from the rest of the segment by a diagonal furrow, or by one which is parallel to the equator of the egg-sphere; and in consequence smaller central segments cut off from the yolk in every direction, and larger peripheral portions still connected with the yolk, may be distinguished (Fig. 120 *C*). From now on, furrows which are radial, and ones which are parallel to the equator, alternately make their appearance, so that the germinal disc becomes more and more split up, the segments being so arranged that the smaller ones are in the centre of the disc, and the larger ones on its circumference (Fig. 121 *C*). Many of the segments which are still attached to the yolk become constricted, so that the nuclear spindle is slanting or vertical, the consequence of which is, that when division occurs one of the daughter nuclei is situated in the yolk-mass. In this manner the yolk-nuclei are produced by partial cleavage; an especially large number of them are embedded in the superficial layers of yolk on the periphery of the segmented germinal disc. Compare the interesting observations of Rückert (VII. 36), and Oppel (VII. 34), from which it appears that in Selachians and Reptiles yolk-nuclei develop in consequence of over-impregnation.

III. So-called Free Cell-Formation.

The peculiarity of this form of multiplication consists in this, that the nucleus in a cell subdivides several times consecutively, whilst the protoplasmic body remains undivided for a considerable time without showing the least inclination towards even a partial cleavage. After bipartition has been repeated several times, the number of nuclei in a single protoplasmic body may amount to several hundreds. These arrange themselves at regular distances from one another. Finally a period arrives when the many-nucleated mother-cell becomes either suddenly or gradually split up into as many daughter-cells as there are nuclei in it.

Free cell-formation occurs chiefly, in both plants and animals, during the development of the sexual products. In order to demonstrate it, I will select three examples: the superficial segmentation of the centrolecithal eggs of *Arthropoda*, the formation of the endosperm in the embryo-sac within the ovule of *Phanerogamia*, and the formation of spores in the sporangia of *Saprolegnia*.

The yolk mass is generally collected in the centre of the egg in Arthropoda, being surrounded by a thin peripheral layer of protoplasm. Hence the eggs are called centrolecithal, *i.e.* eggs with yolk in the centre, in distinction to telolecithal eggs, in which the yolk is situated at the poles (Balfour VI. 3). The cleavage nucleus, surrounded by a protoplasmic envelope, is generally in the centre of the nutritive yolk; here it divides into two daughter-nuclei, whilst the division of the egg itself does not immediately

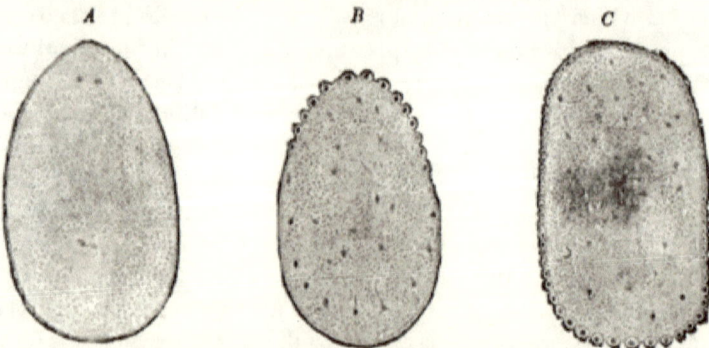

Fig. 122.—Superficial cleavage of the egg of an insect (*Pieris crataegi*) (after Bobretsky; from R. Hertwig, Fig. 100): *A* division of the cleavage nucleus; *B* the nuclei raise themselves and commence to form a germinal layer (blastoderm); *C* formation of blastoderm.

follow. These daughter-nuclei (Fig. 122 *A*) then divide into four, these four into eight, the eight into sixteen, and so on, whilst the egg as a whole remains unsegmented. Later on the nuclei separate from one another, and for the most part move gradually to the surface (Fig. 122 *B*), penetrating into the protoplasmic envelope, where they arrange themselves at equal distances from one another. Not until this has occurred does the egg commence to segment, the peripheral layer splitting up into as many cells as there are nuclei in it, whilst the central yolk remains intact, or is only split up at a much later period. This latter occurs when in the eggs of insects, as in telolecithal eggs, the yolk contains yolk nuclei, or merocytes (Fig. 122 *C*).

The wall of the embryo-sac in *Phanerogamia* is coated with a protoplasmic lining, which at a certain stage of development contains several hundred regularly arranged nuclei; these were formerly considered to develop like crystals in a mother-liquor; but we know now, that they are produced by the repeated bipartition of a mother nucleus, as in the eggs of *Arthropoda* (Fig. 123). The divisions occur almost simultaneously in any one region of the embryo-sac. If the preparation is successful, nuclei in numerous stages of division may be observed at one time in a small space (Fig. 123).

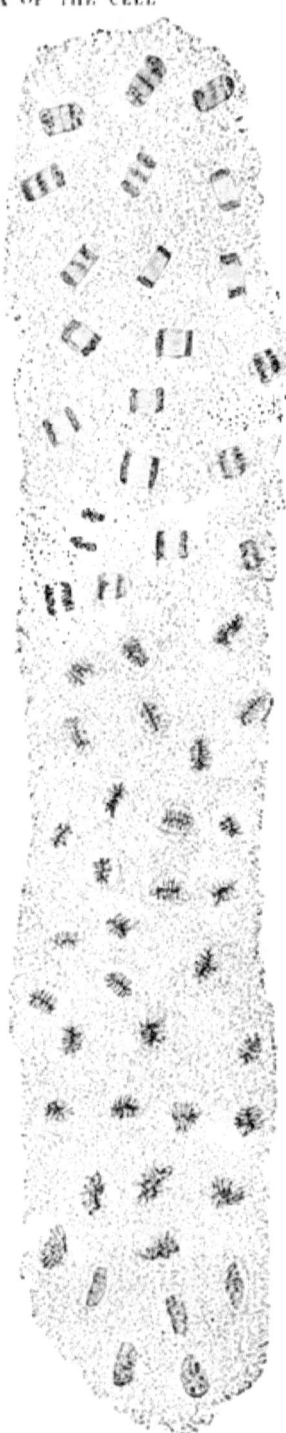

FIG. 123. *Fritillaria imperialis.* Protoplasmic lining from the embryo-sac. A strip showing all the phases of nuclear division. (After Strasburger, *Botan. Praktecum*, Fig. 106.)

After a sufficient number of nuclei have developed, a further stage supervenes, when cells are formed (Fig. 124). Between the nuclei, which are arranged at regular distances from one another, the protoplasm differentiates itself into radial fibrillæ. Further it develops connecting threads in all directions, which thicken at their centres, and form *cell-plates*. In the cell-plates the cellulose walls make their appearance in the manner already described. These swell up easily, and owing to their formation, a portion of the protoplasmic lining becomes encapsuled around each nucleus to form the protoplasm of the cell. Sometimes two nuclei are enclosed in one cell; these subsequently are either separated from one another by a partition wall, or, as in *Corydalis cava*, fuse together to form a single cell.

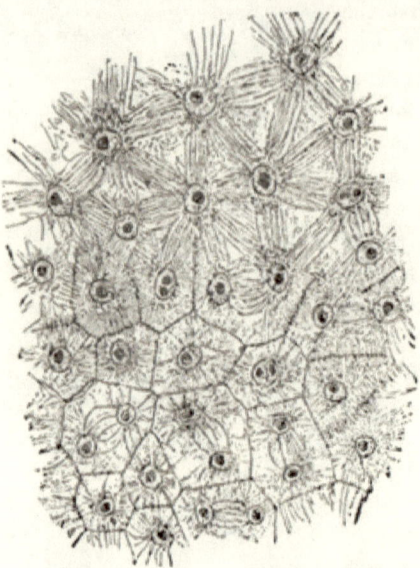

Fig. 124.—*Reseda odorata*. Protoplasmic lining of the embryo-sac at the commencement of free cell-formation. (× 240; after Strasburger, *Botan. Prakticum*, Fig. 182.)

The sporangium of *Saprolegnia* is, to commence with, a long cell filled with protoplasm. Later on the nuclei in it increase very much in number through bipartitions, which for the most part occur simultaneously. After a time they distribute themselves evenly throughout the cell-space. The protoplasm in the neighbourhood of each nucleus then differentiates itself into a small mass, which surrounds itself with a firm glistening envelope; by this means the cell contents split up simultaneously into as many spores as there are small nuclei present in the cell. Later on these are passed to the exterior by the bursting of the mother-cell, the sporangium.

The formation of swarm-spores in *Radiolaria*, which has been already mentioned, affords us another peculiar instance of so-called free cell-formation.

IV. Division with Reduction.

During the final development of ova and spermatozoa, certain peculiar processes of division occur, which have for their function the preparation of the sexual cells. The essential characteristic of this is, that in the double division that occurs the second follows the first so quickly, that the nucleus has no time to enter the resting condition. The result is, that the groups of nuclear segments produced by the first division are immediately split up into two daughter-groups without previously undergoing longitudinal cleavage. Hence, at the end of the second division, the mature egg- and sperm-cells only contain half the number of nuclear segments, and half as much nuclein substance, as are present in the nuclei produced by ordinary cell division in the same animal (Hertwig VI. 34). To this phenomenon the name of "division with reduction" has been given (Weismann VI. 77). Division with reduction is most easily followed in the sperm- and egg-cells of *Ascaris megalocephala*.

In the testis tube a certain number of cells are differentiated off to form the sperm-mother-cells. In the large vesicular nucleus (Fig. 125 *I.*), eight long nuclear threads develop out of the

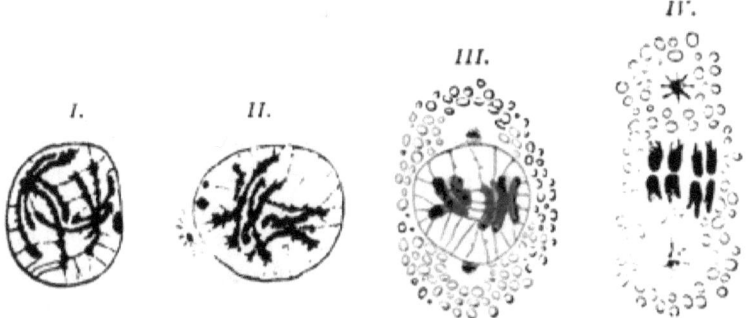

Fig. 125. — Four nuclei of sperm-mother-cells of *Ascaris megalocephala bivalens* at various stages of preparation for division.

chromatin substance. (*Ascaris megalocephala bivalens* has been selected for description.) These are arranged in two bundles, and are connected with the nuclear membrane by linin threads, which stretch out in every direction. Whilst the nucleolus splits up into separate spherules, two centrosomes, surrounded by a small radiation sphere (Fig. 125 *II.*), make their appearance near to one another in the protoplasm, close to the outer surface of the nuclear membrane (Fig. 125 *II.*). The segments then become shorter and thicker (Fig. 125 II., III.). The centrosomes separate from one another, until finally they are situated at opposite sides of, and at some distance from, the vesicular nucleus. By this time, the rest of the nucleolus has disappeared; the nuclear membrane becomes dissolved, and the two bundles, each containing four nuclear segments, arrange themselves in the equator between the centrosomes; then each bundle splits up into two daughter-bundles containing two nuclear segments, which separate and move towards the poles (Figs. 125 IV., 126 I.). The sperm-mother-cell now becomes constricted into

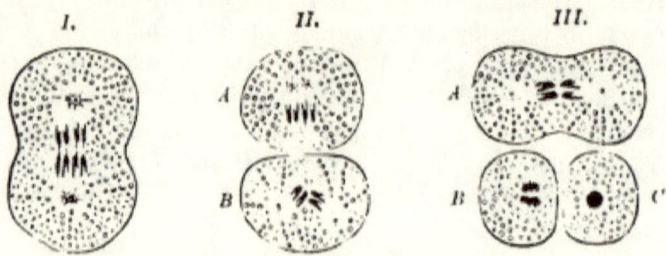

FIG. 126.—Diagram showing the development of sperm-cells from a sperm-mother-cell of *Ascaris megalocephala bivalens*. I. Division of the sperm-mother-cell into two sperm-daughter-cells. II. The two sperm-daughter-cells (A, B) immediately prepare to divide a second time. III. The sperm-daughter-cell A divides into two grand-daughter-cells. B and C are grand-daughter-cells, which have been produced by the division of the daughter-cell B of Fig. II.

two daughter-cells of equal size (Fig. 126 II.). Whilst this process of constriction is taking place, the changes commence which lead up to the second division (Fig. 126 I.), the centrosome of each daughter-cell splits up into two parts which travel, each surrounded by its own radiation sphere, in opposite directions, which are parallel to the first division plane (Fig. 126 A, B). The nuclear segments produced by the first division immediately afford the material for the second division, without passing through the vesicular resting condition. They move until they are situated between the newly-developed poles of the second division figure (Fig. 126 II., B), and then divide into two groups, each of which contains two nuclear segments; these groups then separate, and move towards the poles, after which the second constriction commences (Fig. 126 III., A). Whilst after the first division each daughter-cell contains four of the eight nuclear segments, which have developed beforehand in the resting nucleus, each grand-daughter-cell contains only two. For, in consequence of the second division following so closely on the first that the resting condition was missed, an augmentation of nuclear substance, and an increase in the number of the nuclear segments, through longitudinal cleavage, have been unable to take place. In consequence, the number of segments has been diminished or reduced to half the normal number.

In exactly the same way division with reduction occurs in the egg of *Ascaris megalocephala* during the process of ripening.

The sperm-mother-cell corresponds to the unripe egg, or egg-mother-cell. Here also eight nuclear segments, arranged in two bundles, develop in the germinal vesicle (Fig. 127 *I.*). After the nuclear membrane has been dissolved, they arrange themselves in the equator of the first direction spindle, which rises up to the surface of the yolk (Fig. 127 *II.*), and in the manner already

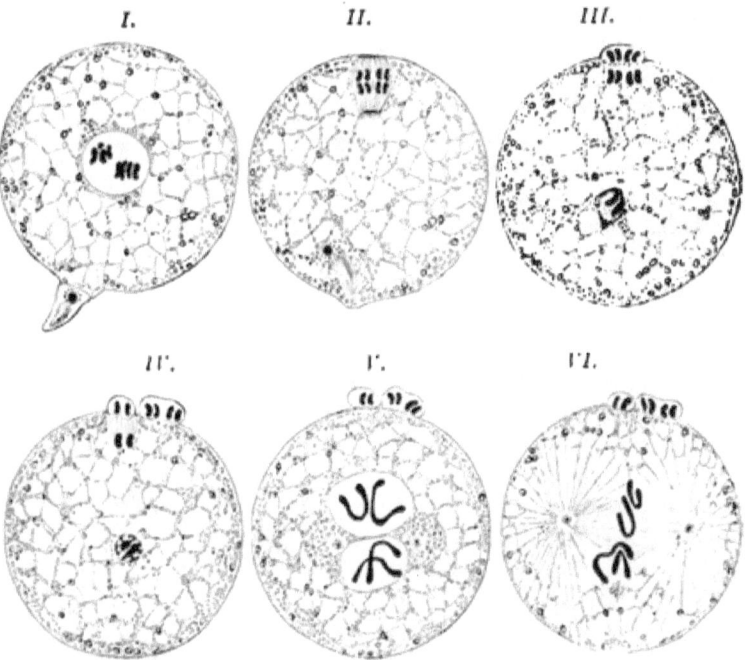

FIG. 127.—Diagram of the development of polar-cells and the fertilisation of the egg of *Ascaris megalocephala bivalens*.

described (p. 228) forms the first polar-cell. This process corresponds to the division of the sperm-mother-cell into two daughter-cells. As before (Fig. 126 *I.*), each of the two unequally large products of division, viz. the egg-daughter-cell and the polar-cell which was produced by budding, receive from the two bundles of four segments two daughter-bundles each containing two segments.

Here also the second division follows the first so closely, that the resting stage is omitted. Out of the material of that half of the spindle which remained behind in the egg-daughter-cell, a second complete spindle develops directly, containing only four segments, arranged in pairs. A second budding produces both the second polar-cell (Fig. 127 *IV.*), and the grand-daughter egg-cell, or the mature egg, each division product containing only two nuclear segments.

If we disregard the fact that the division products, when the egg is ripe, are very unequal in size (budding), the processes which take place resemble so exactly those already described as occurring during sperm formation, that through them some light is thrown upon the *raison d'être* of the polar-cells. Whilst on the one hand four spermatozoa (Fig. 126 *III.*, *A*, *B*, *C*) develop out of a sperm-mother-cell (Fig. 126 *I.*), on the other only one egg capable of being fertilised (Fig. 127 *V.*) and three abortive eggs arise out of an egg-mother-cell. These latter still remain in a rudimentary form, although they play a part in the physiologically important division with reduction.

It has been noticed in many other objects besides Nematodes, that the mature sexual products only possess half as many nuclear segments as the tissue cells of the organism in question; this was observed by Boveri (VI. 6) in the mature egg-cells of the most various classes of the animal kingdom, by Flemming (VI. 13 *II.*), Platner (VI. 52), Henking (VI. 27), Ishikawa (VI. 40), Häcker (VI. 24), vom Rath (VI. 55), in mature spermatozoa of *Salamandra*, *Gryllotalpa*, *Pyrrhocoris*, *Cyclops*, etc., and by Guignard (VI. 23 *b*), in the nuclei of the polar-cells, which are formed during fertilisation, and in the nucleus of the mature egg-cell of *Phanerogamia*.

Maupas (VII. 30) and Richard Hertwig (VII. 21) observed that a reduction of nuclear substance occurs also in *Infusoria* before fertilisation; however, further details on this subject are given later, on p. 269 (Chapter VII.).

In all the above-mentioned cases, the reduction of nuclear substance occurs before the egg-cell is fertilised by the spermatozoon. It appears, however, that the reduction of nuclear substance may occur after fertilisation has taken place, as *a priori* appears quite possible, as a result of the first division. At any rate that is the way in which I explain the interesting observations of Klebahn (VI. 43) upon two species of low Algæ, *Closterium* and *Cosmarium*. A more detailed account is given in the chapter on the process of fertilisation, p. 279.

IV. Influence of the Environment upon Cell-Division.

The complex play of forces, exhibited to the spectator at each cell-division, can, just like the phenomena of protoplasmic movements, which have been already described, be influenced to a considerable extent by external agencies. Only here, for obvious reasons, the conditions are more complicated than with the protoplasmic movements, because bodies differing in structure, such as protoplasm, nuclear segments, spindle threads, centrosomes, etc., are concerned, and these can be altered in very various ways. As yet very little experimental work has been done upon the subject. If the question is raised as to how the processes of nuclear division are affected at any individual stage by thermal, mechanical, electrical or chemical stimuli, the answer is but unsatisfactory. The most complete experiments that have been made at present have been upon Echinoderm eggs, whose reactions during division to thermal and chemical stimuli have been carefully observed.

It is generally accepted that the rate of cell-division is affected by the temperature, but what are the exact maximum and minimum temperatures, and what changes in the nuclear figure are produced by temperatures exceeding the maximum, have not yet been accurately determined.

I (VI. 32, 33) have conducted a series of experiments upon the influence of temperature from $1°$ to $4°$ Celsius below zero.

If dividing Echinoderm eggs are cooled down for about 15 to 20 minutes from $1°$ to $4°$ Celsius below zero, after a few minutes the whole achromatin portion of the nuclear figure becomes disintegrated and destroyed, whilst the chromatin portion forming the nuclear segments experiences only small or unimportant changes. The most instructive figures are seen when the nuclear segments are arranged in the equator (Fig. 128 *A*), or when they have already migrated to the two poles, as can be seen from Fig. 128 *B*; the protoplasmic radiations and the spindle threads have absolutely disappeared, whilst the radiation spheres in the neighbourhood of the centrosomes are marked by bright portions in the yolk. The nuclear segments alone are unaltered in appearance and position.

As long as the eggs are under the influence of the cold, the nuclear figures remain in this condition; however, the rigidity gradually disappears when the eggs are placed in a drop of water upon an object glass, and gradually warmed up to the temperature of the room. After 5 or 10 minutes the two polar radiations develop again at the same places as before, at first being only faintly seen, but finally being as distinct as ever; the spindle threads reappear between the two poles, and division proceeds in the usual manner. In such cases the cold has acted only as a check, the process of division simply going on from the point at which it was arrested by the cold.

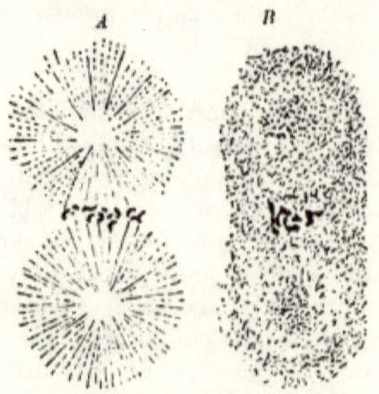

Fig. 128.—*A* Nuclear figure of an egg of *Strongylocentrotus*, one hour and twenty minutes after fertilisation. *B* Nuclear figure of an egg of *Strongylocentrotus*; this was killed after having been kept for two hours and fifteen minutes in a freezing mixture, with a temperature of −2°, in which it was placed one and a half hours after the occurrence of fertilisation.

A greater effect is produced if the eggs are subjected for about 2 to 3 hours to a temperature of from 2° to 3° Celsius below zero. The whole nuclear figure is then fundamentally altered, and hence, when the cold rigor is over, it is obliged to reconstruct itself entirely, on which account a longer period of recuperation is necessary. The nuclear segments either become fused together to form an irregularly-lobulated body, or they develop into a small vesicular nucleus, such as is formed during the reconstruction process after division. Then changes begin anew, which result in the formation of polar radiations, and frequently of more or less abnormal nuclear division figures. In fact the division of the egg-body is not only considerably delayed, but even pathologically altered.

Similarly certain chemical substances exert a marked effect upon the process of division (·05 solution of sulphate of quinine and 5 per cent. chloral hydrate). If eggs which have developed spindles, and which exhibit the equatorial arrangement of the nuclear segments, are subjected for about 5 to 10 minutes to the action of the above-mentioned substances, the pole radiations soon commence to disappear completely. However, after a short period of rest, they reappear, and division proceeds as usual. If, however, the substances are allowed to act upon the eggs for from 10 to 20 minutes, a still greater disturbance is produced, resulting in many cases in a very peculiar and, in its way, typical course of the division process. Not only are the pole radiations completely destroyed, but the nuclear segments become gradually transformed into the vesicular resting condition of the nucleus (Fig. 129 *A*). This constitutes the starting point of a new but considerably modified process of division (O. and R. Hertwig VI. 38).

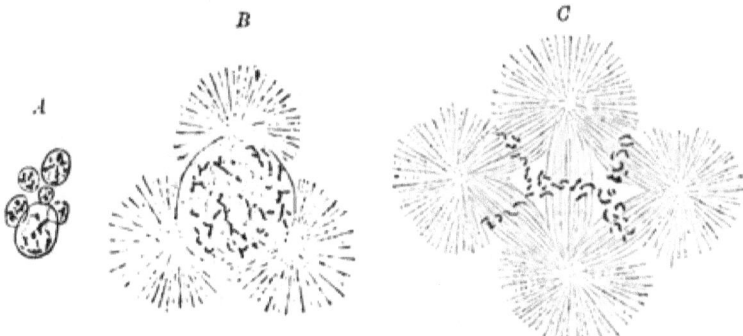

FIG. 129.—Nuclei of eggs of *Strongylocentrotus* which, one and a half hours after the act of fertilisation has occurred, have been placed in ·025 per cent. solution of quinine sulphate, where they remained for twenty minutes. *A* Nuclear figure of an egg, which was killed one hour after it was removed from the quinine solution; *B* nuclear figure of an egg, killed somewhat later; *C* nuclear figure of an egg, killed two hours after it was removed from the quinine sulphate solution.

Instead of two radiations, four develop immediately upon the surface of the nuclear vesicle (Fig. 129 *B*, in which one radiation is obscured). If treated with quinine, these soon become sharply defined; when, however, chloral is used, they remain permanently faint, and confined to the immediate neighbourhood of the nucleus. The nuclear membrane next becomes dissolved; five spindles develop between the four poles, and upon these the nuclear segments distribute themselves equatorially, thus producing a characteristic figure (Fig. 129 *C*). The nuclear segments then move towards the four poles, and form the basis for four vesicular nuclei, which separate from one another and travel towards the surface of the yolk. The egg then begins, by means of two cross furrows, to become constricted into four corresponding segments.

However, as a rule, this division into four portions is not completed until after the four nuclei have begun to make preparations for dividing again by forming spindles with two pole radiations

At the same time, the furrows already mentioned deepen, so that each spindle comes to lie in a protuberance or bud. Now the splitting up becomes either pretty well completed, or the four spindles, before the furrows have penetrated far into the yolk, commence to divide, the nuclear segments travelling towards the poles. The result of this is that the four first protuberances begin to become constricted a second time and to separate from one another (cell-budding, bud formation).

The most striking of the phenomena described above is the sudden appearance of the four pole radiations, for which, according to our present knowledge, an equal number of centrosomes must have served as bases. An explanation of this is afforded us by the processes connected with the fertilisation of the Echinoderm egg, which are discussed on p. 259.

Modifications of the form of nuclear transformation, shown in Fig. 129 *C*, occur not infrequently; these are due to one of the radiations being somewhat separated from the three others (Fig. 130). In this case the three that are situated close to one another

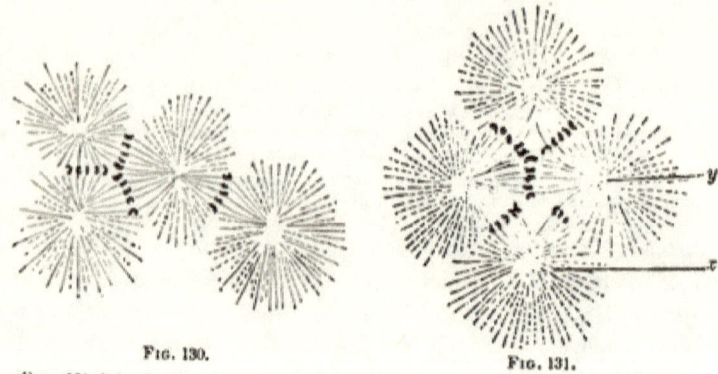

Fig. 130.   Fig. 131.

Figs. 130, 131.—Nuclear figures with four poles from *Strongylocentrotus* eggs, which, one and a half hours after the act of fertilisation, have been placed for twenty minutes in 0.5 per cent. solution of quinine, and which have been killed two hours after their removal from the quinine solution.

are united by the three spindles to form a triaster. In the centre of the equilateral triangle thus formed, the three nuclear planes intersect, producing another regular figure. The fourth radiation, which is situated at one side, is connected by a single spindle with the radiation nearest to it.

Fig. 131 may be regarded as an intermediate stage between Figs. 129 and 130. Here the radiation x, which lies somewhat apart, is connected by means of two spindles to the remaining portion of the figure, which forms a triaster. Of these two spindles one is only faintly and imperfectly developed, and is further remarkable for the small number of its nuclear segments. Apparently it would never have made its appearance if radiation x had been at a somewhat greater distance from radiation y.

Nuclear figures with three, four or more poles (triaster, tetraster, polyaster, multipolar mitoses), have been frequently observed by pathological anatomists in tissues altered by disease (Arnold, Hansemann, Schottländer, Cornil, Denys, etc., VI. 1, 10, 11, 25, 67); they occur with especial frequency in malignant tumours, such as carcinoma, and resemble to a remarkable extent those produced artificially in egg-cells, such as are represented in Figs. 129 to 131. Apparently the cause for the abnormal appearances may be traced to chemical stimuli. Thus Schottländer (VI. 67) was able to excite pathological nuclear division in the endothelium of Descemet's membrane by cauterising the transparent cornea of the Frog's eye with chloride of zinc solution of a certain strength, and thus inducing inflammation. It is remarkable how much the number of nuclear segments may vary in individual spindles. For instance, Schottländer found as many as twelve segments in some spindles, and in others only six or even three; the same was observed in Echinoderm eggs.

Further, multipolar nuclear figures may apparently be due to other causes, about which at present extremely little is known to us. For instance, a very common cause is the presence of several nuclei in one cell. Such a condition can be easily produced artificially by injuring egg-cells in some suitable way, and by subsequently fertilising them (Fol VI. 19 b; Hertwig VI. 30 a, 32, 33, 38). Under these circumstances instead of one single spermatozoon entering in the usual manner, two, three, or more make their way into the yolk. The consequence of this kind of over-fertilisation (polyspermia) is the formation of several sperm nuclei, corresponding in number to the spermatozoa which entered. Some of these approach the egg nucleus, and since each of them has brought a centrosome with it into the egg, a corresponding number of pole radiations develop around the egg nucleus. And thus, according to the number of spermatozoa, the egg nucleus becomes transformed into a nuclear division figure with three, four, or more radiations.

Further, those sperm nuclei which are not in contact with the egg nucleus, but which remain isolated in the yolk, very frequently give rise to peculiar, multipolar nuclear figures. They next become transformed into small sperm spindles. Neighbouring spindles then frequently approach each other, so that two pole radiations, and consequently the centrosomes which they contain, are fused together to form one. In this manner the most various collections of spindles may be produced according to the amount of coalescence which occurs, especially when over-fertilisation has taken place to a high degree. Further the multi-radiated figure, proceeding from the over-fertilised egg nucleus, may become yet still more complicated in structure by the formation of male nuclear spindles.

The interesting discoveries of Denys on the giant cells of bone marrow, and of Kostanecki (VI. 46) on those in the embryonic livers of mammals, may be explained in a similar manner. Several centrosomes, proportionate in number to the nuclei, are present in the cell. Hence when the whole cell contents commence to divide, several centrosomic radiations have to develop, and amongst these the nuclear segments, which under certain circumstances may number several hundreds, arrange themselves in peculiarly branched nuclear plates, such as have been depicted by Kostanecki in Fig. 132. When subsequently the mother-segments split up into daughter-segments, these move off in groups towards the

Fig. 132.

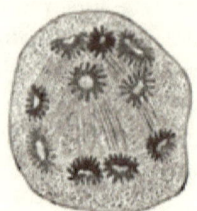
Fig. 133.

Fig. 132.—Multicentrosomic nuclear division figure, with several groups of mother-segments, from a giant cell from the liver of a mammalian embryo. (After Kostanecki.)

Fig. 133.—Multicentrosomic nuclear division figure of a giant cell from the liver of a mammalian embryo; the daughter-segments form several groups, which have travelled towards the numerous centrosomes. (After Kostanecki.)

poles of the complicated nuclear division figure, where they form a large number of small spheres (Fig. 133). Later on, a nucleus develops out of each sphere; finally the giant cell splits up into as many portions as there were nuclei—that is to say, spheres consisting of daughter-segments—present in the cell.

The observations of Henneguy (VI. 28) on Trout eggs belong to the same category. It is well known, that a large number of nuclei (merocytes) are scattered throughout the yolk layer; this is situated below the germinating cells in eggs, which are partially segmented by furrows. Occasionally some of them collect together to form small spindle aggregations, whilst at the same time they are making preparations for division. Hence it is very instructive to see, that in the following case, described by Hennegny (Fig. 134), the centrosomes act as attraction centres. Two merocytes, which are in the act of dividing, lie close together in the common mass of yolk, so that the longitudinal axis of spindle B would, if produced, cut spindle A in its equator; we see also that the one centrosome b is very near to spindle A. In con-

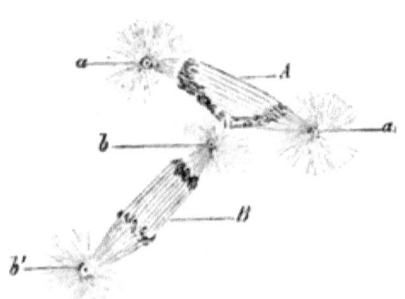

FIG. 134.—Two nuclear spindles from the yolk of the germinal disc of a Trout's egg: the centrosome is exerting a disturbing influence upon the arrangement and distribution of the daughter-segments of the second spindle. (After Henneguy.)

sequence, the arrangement of the daughter-segments of spindle A has been disturbed to a considerable extent. Instead of their being arranged in two groups near the centrosomes, a, a, as would occur normally, a number of those which are within the attraction sphere of the centrosome b of the neighbouring foreign spindle have been drawn towards it. In a word: the centrosome of the one spindle has evidently exerted a disturbing influence upon the arrangement and distribution of the daughter-segments of the other spindle.

Henneguy has observed triasters, such as the one depicted in Fig. 135, and also tetrasters, in the germinal cells of the same object; these gradually separated themselves from the layer of merocytes.

At the close of this fourth section we may mention the degeneration processes, which sometimes occur in cell nuclei, apparently as the result of injurious influences. Especially in the sexual organs, individual germ cells, or groups of them, appear to degenerate before

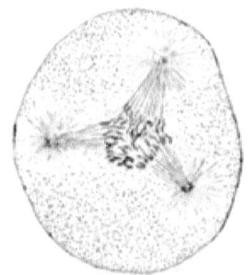

FIG. 135.—Cell with a tri-centrosomic nuclear figure: from a Trout embryo. (After Henneguy.)

they have reached maturity, as has been observed by Flemming and Hermann in *Salamandra maculata*, and by myself in *Ascaris megalocephala*. The framework of the nuclei disintegrates, and the nuclein collects together into a compact mass, which is remarkable for its strong affinity for the most various stains. The protoplasm diminishes in quantity, in proportion to that present in similar normal germ cells. Such a stunted cell with a com-

FIG. 136.—*A* Sperm cell with a degenerated nucleus from the testis of a *Salamandra maculata* (from Flemming, Pl. 25, Fig. 51 a). *B* Residuary body (*corps résiduel*) from the testis of *Ascaris megalocephala*. Nuclear degeneration.

pletely disorganised nucleus is depicted in Fig. 136. *A* is a germinal cell from the testis of *Salamandra*; *B*, a germinal cell of *Ascaris*, such as is found both in the testis and ovary, and which is known by the name of *corps résiduel*, or residuary body. Wasielewski, by injecting turpentine into the testes of mammals, has succeeded in inducing experimentally a similarly degenerated condition in the nuclei of germ cells.

Concerning the physiological importance of the nuclear division processes, compare Chapter IX., section 3, especially that portion dealing with the equal distribution of the multiplying inherited mass amongst the cells proceeding from the fertilised egg.

Literature VI.

1. JULIUS ARNOLD. *Ueber die Theilungsvorgänge an den Wanderzellen.* Archiv für mikroskop. Anatomie. Bd. XXX. Ferner mehrere Aufsätze in Virchow's Archiv. Bd. XCIII., XCVIII., CIII.
2A. AUERBACH. *Organologische Studien.* Zweites Heft. Ueber Neubildung und Vermehrung der Zellkerne.
2B. AUERBACH. *Zur Kenntniss der thierischen Zellen.* Sitzungsber. d. kgl. preuss. Akademie der Wissenschaften zu Berlin. 1890.
3. BALFOUR. *A Treatise on Embryology.* London. 1880.
4A. VAN BENEDEN. *Recherches sur la maturation de l'œuf, la fécondation, et la division cellulaire.* Archives de biologie. Vol. IV. 1883. (Trans. by Cunningham, Q.J.M.S., Jan., 1885.)
4B. VAN BENEDEN u. NEYT. *Nouvelles recherches sur la fécondation et la division mitosique chez l'ascaride mégalocéphale.* Leipzig. 1887.

5. BORN. *Ueber den Einfluss der Schwere auf das Froschei.* Archiv für mikroskop. Anatomie. Bd. XXIV.
6. BOVERI. *Zellenstudien.* Jenaische Zeitschrift. 1887, 1888, 1890.
7. BOVERI. *Ueber den Antheil des Spermatozoons an der Theilung der Eier.* Sitzungsber. der Gesellsch. f. Morph. u. Physiol. in München. 1887.
8. BRANDT. *Neue Radiolarienstudien.* Mittheil. des Vereins Schleswig-Holstein. Aerzte. Januar 1890.
9. CARNOY. See Literature IV.
10. CORNIL. *Sur la multiplication des cellules de la moelle des os par division indirecte dans l'inflammation.* Arch. de phys. norm. et patholog. 1887.
11. CORNIL. *Sur le procédé de division indirecte des noyaux et des cellules epitheliales dans les tumeurs.* Arch. de phys. norm. et path. 3. sér. T. VIII.
12. W. FLEMMING. *Zellsubstanz, Kern und Zelltheilung.* Leipzig. 1882.
13. W. FLEMMING. *Neue Beiträge zur Kenntniss der Zelle.* I. Theil. Archiv für mikrosk. Anatomie. Bd. XXIX. 1887. II. Theil: Ebenda. Bd. XXXVII. 1891.
14. W. FLEMMING. *Ueber Zelltheilung.* Verhandl. der anat. Gesellschaft zu München. 1891, p. 125.
15. W. FLEMMING. *Ueber Theilung und Kernformen bei Leukocyten u. über deren Attractions-sphären.* Archiv f. mikrosk. Anatomie. Bd. XXXVII. 1891, p. 249.
16. W. FLEMMING. *Amitotische Kerntheilung im Blasenepithel des Salamanders.* Archiv für mikrosk. Anatomie. Bd. XXXIV.
17. W. FLEMMING. *Attractionssphäre u. Centralkörper in Gewebszellen u. Wanderzellen.* Anat. Anzeiger. 1891.
18. FOL. *Die erste Entwicklung des Geryonideneies.* Jenaische Zeitschr. Vol. VII. 1873.
19A. FOL. *Sur le commencement de l'hénogénie.* Archives des sciences phys. et natur. Genève. 1877.
19B. FOL. *Archives des sciences physiques et naturelles.* Genève, 15. Oct. 1883.
20. FOL. *Sur l'œuf et ses enveloppes chez les Tuniciers.* Recueil zoologique Suisse.
21. FRENZEL. *Die nucleoläre Kernhalbirung etc.* Archiv für mikroskop. Anatomie. Bd. XXXIX. 1892.
22. GÖPPERT. *Kerntheilung durch indirecte Fragmentirung in der lymphatischen Randschicht der Salamanderleber.* Archiv f. mikrosk. Anatomie. Bd. XXXVII. 1891.
23A. GUIGNARD. *Recherches sur la structure et la division du noyau cellulaire.* Annales des scienc. nat. 6. sér. T. XVII. 1884.
23B. GUIGNARD. *Nouvelles études sur la fécondation, comparaison, etc.* Annales des scienc. nat. T. XIV. Botanique. 1891.
24. V. HÄCKER. *Die Eibildung bei Cyclops u. Canthocamptus.* Zool. Jahrbücher. Abth. f. Anat. u. Ontogenie. Bd. V.
25. DAVID HANSEMANN. *Ueber pathologische Mitosen.* Virchow's Archiv. Bd. CXXIII. 1891.

26. DAVID HANSEMANN. *Ueber asymmetrische Zelltheilung in Epithelkrebsen und deren biologische Bedeutung.* Virchow's Archiv. Bd. CXIX.
27. HENKING. *Untersuchungen über die ersten Entwicklungsvorgänge in den Eiern der Insecten.* Theil 1–3. Zeitschr. f. wissenschaftl. Zoologie. Bd. XLIX., LI., LIV.
28. HENNEGUY. *Nouvelles recherches sur la division cellulaire indirecte.* Journal de l'anatomie. Bd. XXVII. 1891.
29. F. HERMANN. *Beitrag zur Lehre von der Entstehung der karyokinetischen Spindel.* Archiv f. mikrosk. Anatomie. Bd. XXXVII., p. 569.
30A. O. HERTWIG. *Beiträge zur Kenntniss der Bildung, Befruchtung und Theilung des thierischen Eies.* Morphol. Jahrbücher. Bd. I., III. u. IV. 1875, 1877, 1878.
30B. O. HERTWIG. *Die Chaetognathen, eine Monographie.* 1880.
31. O. HERTWIG. *Welchen Einfluss übt die Schwerkraft auf die Theilung der Zellen?* Jena. 1884.
32. O. HERTWIG. *Experimentelle Studien am thierischen Ei vor, während und nach der Befruchtung.*
33. O. HERTWIG. *Ueber pathologische Veränderung des Kerntheilungsprocesses in Folge experimenteller Eingriffe.* Internationale Beiträge zur wissenschaftl. Medicin.
34. O. HERTWIG. *Vergleich der Ei- und Samenbildung bei Nematoden. Eine Grundlage für celluläre Streitfragen.* Archiv f. mikroskop. Anatomie. Bd. XXXVI. 1890.
35. R. HERTWIG. *Beiträge zur Kenntniss des Acineten.* Morphol. Jahrbücher. Bd. I. 1875.
36. R. HERTWIG. *Zur Histologie der Radiolarien.* Leipzig. 1876.
37. R. HERTWIG. *Ueber den Bau und die Entwicklung der Spirogona gemmipara.* Jenaische Zeitschrift. Bd. XI. 1877.
38. O. u. R. HERTWIG. *Ueber den Befruchtungs- und Theilungsvorgang des thierischen Eies unter dem Einfluss äusserer Agentien.* Jena. 1887.
39. E. HEUSER. *Beobachtungen über Zelltheilung.* Botanisches Centralblatt. 1884.
40. ISHIKAWA. *Studies of Reproductive Elements.* 1. Spermatogenesis, Ovogenesis and Fertilisation in Diaptomus. Journal of the College of Science, Imperial University. Japan. Vol. V. 1891.
41. JOHNSON. *Amitosis in the Embryonal Envelopes of the Scorpion.* Bulletin of the Museum of Comparative Zoology at Harvard College. Vol. XXII. 1892.
42. JOHOW. *Die Zellkerne von Chara foetida.* Botanische Zeitung. 1881.
43. KLEBAHN. *Die Keimung von Closterium und Cosmarium.* Pringsheims Jahrbücher f. wissenschaftl. Botanik. Bd. XXII.
44. KÖLLIKER. *Entwicklungsgeschichte der Cephalopoden.* 1844.
45. KÖLLIKER. *Die Lehre von der thierischen Zelle.* In Schleiden u. Nägeli's wissenschaftl. Botanik. Heft 2.
46. v. KOSTANECKI. *Ueber Kerntheilung bei Riesenzellen nach Beobachtungen aus der embryonalen Säugethierleber.* Anatomische Hefte. 1892.
47. H. v. MOHL. *Ueber die Vermehrung der Pflanzenzellen durch Theilung.* Dissertation. Tübingen, 1835. Flora, 1837.

48. NÄGELI. *Zellkern, Zellbildung und Zellenwachsthum bei den Pflanzen. In Schleiden und Nägeli's Zeitschr. f. wissenschaftl. Botanik. Bd. II. u. III.*
49. PFLÜGER. *Ueber den Einfluss der Schwerkraft auf die Thielung der Zellen. Archiv f. die gesammte Physiologie. Bd. XXXI. u. XXXII.* 1883.
50. PFLÜGER. *Ueber die Einwirkung der Schwerkraft u. anderer Bedingungen auf die Richtung der Zelltheilung. 3. Abh. Archiv f. d. gesammte Physiologie. Bd. XXXIV.* 1884.
51. PLATNER. *Die Karyokinese bei den Lepidopteren als Grundlage für eine Theorie der Zelltheilung. Internationale Monatsschrift. Bd. III.* 1885.
52. PLATNER. *Beiträge zur Kenntniss der Zelle u. ihrer Theilungserscheinungen. Archiv f. mikroskop. Anatomie. Bd. XXXIII.* 1889.
53. RABL. *Ueber Zelltheilung. Morpholog. Jahrb. Bd. X.* 1885, *und Anat. Anzeiger. Bd. IV.* 1889.
54. RANVIER. *Technisches Lehrbuch der Histologie. Leipzig.* 1888.
55. O. VOM RATH. *Zur Kenntniss der Spermatogenese von Gryllotalpa vulg. mit besonderer Berücksichtigung der Frage nach der Reductionstheilung. Archiv f. mikrosk. Anatomie. Bd. XL.* 1892.
56. RAUBER. *Formbildung u. Cellularmechanik. Morpholog. Jahrbüch. Bd. VI.*
57. RAUBER. *Thier u. Pflanze. Akademisches Programm. Zoolog. Anzeiger.* 1881.
58. REICHERT. *Beitrag zur Entwicklungsgeschichte der Samenkörperchen bei den Nematoden. Müller's Archiv f. Anat. u. Physiol. etc.* 1847.
59. REICHERT. *Der Furchungsprocess u. die sogenannte Zellenbildung um Inhaltsportionen. Müller's Archiv.* 1846.
60. REMAK. *Ueber extracellulare Entstehung thierischer Zellen u. über Vermehrung derselben durch Theilung. Müller's Archiv.* 1852.
61. REMAK. *Untersuchungen über die Entwicklung der Wirbelthiere.* 1855.
62. RETZIUS. *Studien über die Zellentheilung. Biolog. Untersuchungen. Jahrgang* 1881.
63. ROUX. *Ueber die Bedeutung der Kerntheilungsfiguren. Leipzig.* 1883.
64. SACHS. *Die Anordnung der Zellen in jüngsten Pflanzentheilen. Arbeiten des botan. Instituts in Würzburg. Bd. II.*
65A. SCHÄFER. *On the Structure of the immature Ovarian Ovum in the common Fowl and in the Rabbit. Proceedings of the Royal Society. London.* 1880.
65B. SCHEWIAKOFF. *Ueber die karyokinetische Kerntheilung der Euglypha alveolata. Morpholog. Jahrbüch. Bd. XIII.* 1888.
66. SCHNEIDER. *Untersuchungen über Platkelminthen. Jahrb. d. oberhessischen Gesellsch. f. Natur- u. Heilkunde.* 1873.
67. SCHOTTLÄNDER. *Ueber Kern- und Zelltheilungsvorgänge in dem Endothel der entzündeten Hornhaut. Archiv f. mikroskop. Anatomie. Bd. XXXI.* 1888.
68. MAX SCHULTZE. *De ovorum ranarum segmentatione, quae Furchungsprocess dicitur. Bonn.* 1863.
69. MAX SCHULTZE. *Untersuchungen über die Reifung und Befruchtung des Amphibieneies. Zeitschrift f. wissenschaftl. Zoologie. Bd. XLV.* 1887.

70. Solger. *Zur Kenntniss der Pigmentzellen.* Anatom. Anzeiger. 1891, p. 162.
71. Ed. Strasburger. *Zellbildung u. Zelltheilung.* 3 Aufl. 1880.
72. Ed. Strasburger. *Die Controversen der indirecten Kerntheilung.* Archiv für mikroskop. Anatomie. Bd. XXIII. Bonn. 1884.
73. Ed. Strasburger. *Histologische Beiträge. Heft I.: Ueber Kern- u. Zelltheilung im Pflanzenreiche, etc.* Jena. 1888.
74. Vejdovsky. *Entwicklungsgeschichtliche Untersuchungen.* Prag. 1888.
75. Vialleton. *Recherches sur les premières phases du développement de la seiche.* Paris. 1888.
76. W. Waldeyer. *Karyokinesis and its Relation to the Process of Fertilisation.* Quart. Journ. of Micr. Science. Vol. XXX. 1889 (Trans. from the Arch. f. mikr. Anat., Vol. XXXII., 1888), containing a complete bibliography up to 1888.
77. Weismann. *Ueber die Zahl der Richtungskörper und über ihre Bedeutung für die Vererbung.* Jena. 1887.
78. R. Zander. *Ueber den gegenwärtigen Stand der Lehre von der Zelltheilung.* Biolog. Centralblatt. Bd. XII. 1892.
79. H. E. Ziegler. *Die biologische Bedeutung der amitotischen Kerntheilung im Thierreich.* Biolog. Centralblatt. Bd. XI. 1891.
80. Ziegler u. vom Rath. *Die amitotische Kerntheilung bei den Arthropoden.* Biolog. Centralblatt. Bd. XI. 1891.
81. Bütschli. *Studien über die ersten Entwicklungsvorgänge der Eizelle, Zelltheilung und Conjugation der Infusorien.* Abhandl. d. Senkenberg. naturf. Gesellsch. 1876.
82. Rich. Hertwig. *Ueber die Kerntheilung bei Actinosphärium.* Jenaische Zeitschr. f. Naturw. 1884.
83. Rich. Hertwig. *Ueber die Gleichwerthigkeit der Geschlechtskerne bei den Seeigeln.* Sitzungsberichte d. Gesellsch. f. Morph. u. Phys. in München. Bd. IV. 1888.
84. Rich. Hertwig. *Ueber Kernstructur und ihre Bedeutung für Zelltheilung u. Befruchtung.* Ebenda. Bd. IV. 1888.
85. Oscar Hertwig. *Das Problem der Befruchtung und der Isotropie des Eies, eine Theorie der Vererbung.* Jenaische Zeitschrift. 1884.
86. Blochmann. *Ueber directe Kerntheilung in der Embryonalhülle der Skorpione.* Morphol. Jahrb. Bd. X. 1885.
87. v. Davidoff. *Untersuchungen zur Entwicklungsgeschichte der Distaplia magnilarva, einer zusammengesetzten Ascidie.* Mittheil. aus d. zoolog. Station zu Neapel. Bd. IX.
88. L. Sheldon. *The Maturation of the Ovum in Peripatus.* Quart. Journ. of Microsc. Science. Vol. XXX. 1889.
89. A. Thomson. *Recent Researches on Oogenesis.* Quart. Journ. of Microsc. Science. Vol. XXVI. June, 1886.
90. J. G. McKendrick. *The Modern Cell Theory.* Proc. of the Glasgow Phil. Soc. 1888.
91. J. S. Burdon Sanderson. *Address to the Biological Section of the British Association.* Report for 1889.
92. G. S. Woodhead and G. E. C. Wood. *The Physiology of the Cell Considered in its Relation to Pathology.* Edinburgh Med. Journal, 1890.

93. W. H. CALDWELL. *The Embryology of Monotremata and Marsupialia, Part I., Philosophical Transactions of the Royal Society of London for the year* 1887.
94. G. L. GALLAND. *The Nature and Varieties of Leucocytes, Laboratory Reports of R. Coll. of Physicians of Edinburgh.* Vol. III. 1891.
95. ROBINSON. *On the Early Developmental History of Mammalia, especially as regards the Inversion of the Layers.* Quart. Journ. Mic. Sci., XXXIII., 369.
96. SEDGWICK. *On Elasmobranchs.* Quart. Journ. Micr. Sci., XXXIII., 559.
97. J. B. CARNOY. *Some Remarks on the recent Researches of Zacharias and Boveri upon the Fecundation of Ascaris megalocephala. Report on the 57th meeting of the Brit. Ass. for the Advancement of Science at Manchester,* 1887.
98. A. WEISMANN. *Essays upon Heredity and kindred Biological Problems (translations).* Oxford, Clarendon Press. 1889.

CHAPTER VII.

THE VITAL PROPERTIES OF THE CELL.

V. Phenomena and Methods of Fertilisation. Cell reproduction by means of cell division, such as is described in Chapter VI., does not, at least for the majority of organisms, appear to be able to continue for an indefinite period; the process of multiplication, after a shorter or longer period, comes to a standstill, unless it is stimulated afresh by the excitatory processes, which are grouped together under the name of fertilisation. Only the very lowest organisms, such as fission fungi, appear to be able to multiply indefinitely by repeated divisions; for the greater part of the animal and vegetable kingdoms the general law may be laid down, that after a period of increase of mass through cell division a time arrives when two cells of different origin must fuse together, producing by their coalescence an elementary organism which affords the starting-point for a new series of multiplications by division.

Hence the multiplication of the elementary organism, and with it life itself, resolves itself into a cyclic process. After generations of cells have been produced by division, the life-cycle returns to the same starting-point, when two cells must unite in the act of fertilisation, and thus constitute themselves the foundation of a new series of generations. Such cycles are termed generation cycles. They occur in the whole organic kingdom in the most various forms.

In unicellular organisms, for instance, the generation cycle consists of a large number of independent individuals, which sometimes amount to thousands. The fertilised elementary organism multiplies by repeated divisions, producing descendants, which do not require fertilisation, until a period arrives when a new generative act occurs between the generations which have been produced asexually. These phenomena have been most carefully observed in Infusoria. Thus Manpas (VII. 30, p. 407) has proved by a number of experiments upon *Leucophrys patula*, a species of Infusorian, that only after 300 generations have been produced from a fertilised individual does the generation cycle close, the descendants now showing for the first time the inclination and capacity for sexual conjugation. In *Onychodromus grandis* this

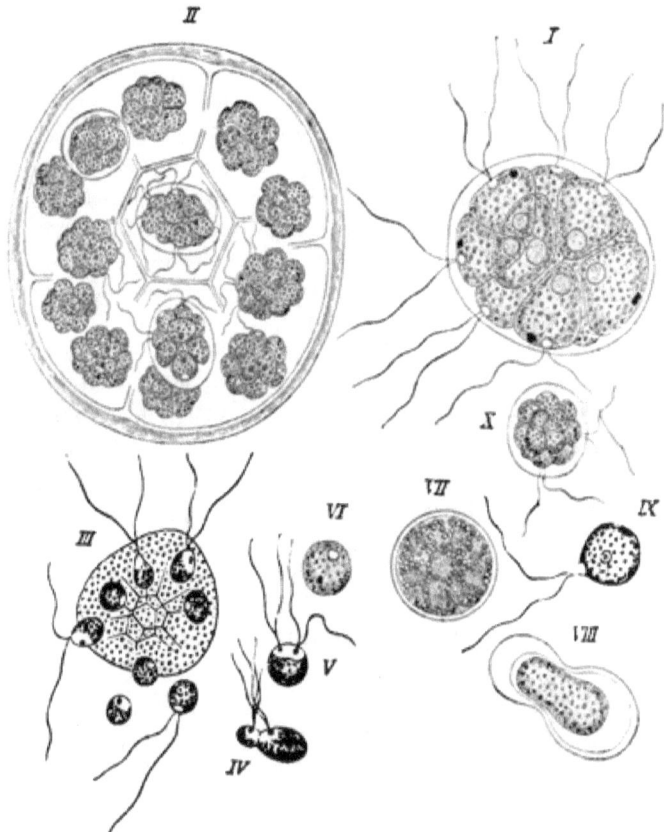

FIG. 137.—Development of *Pandorina morum* (after Pringsheim; from Sachs, Fig. 411): *I* a swarming colony; *II* the same, split up into sixteen daughter-colonies; *III* a sexual family, through the gelatinous envelope of which the individual cells are passing out; *IV*, *V* conjugation of the swarm-spores; *VI* a newly-formed zygote; *VII* a full-grown zygote; *VIII* transformation of the contents of a zygote into a large swarm-cell; *IX* the same, after having been set free; *X* the young colony derived from the swarm-cell.

condition occurs after the 140th generation, and in *Stylonichia pustulata*, after the 130th generation.

In multicellular organisms the cells, which are produced by the division of a fertilised egg, remain associated together, forming a colony of cells or an organic individual of a higher order. Regarded from the common point of view, from which we here treat the sexual question, they may be compared to the collection of cell individuals, multiplying asexually by division, which are derived from a fertilised mother Infusorian. The generation cycle closes here, when in the multicellular organism sexual cells, which have become mature, unite after the processes of fertilisation have occurred, and thus form the starting-point for new generations of dividing cells. The generation cycle may, in this case, present a very different picture, being sometimes very complicated in character. The simplest form is seen in many of the lower multicellular Algæ, such as *Eudorina*, or *Pandorina*. A cell colony (Fig. 137) is produced by the repeated division of the fertilised cell. After having lived for a definite period, all the cells become sexual cells. In order that conjugation may occur, the whole colony produced by cell division splits up into individuals, which serve as starting-points for new generation cycles.

The capacity, which each cell thus exhibits of reproducing the whole multicellular organism, is not seen when the organism is somewhat more highly developed. The cell substance, which has been derived from a fertilised egg, and which has multiplied by division to an immeasurable extent, then separates itself into two masses, one of which consists of cells which serve to build up the tissues and organs of the plant or animals, and the other of those destined to function in reproduction. In consequence the organism generally remains unaffected in itself when it reaches sexual maturity; it continues to detach the sexual elements from itself, so that they may start new generation cycles, until in consequence of the deterioration of the cells of its own body, or from any other cause, it succumbs to death (Nussbaum VII. 33; Weismann VII. 48).

In its purest form, a fixed and definite cycle is only to be met with in the higher animals, in which multiplication of individuals is only possible through sexual reproduction. In many species of the animal and vegetable kingdoms sexual and asexual multiplication take place simultaneously. In addition to the cells which require fertilisation, there are others which do not need it, and which, having detached themselves from the organism in the form of spores or pseud-ova, or as large groups of cells (buds, shoots, etc.), give rise to new organisms solely by repeated divisions, without sexual intercourse (vegetative reproduction). Or, to speak generally, between two acts of fertilisation a large number of events, which are the result of cell division, are introduced; these, however, need not belong to a single highly developed physiological individual, but may be shared by numerous individuals. This may occur in one of two ways.

In the one case the organism proceeding from the fertilised egg is unable itself to form sexual cells; it is only able to multiply by means of buds, spores, or parthenogenetic ova. These, or their asexually produced descendants, then become sexually mature, and develop the capacity of producing ova and spermatozoa. Such a cycle of events is called a regular alternation of generations (Hydroid polyps, Trematodes, Cestodes, parthenogenesis of Aphides, Daphnids, etc. Higher Cryptogams).

In the second case the organism derived from the fertilised egg multiplies both sexually and asexually. The consequence of this is, that even in the same species of plant or animal the generation cycle must vary considerably. Between the completion of the first and the commencement of the second act of fertilisation, either, on the one hand, only cell descendants arise which belong to the single individual from which the fertilised egg was derived, or one or more generations, the number in some cases being very large, intervene, until finally the eggs of an individual, produced by budding, become fertilised. In consequence, fertilisation here assumes the character of a facultative process, which is not absolutely necessary for the continuation of the species, at any rate, so long as it has not been proved that there are definite limits to vegetative multiplication. At present this cannot be demonstrated in numerous plants, which appear to be able to multiply indefinitely by means of runners, tubers, etc.

When we consider such cases, we must admit that the vital processes may continue indefinitely simply by repeated division of the cells themselves, without the intervention of the act of fertilisation; still, on the other hand, we are bound to conclude, on account of the wide distribution throughout the whole organic kingdom of the phenomenon of fertilisation, that this institution is of essential importance amongst the vital processes, and that it is fundamentally connected with the life of the cell. Fertilisation is in fact a cellular problem.

Our present subject is most closely connected with the study of the cell, especially as concerns its irritability and divisibility.

Hence this chapter may be divided into two sections: the Morphology and the Physiology of the process of fertilisation.

I. **The Morphology of the Process of Fertilisation.**
Up till now the process of fertilisation has been thoroughly worked out to the most minute details in three objects: in the animal egg, in the embryo-sac of Phanerogams, and in Infusoria. Although these three objects belong to different kingdoms of the organic world, they show a marked similarity in all the processes peculiar to fertilisation. It is, therefore, most suitable to commence this section by investigating these three objects. We will then occupy ourselves with the more general points of view provided by a study of comparative morphology, discussing:—

1. The different forms of sexual cells, the relative importance

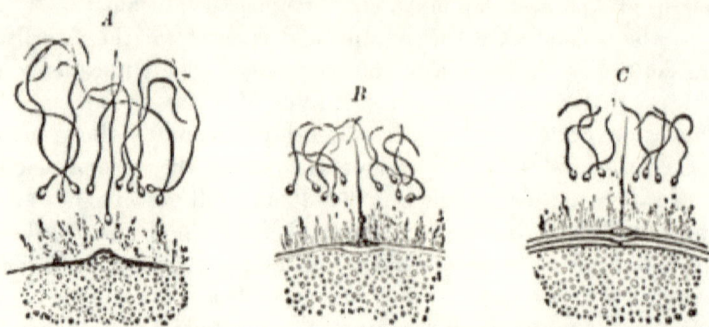

FIG. 138.—*A, B, C* small sections from the eggs of *Asterias glacialis* (after Fol). The spermatozoa have already penetrated into the gelatinous sheath covering the ova. In *A* a protuberance is commencing to raise itself to meet the most advanced spermatozoon. In *B* the protuberance and spermatozoon have met. In *C* the spermatozoon has entered the ovum. By this time a yolk membrane with a funnel-shaped opening has developed.

of the cell-substances, which are concerned in the generative act, and the idea of "male and female sexual cells."

2. The original and fundamental forms of sexual generation, and the derivation of sexual differences in the animal and vegetable kingdom.

1. **Fertilisation of the Animal Egg.** Echinoderm ova (Hertwig VI. 30; Fol. VI. 19, VII. 14) are classical subjects for the study of the processes of fertilisation, as also are the eggs of *Ascaris megalocephala* (van Beneden VI. 4 a, 4 b; Boveri VI. 6, etc.). They complement each other, for some phases of the process are more easily to be demonstrated in the one, whilst others are more plainly to be seen in the other.

a. **Echinoderm Eggs.** In most Echinoderms, the minute transparent ova are laid in sea-water, in a completely mature condition, having already budded off the pole cells (p. 229), and developed a small egg nucleus. They are surrounded by a soft gelatinous sheath, which can be easily penetrated by the spermatozoa (Fig. 138 *A*).

The spermatozoa are exceptionally small, and consist, as is the case in most animals, of (1) a head resembling a conical bullet; (2) a small spherule, the middle portion or neck; and (3) a delicate, contractile, thread-like tail. The head contains nuclein, the middle portion paranuclein, whilst the tail consists of modified protoplasm, and may be compared to a flagellum.

If ova and spermatozoa are brought together in sea-water, several of the latter immediately attach themselves to the gelatinous envelope of each ovum. Of these, however, only one normally fertilises each egg, namely, that one which, by means of the undulating movements of its tail, was the first to approach its surface (Fig. 138 *A–C*). At the spot where the apex of the head impinged, the hyaline protoplasm constituting the peripheral layer of the ovum raises itself up to form a small protuberance, the receptive protuberance. Here the head, impelled by the undulating movements of the tail, bores its way into the ovum, which at this moment, excited by the stimulus, excretes a delicate membrane, the vitelline membrane, upon its surface (Fig. 138 *C*), and, apparently by means of the contraction of its contents, presses some fluid out of the yolk. In consequence, a gradually increasing intervening space, which commences at the receptive protuberance, develops between the yolk and the yolk membrane. By this means the entrance of another spermatozoon is prevented.

Processes occurring in the interior of the yolk follow the external union of the two cells; these may be grouped together under the common name of internal fertilisation.

The tail ceases to move, and soon disappears from view; the head, however, slowly pushes its way into the yolk (Fig. 139 *A*); meanwhile, it absorbs fluid (Fig. 139 *B*), and swells up to form a small vesicle, which may be called the sperm-nucleus, or male pro-nucleus, since its essential constituent is the nuclein of the head of the spermatozoon; hence it becomes intensely stained by carmine, etc. Fol has lately shown that immediately in front of it, on the side which is directed to the centre of the egg (Fig.

139 *A, B*), there is a much smaller spherule, around which the yolk commences to arrange itself in radial striæ (Fig. 140 *A*), forming a radiated figure (a star); this star grows gradually more distinct, and at the same time extends itself farther away from the spherule. Since it seems to be derived from the neck of the

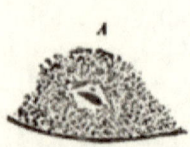 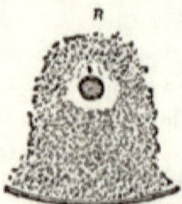

Fig. 139.—*A* and *B* represent portions of a section of a fertilised egg of *Asteracanthion*. A centrosome (sperm-centrum) has moved out a little in advance of the sperm-nucleus. (After Fol.)

spermatozoon, Fol has called it the sperm-centrum (male centrosome). A corresponding spherule can be seen close to the egg-nucleus, on that side which is turned away from the sperm-nucleus; Fol has called this the ovo-centrum (female centrosome).

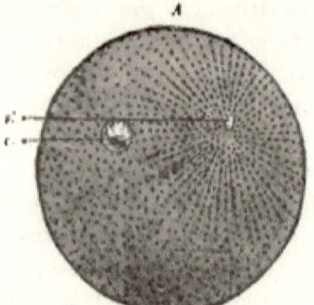 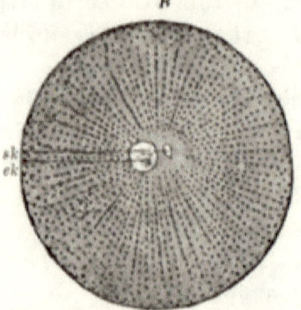

Fig. 140.—*A* Fertilised egg of a Sea-urchin (O. Hertwig, *Embryology*, Fig. 18). The head of the spermatozoon, which has penetrated into the egg, has been converted into a sperm-nucleus (*sk*) surrounded by a protoplasmic radiation, and has approached the egg-nucleus (*ek*). *B* Fertilised egg of a Sea-urchin (O. Hertwig, *Embryology*, Fig. 19). The sperm-nucleus (*sk*) and the egg-nucleus (*ek*) have approached each other, and are both surrounded by a protoplasmic radiation.

An interesting phenomenon now commences to attract attention (Fig. 140 *A, B*). The egg- and sperm-nuclei (male and female pro-nuclei) mutually attract each other, as it were simultaneously, and travel through the yolk towards each other with increasing velocity; the sperm-nucleus (*sk*) with its radiation containing the centrosome always moving in front of it, travels more quickly than the egg-nucleus (*ek*) with its ovo-centrum. Soon they meet in the centre of the egg, to become surrounded by an aureole of non-granular protoplasm, outside of which there is a radiation sphere, common to them both (sun-like figure and aureole of Fol).

During the course of the next twenty minutes the egg-nucleus and the sperm-nucleus fuse together to form a single germinal or cleavage nucleus (Fig. 141 *I-IV*); at first they lie close to one another, flattening their contingent surfaces (Fig. 141 *II*), until finally the lines of demarcation disappear, so that they unite to form a common nuclear vesicle. In this the substance derived from the spermatozoon may be distinguished for a considerable time as a distinct granular mass of nuclein, which eagerly absorbs staining solutions.

The fusion of the centrosomes (Fig. 141 *I*) follows closely on the union of the nuclei. They lie, surrounded by the homogeneous protoplasmic area, on opposite sides of the cleavage nucleus (Fig. 141 *II*); they then spread themselves out tangentially upon its surface, assuming the shape of a dumb-bell, and finally divide into halves, which move off in opposite directions (Fig. 141 *III*), and travel over one quarter of the circumference of the cleavage nucleus. By means of these circular movements (Fol's quadrille), half of each male centrosome approaches a corresponding half of a female centrosome; the plane in which they meet finally intersects at right angles the one in which they were first represented as lying (Fig. 141 *IV*). Here they fuse together to form the centrosomes of the first division figure. This concludes the process of fertilisation, since all further changes are connected with the division of the nucleus.

b. **Eggs of Ascaris megalocephala.** Further knowledge of the process of fertilisation may be gained from the study of the eggs of *Ascaris megalocephala*. Here the spermatozoon penetrates into the egg before the development of the pole-cells (*cf.* Fig. 127, and the text on p. 237), arriving finally at the centre (Fig. 142 *I*); meanwhile the germinal vesicle, after changing itself, in the manner already described, into the pole spindle, mounts up to the surface of the yolk, and gives rise to several pole cells. Two vesicular nuclei develop, one derived from the nuclear substance of the spermatozoon, which has entered, and the other from one half of the second polar spindle (Fig. 142 *I*). Egg-nucleus and sperm-nucleus (Fig. 142 *II*) then approach each other; in this case, however, the male nucleus is in the centre, whilst the female

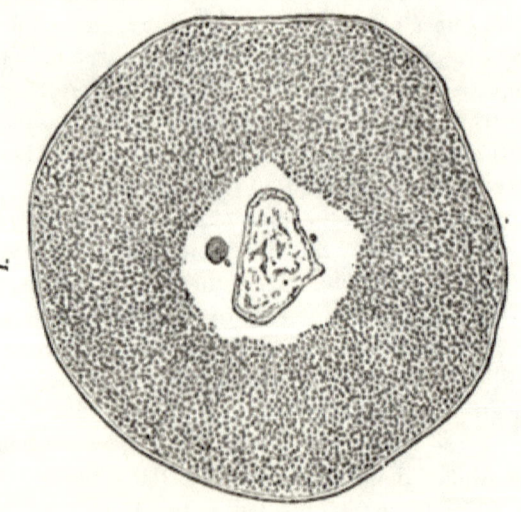

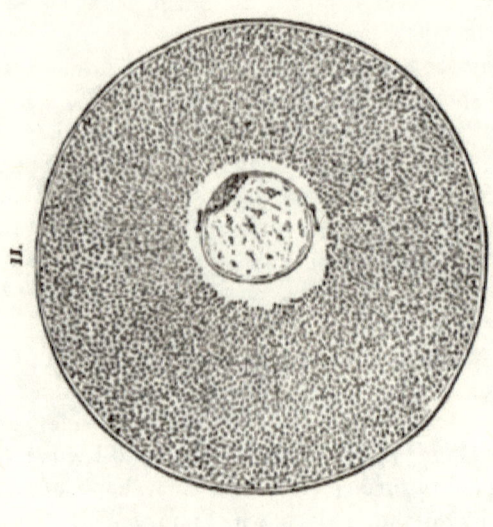

THE VITAL PROPERTIES OF THE CELL 261

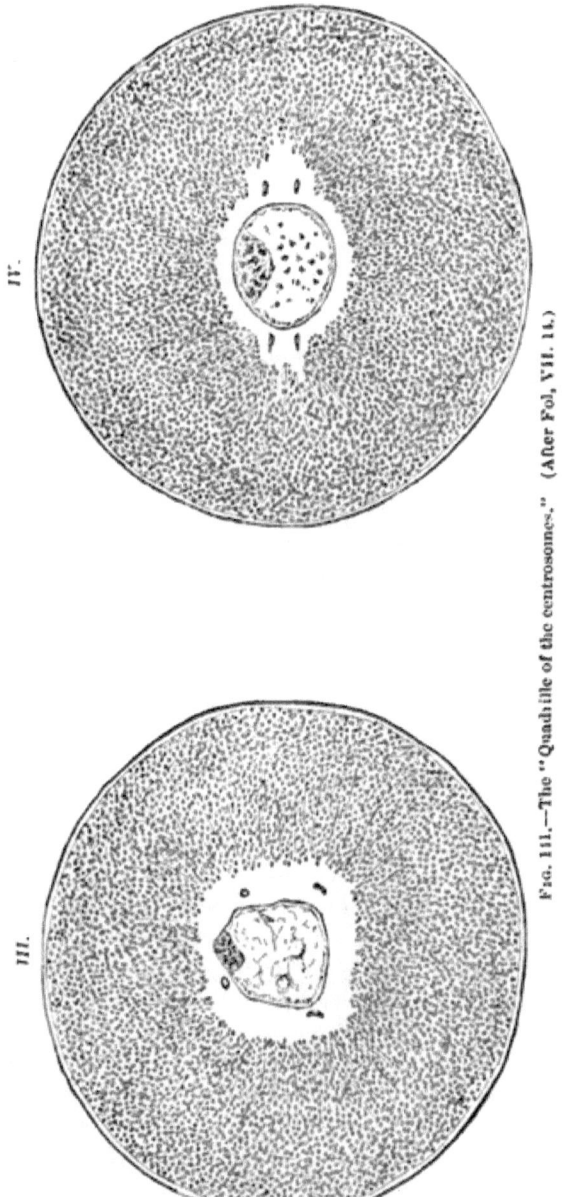

Fig. 111.—The "Quadrille of the centrosomes." (After Fol, VII. 14.)

one makes its way in from the surface, whereas just the reverse occurs in Echinoderm eggs; further, both nuclei are approximately of the same size, and lie close together, although for a time they do not coalesce, but pass through a period of rest. Indeed, even after they have begun to prepare for the formation of the first division spindle, they do not commence to fuse. In consequence of this, and of the further circumstance, that in *Ascaris megalocephala* during nuclear division there develop only a few nuclear segments, which are of considerable size, and hence are easy to count, van Beneden (VI. 4a, 4b) was able to supple-

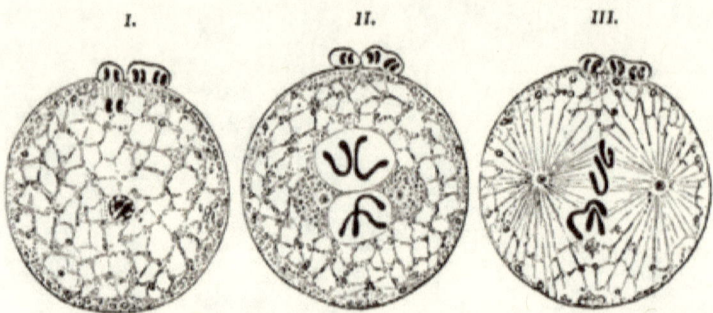

FIG. 142.—(I–III). Three diagrams depicting the course of the processes of fertilisation in *Ascaris megalocephala bivalens*.

ment our knowledge of the process of fertilisation by the following fundamental discovery:—

During the preparation for the first division spindle, the nuclein in the egg- and sperm-nuclei, whilst these are still separated from one another, becomes transformed into a delicate thread which spreads itself out in many coils in the nuclear space. Each thread then divides into two twisted loops of equal size, the nuclear segments (Fig. 142 *II*). On either side of the pair of nuclei a centrosome makes its appearance; however, up till now, no one has been so fortunate as to observe whence these are derived. The line of demarcation between the two nuclei and the surrounding yolk now disappears.

Between the two centrosomes (Fig. 142 *III*), which are surrounded by a radiation sphere, spindle fibrils develop; these are at first faint, but later on are distinctly visible; they arrange themselves about the four nuclear segments, which have been set free by the breaking up of the nuclear vesicles, so that they rest externally upon the middle of the spindle.

Thus in the egg of the round worm of the horse the union of the two sexual nuclei, which is the final stage of fertilisation, only occurs during the formation of the first division spindle, in which process they take an important part. The important principle enunciated by van Beneden is as follows: Half of the nuclear segments of the first division are derived from the egg-nucleus, and half from the sperm-nucleus, hence they may be distinguished as male and female. Now since in this case, as before in nuclear division, the four segments split longitudinally, and then separate, and move towards the two centrosomes, two groups of four daughter-loops are formed, of which two are of male and two of female origin. Each group then transforms itself into the resting nucleus of the daughter-cell. Thus it is indisputably proved, that each daughter-nucleus in each half of the egg produced by the first division process contains two equal quantities of nuclein, one of which is derived from the egg-nucleus, and the other from the sperm-nucleus.

2. The Fertilisation of Phanerogamia.

The discoveries which have been made concerning the processes of fertilisation in Phanerogamia correspond most completely with those which have been observed in the animal kingdom. Strasburger (VII. 38) and Guignard (VII. 15) stand in the first rank of investigators. The most suitable objects for examination are the Liliaceæ, especially *Lilium martagon* and *Fritillaria imperialis*. One of the cells, into which the pollen grain divides in Phanerogams, corresponds to the spermatozoon, whilst the vegetable egg-cell, which in the ovule is enclosed in the ovary of the gynæcium, forms the most important portion of the embryo-sac, and corresponds to the animal egg.

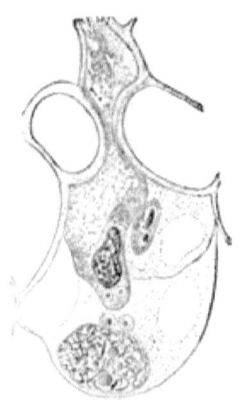

Fig. 143.—Section through the embryo-sac of *Lilium martagon* (after Guignard XV., Fig. 75). At the end of the pollen-tube, whose weakened wall is allowing its contents to escape, the sperm-nucleus may be seen with its two centrosomes. The egg-nucleus is also provided with two centrosomes. On the right, at the end of the pollen tube, a synergida may be distinguished which has commenced to disintegrate.

When the pollen grain has reached the stigma of the style, its contents commence to emerge through a weakened portion of the membrane, and to develop into a long tube (Fig. 143), which penetrates into the style until it reaches an embryo-sac. Here it presses between the two synergidæ right into the egg-cell. The pollen grain and the pollen tube contain two nuclei, the vegetative one, which takes no part in fertilisation, and the sperm-nucleus. This latter comes to lie at the end of the pollen tube, after this has made its way to the egg-cell; thence it passes through the weakened cellulose wall into the protoplasm of the egg, whilst two centrosomes advance in front of it; these latter were discovered by the French investigator, Gnignard (Fig. 143). It soon meets the egg-nucleus, which is somewhat larger, and on whose surface also a pair of centrosomes may be distinguished.

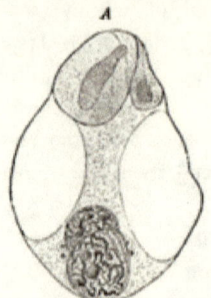 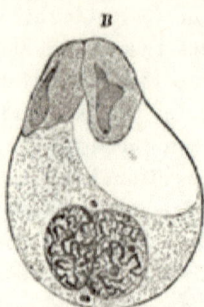

FIG. 144.—Egg from *Lilium martagon* (after Guignard XVI., Figs. 80 and 81): *A* a short time after the union of the egg- and sperm-nuclei; *B* a later stage. The fusing of the centrosomes is nearly completed.

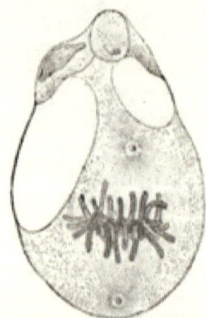

FIG. 145. — Egg - cell from the embryo-sac of *Lilium martagon*, with its nucleus undergoing division. The nuclear plate consists of twenty-four nuclear segments. (After Guignard XVI. Fig. 83.)

The two nuclei (Fig. 144) then coalesce, as do also the four centrosomes; these latter unite so as to form two new pairs, of which each is composed of one element of male and one of female origin. The new pairs are situated on opposite sides of the cleavage nucleus, and there develop into the two centrosomes of the first nuclear spindle (Fig. 145).

In the same way as in animal sexual cells, the nuclein and the number of nuclear segments derived from it are decreased during the formation of the pollen-cell and of the egg-cell to one half of the quantity present in a normal nucleus. For instance, whilst in *Lilium martagon* the normal nucleus develops during its division 24 nuclear segments which split up into 48 daughter-segments, in the nuclei of egg- and sperm-cells there are but 12. It is only when the two nuclei unite that they form a complete nucleus, from which arises the first division spindle with its 24 mother-segments, 12 being of male and 12 of female origin.

As concerns the centrosomes, a slight difference is shown by Echinoderms and Phanerogams. In the former, the centrosome at the beginning is single in both egg- and sperm-nuclei, and only becomes doubled through division; in the latter, on the other hand, two centrosomes are seen at a very early period both in the pollen-tube and in the egg-cell.

If we compare the results mentioned on the preceding pages (256-264), we may lay down the following fundamental laws referring to the process of fertilisation as it occurs in animals and phanerogamous plants:—

During fertilisation morphological processes, plainly to be demonstrated, occur. The most important and essential of these is the coalescence of the two nuclei which are derived from different sexual cells, that is to say, the coalescence of the egg- and the sperm-nuclei.

During the act of fertilisation two important processes of coalescence occur:—

1. Equivalent quantities of male and female stainable nuclear substance (nuclein) unite together.

2. Each of the halves obtained by the division of a male centrosome unites with a corresponding half of a female centrosome, by means of which the two centrosomes of the first nuclear division figure are obtained.

In the male and female alike, the stainable nuclear substance has been reduced to one half of the normal quantity, both as regards mass and the number of nuclear segments which it contains. Hence it is only after they have fused together that the full amount of substance and the complete number of segments contained by a normal nucleus are again present.

3. **The Fertilisation of Infusoria.** Certain Infusoria afford us especially important objects for the investigation of the subject of fertilisation. The sexual processes occurring in them were discovered by Balbiani and Bütschli (VII. 6), who were pioneers in this work, and they have lately been rendered much clearer by the classical labours of Maupas (VII. 30) and of Richard Hertwig (VII. 21).

Infusoria, as it is well known, differ from other lower organisms in one very interesting peculiarity, namely, that their nuclear apparatus has split up into two kinds of nuclei, which differ physiologically, *i.e.* into the chief nucleus (macro-nucleus) (Fig. 146 *k*), and into one or more sub-nuclei or sexual nuclei (*n, k*) (micro-nuclei). If plenty of nourishment be present, the Infusoria, which may be cultivated for observation in a small drop of water, multiply by means of the usual transverse division (Fig.

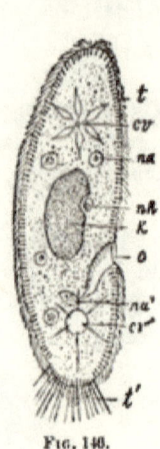

Fig. 146.

Fig. 147.

Fig. 146.—*Paramœcium caudatum* (semi-diagrammatic) (R. Hertwig, Zool., Fig. 130) : *k* nucleus ; *nk* paranucleus ; *o* mouth aperture (cytostom) ; *na'* food vacuole during process of formation ; *na* food vacuole ; *cv* contractile vacuole in contracted condition ; *cv'* contractile vacuole in extended condition ; *t* trichocysts ; *t'* the same extended.

Fig. 147.—*Paramœcium aurelia*, undergoing process of division. Fig. 2 shows how at an earlier stage the cytostom of the lower animal is formed by means of constriction from the upper one (R. Hertwig, Zool., Fig. 140) : *k*, *nk*, *o*, nucleus, paranucleus, and mouth aperture of upper portion ; *k'*, *nk'*, *o'*, nucleus, paranucleus, and mouth aperture of lower portion.

147), when the macro- and micro-nuclei extend themselves simultaneously in a longitudinal direction and divide.

This asexual multiplication is so energetic under favourable conditions that a single individual may, during the period of six days, divide thirteen times, and thus produce about 7,000 or 8,000 descendants.

However, it has been shown, especially by the culture experiments of Maupas and Richard Hertwig, that an Infusorian is unable to maintain the species for any length of time, and to continue to multiply by simple division, even if nourishment be supplied to it. The individuals undergo changes with regard to the nuclear apparatus ; they may even completely lose it, when they no longer divide, but die, as a result of the changes induced by age, or, as Maupas has expressed it, of *senile degeneration*. In order to maintain the species, it seems to be absolutely necessary that after definite periods two individuals should unite together in a sexual act. In cultures such acts occur simultaneously throughout the colony, so that a conjugation epidemic may be said to occur occasionally.

During an epidemic, which lasts for several days, the observer sees hardly any isolated Infusoria in the culture glass, for they are nearly all joined together in pairs. Maupas states that conjugation occurs in *Leucophrys patula* in the 300th generation, in *Onychodromus* in the 140th, and in *Stylonichia* in the 120th generation. By a diminution of the amount of nourishment, the onset of an epidemic may be hastened; by an increase it may be postponed, or even permanently prevented, in which case the individuals perish from senile degeneration.

If, after these preliminary remarks, we examine more closely the process of fertilisation, we find that, during a period of several days, the following peculiar and interesting changes take place in the couples of Infusoria. We will take as the basis of our description the *Paramœcium caudatum*, for, since it possesses but one nucleus and one single paranucleus, it presents simpler conditions than those seen in most other species (Fig. 148).

When the inclination for conjugation arises, " two paramœcia come close together; at first only their anterior ends touch, but later on their whole ventral surfaces are in contact, their mouth openings being opposite to one another " (Fig. 148 *I, o*). An irregular thickening develops over a small area in the neighbourhood of these latter, if conjugation lasts for any considerable period. Meanwhile the nuclear apparatus, including both the chief nucleus and the paranucleus, has undergone fundamental changes.

The chief nucleus becomes somewhat enlarged, its surface being at first covered with protuberances and depressions (Fig. 148 *II–IV, k*); these protuberances extend themselves into longer processes, which later on become separated off, and then gradually split up into still smaller pieces (*V, VI, k*). Thus the whole chief nucleus becomes broken up into several small segments, which distribute themselves all over the body of the Infusorian (*VII*), and finally become dissolved and absorbed. In a word, the main nucleus, having played its part, becomes completely disintegrated, during and after conjugation.

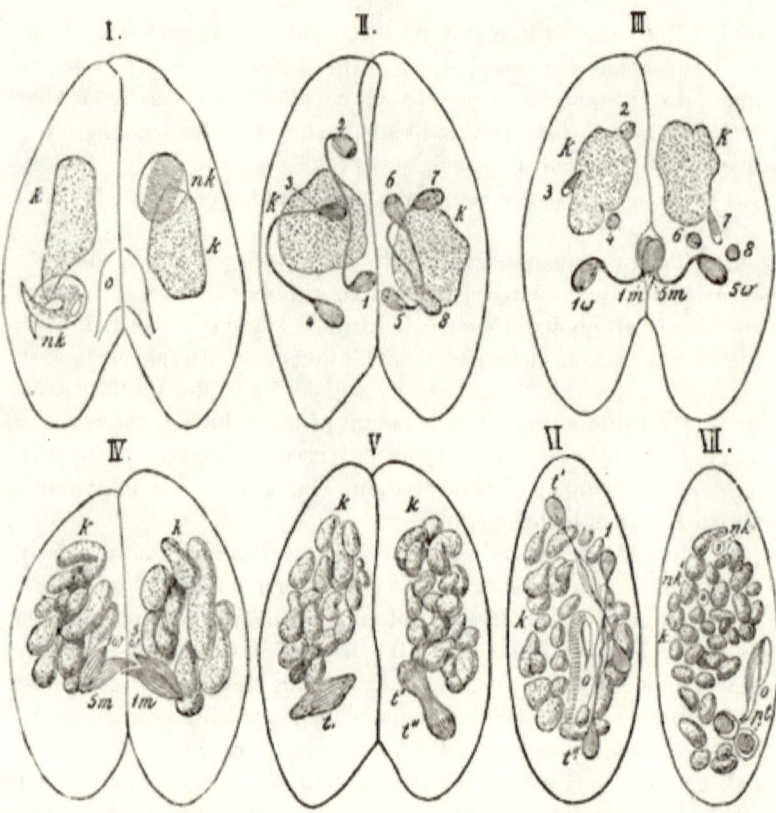

Fig. 148.—Conjugation of *Paramœcium* (R. Hertwig, Zool., Fig. 141): nk paranuclei; k nuclei of conjugating animals. *I* The paranucleus transforms itself into a spindle; in left-hand animal the sickle-stage, in right-hand animal the spindle-stage, are represented. *II* Second division of paranucleus into chief spindle (marked 1 in left, and 5 in right) and subsidiary spindles (2, 3, 4 in left, and 6, 7, 8 in right). *III* Subsidiary spindles show degeneration (2, 3, 4 in left, 6, 7, 8 in right), the chief spindles divide into male and female spindles (1 into 1 m and 1 w in left, and 5 into 5 m and 5 w in right). *IV* Transmigration of male spindles nearly completed (fertilisation). One end remains in the mother animal, whilst the other has united itself with the female spindle of the other animal (1 m with 5 w, and 5 m with 1 w). The main chief nucleus has become converted into segments. *V* The primary division spindle resulting from the union of the male and female spindles divides into secondary division spindles t' and t". *VI, VII* After the termination of conjugation. The secondary division spindles separate from one another, and come to lie amongst the rudiments of the new paranucleus (nk'), and of the new chief nucleus (pt, placentæ). The degenerated original nucleus commences to disintegrate. Since the *Paramœcium caudatum* has been selected to demonstrate the initial stages, and *P. aurelia* the final stages, *I-III* represent the former, and *IV-VII* the latter. The difference between the two consists in this, that *P. caudatum* has only one paranucleus, whilst *P. aurelia* has two, and also that in the latter, nuclear disintegration commences even in the first stage (stage 1).

During the retrogressive metamorphosis of the chief nucleus, the small paranucleus undergoes most important changes, which always recur in the same manner, and which may be compared to the phenomena of maturation and fertilisation seen in animal eggs. It enlarges itself by taking up fluid from the protoplasm, its contents assume a filiform appearance, until finally it transforms itself into a little spindle (Fig. 148 *I*, *nk*). This spindle divides into two parts, which soon develop into two new spindles; these in their turn become constricted and divide into two, so that finally four spindles, which have developed out of the paranucleus, are present in the neighbourhood of the main nucleus, which is undergoing transformation (Fig. 148 *II*, 1-4, 5-8).

During the further course of development, three of these four paranuclear spindles disintegrate (*III*, 2, 3, 4, 6, 7, 8). They become transformed into globules, which finally cannot be distinguished from the segments of the chief nucleus, whose fate they share. They strikingly recall the formation of the pole cells during the maturing of animal eggs, and in consequence have been compared to them by many investigators.

The fourth or chief spindle alone persists (*II*, 1 and 5); it takes part in the process of fertilisation, and serves as the foundation for the new formation of the whole nuclear apparatus in the body of the Infusorian. Which of these four spindles, derived from the original paranucleus, eventually becomes the chief spindle, depends, according to Maupas, solely and entirely upon its position. They are all four precisely alike as regards structure. The one which happens to be nearest to the above-mentioned zone of irregular thickening becomes the chief spindle (*II*, 1 and 5). Here it places itself at right angles to the surface of the body, extends itself longitudinally, and again divides into two (*III*, 1*w*, 1*m*; 5*w*, 5*m*).

Each of the halves contains apparently only about half as many spindle fibrils, and half as many chromatic elements as one of the earlier spindles. According to the observations made by Richard Hertwig, during the division of the chief spindle the number of spindle fibrils has been reduced to one half, a process similar to that occurring in the nuclei of animal and plant sexual cells. Hence these very characteristic nuclei play the same part as those of ova and spermatozoa, and may be distinguished as male and female, or as migratory and stationary nuclei.

Further, which of the two nuclei is to be migratory and which stationary cannot be foretold from their structure, for it depends solely and entirely upon their position and their consequent rôle during the process of fertilisation. Thus the portions which are situated nearest to the zone of thickening become the migratory nuclei (III, $1m$, $5m$); the two conjugating bodies exchange these migratory nuclei; these pass each other across the protoplasmic bridge, which has been formed for this purpose. During this exchange, the male migratory nuclei possess the structure of spindles (IV, $5m$, $1m$). After the exchange has been completed, each male nucleus coalesces with a stationary or female nucleus, which is also in the form of a spindle (IV, $1w$, $5w$), so that now each animal possesses only one spindle—the division spindle (vt)—if we disregard the segments of the chief nucleus, and the paranucleus, which are gradually undergoing disintegration.

The similarity to the process of fertilisation, as it occurs in Phanerogamia and animals, is striking. In Paramæcia, the stationary and migratory nuclei unite to form a division spindle, just as in plants and animals the egg- and sperm-nuclei unite to form the germinal nucleus. The division spindle serves to replace the old nuclear apparatus, which is becoming dissolved. It increases considerably in size (Fig. 148 V, t). The chromatin elements inside it arrange themselves into a plate; they then divide and move apart towards opposite ends of the spindle, almost up to the poles, thus forming the daughter-plates (V, right t' t''). The two halves remain united for a considerable time by a connecting strand. They then develop in a roundabout fashion into chief nucleus and paranucleus; in *Paramæcium aurelia* (Fig. 148 VI) for example, the daughter-spindles (t', t''), which have been formed out of the primary division spindles, re-divide, and so produce four spindles (VI), two of which develop into paranuclei (nk', nk'), whilst the other two coalesce to form the chief nucleus (pt). Thus, in Infusoria, "fertilisation brings about a complete re-organisation of the nuclear apparatus, and at the same time of the Infusorian" (Richard Hertwig).

Sooner or later, after the exchange of migratory nuclei, the two individuals separate from one another (Fig. 148 VI, VII). A longer period is necessary for the reabsorption of the useless portions of the nucleus, and for their replacement by new formations. The individuals, which have thus become rejuvenated, have regained the capacity of multiplying enormously by means of division, until again the necessity for a new "conjugation epidemic" arises.

The conjugation period at the same time causes a somewhat lengthy cessation of multiplication in the life of the Infusorian, as Maupas, for instance, has plainly shown in the case of *Onychrodromus grandis*, where, if the temperature is kept at from 17° to 18°, an interval of six and a half days occurs between the commencement of conjugation and the first subsequent division. During this period, if conjugation is not taking place, a single individual, when provided with sufficient nourishment, divides thirteen times; that is to say, it produces from 7,000 to 8,000 descendants.

In most Infusoria, as in the cases described here, both conjugating individuals behave in the same way, each functioning towards the other as male and female, that is to say, both imparting and receiving. Fixed forms of Infusoria, however, such as *Vorticellæ*, etc., behave in an interesting and somewhat different fashion.

The *Epistylis umbellaria* (Fig. 149) may serve as an example. When a conjugation period is approaching, several individuals of the colony of Vorticellæ divide rapidly and repeatedly, thus producing a generation of individuals (*r*) very inferior in size to the mother organism. Other individuals of the colony remain undivided and of normal size. The former are called microgametes, and the latter macrogametes; they differ from one another sexually.

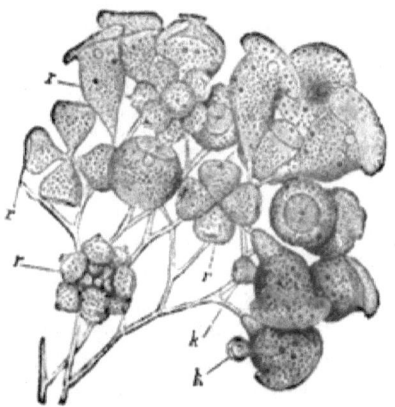

FIG. 149.—*Epistylis umbellaria* (after Graeff; from R. Hertwig, Fig. 142); portion of a colony in the act of conjugation; *r* microzoids produced by division; *k* microgametes in conjugation with macrogametes.

Each microgamete detaches itself from its stalk, swims round in the water, and after a short time attaches itself to a macrogamete in order to conjugate with it (Fig. 149 *k*). Changes occur in the nuclear apparatus similar to those described in detail above in the *Paramœcium*, and migratory nuclei are exchanged here also. However, the macrogamete alone continues to develop, the migratory and stationary nuclei of the primary division spindle coalescing, whilst the corresponding structures in the microgamete are, as it were, paralysed, and, instead of fusing and developing further, degenerate and become dissolved, like the fragments of the chief nucleus and the subsidiary spindles.

In this manner the microgamete loses its independence and individuality, and becomes gradually absorbed into the macrogamete, increasing the size of the latter.

Thus, in consequence of the stationary mode of life of *Vorticella*, a peculiar sexual dimorphism has developed, resulting in the absorption of the smaller of the conjugating individuals, after it has functioned to a certain extent as a male element in fertilising the macrogamete. However, the resemblance to ova and spermatozoa is not complete, although both in *Vorticella* and *Paramæcium* fertilisation commences with the interchange of nuclear material, and only results later on in the formation of a single effective individual.

4. **The various forms of sexual cells; equivalence of participating substances during the act of fertilisation; conception of male and female sexual cells.** Having shown in various instances, that the course of the process of fertilisation, and especially the behaviour of the nucleus during the process, is essentially uniform in animals, plants, and Protozoa, we will now proceed to state more clearly a difference which can be perceived in the cells participating in the act of fertilisation in most organisms, and to point out the importance of this difference. It consists in the unequal size and form of male and female germinal cells. The larger, stationary, and hence receptive cell, is called the female; the male cell, on the contrary, is much smaller, often extremely minute; it is either motile, approaching the egg-cell actively by amœboid movements or by means of flagella, or so small that it is conveyed passively through the air or water to the egg-cell.

What is the importance of this difference? Is it an essential product of the process of fertilisation, or is it brought about by causes of a subsidiary and secondary nature, due to incidental and secondary causes? It is of the greatest importance, in order to decide this question, to determine in what substance and in what portion of the two sexual cells this variation manifests itself.

Each cell consists of protoplasm and nuclear substance. Of these the amount of protoplasm present in the sexual cells may vary considerably, as may be immediately recognised by their appearance; the spermatozoon often contains less than $\frac{1}{100000}$ of the protoplasm present in the ovum. Thus, according to Thuret's computation, the ovum of *Fucus* is as large as from 30,000 to 60,000 antherozoids. In animal sexual cells, the difference is usually still greater, especially when the egg-cells are copiously laden with reserve materials, such as fat-globules, yolk-granules, etc. Indeed, in typically developed spermatozoa the presence of protoplasm at all may be doubted; for the tail, which is attached to the middle portion, consists of contractile substance, which, like muscle fibres, is a differentiation product of the protoplasm of the sperm-cell. In immature spermatozoa, protoplasm is present in the form of drops of various sizes, which, having served their purpose during development, eventually disappear.

Nuclear substance behaves in quite a different way. However much the ovum and spermatozoon may vary as to size, they still invariably contain equal quantities of active nuclear substance. The truth of the above statement cannot be proved by a simple comparison of the two sexual cells, but if the course of the process of fertilisation and of the development of the mature ovum and sperm-cell be watched, it will be seen that they both contain an equal quantity of nuclein, and that during the process of maturation they develop an equal number of nuclear segments. For example, the sperm-nucleus of *Ascaris megalocephala bivalens* consists, like the egg-nucleus, of two nuclear segments of the mother cell; each during fertilisation contributes similar elements, which are utilised in the formation of the germinal nucleus (Fig. 142 *II*). In the same way each nucleus contributes the same amount of polar substance, the male and female centrosome both of which, in the manner described on p. 262, take part in the process of fertilisation (Fig. 141).

In opposition to these conclusions, it might be stated, that the nuclear portions of both egg and sperm-cells before their union are usually very different in appearance, and vary more or less in size. This, however, is easily explained by the fact, that the passive fluid substances may be mixed in greater or less quantities with the active nuclear substance. The minute head of the spermatozoon consists of fairly compact, and hence strongly stainable, nuclein. In the egg-nucleus, which is much larger, the same amount of nuclein is saturated with a quantity of nuclear sap, throughout which it is distributed in the form of minute granules and threads, the result being that the egg-nucleus as a whole is

T

less dense and does not become so strongly stained as the head of the spermatozoon.

This difference in size and consistency soon disappears during the course of the process of internal fertilisation; for the sperm-nucleus, which was at first small, whilst on its way to the egg-nucleus, soon swells up to the same size as the latter by absorbing fluid out of the yolk (Fig. 142 *II*), as is seen in the eggs of most Worms, Molluscs, and Vertebrates. It is true that in isolated cases, as in the eggs of the Sea-urchin (Fig. 141), the nuclei are of different sizes, when they unite; under these circumstances the sperm-nucleus has taken up a smaller quantity of sap than usual, and is consequently somewhat denser in consistency; so that, in spite of the difference in size, we may still assume that an equal amount of solid active constituents is present in both.

It may be demonstrated in suitable objects, that the relative size of egg and sperm-nuclei depends chiefly upon the time at which the egg-cell was fertilised, whether before, during, or after the formation of the polar cells. For instance, if spermatozoa be brought into contact with an egg of *Asteracanthion* whilst the polar cells are developing, the sperm-nucleus must remain for a considerable time in the yolk before fusion commences, and in consequence it swells up during this period by absorbing nuclear sap, until it is of the same size as the egg-nucleus, which develops after the second polar cell has separated off. On the other hand, if fertilisation occurs after the egg-cell is provided with both the polar cells and the egg-nucleus, the sperm-nucleus remains for only a few minutes as an independent body in the yolk, commencing almost immediately after its entrance to fuse with the egg-nucleus. Under these circumstances it keeps small in size, for it is not able to saturate itself in the same way with nuclear sap.

Thus we may consider the following important law as proved, *i.e.* that the two sexual cells, in spite of the fact that frequently they vary considerably in appearance and contain such unequal quantities of protoplasm, contribute equal amounts of nuclear substance (nuclein, in a definite number of nuclear segments, paranuclein, in the ovocentrum and spermcentrum) during the process of fertilisation, and in so far are equivalent.

From this law I deduce the following: the nuclear substances which are derived in equal quantities from two different individuals are invariably the only active substances, upon whose union the act of fertilisation depends; they are the true fertilisation substances. All other substances (protoplasm, yolk, nuclear sap, etc.) are not concerned in fertilisation as such.

This proposition is supported by two important facts:—

Firstly, the complicated processes of preparation and maturation which the sexual cells must undergo. As follows from the statements given on pp. 235–239, the chief result of these processes is not that the nuclear substances are increased through fertilisation, but that they remain constant in amount for the species of plant or animal in question.

Secondly, the phenomena of fertilisation seen in Infusoria. Here, as Maupas and Richard Hertwig both assert, similar individuals remain in contact for a sufficient period in order to exchange halves of equal nuclei. When this exchange of migratory nuclei has been effected, the process of fertilisation is completed, and the two animals separate. Hence it is evident, that the ultimate result of the complicated processes consists in this, that after the fusion of the migratory and stationary nuclei the nucleus in each fertilised individual is composed of nuclear substance derived from two different sources.

If the important substance of fertilisation is contained in the nucleus, the question arises whether the nuclear substance of the spermatozoon differs from that of the egg-cell. This question has been answered in very different ways. Formerly it was generally considered, as Sachs expressed it, that the male element introduced into the ovum a substance which it did not contain before. One view especially has obtained many adherents; it may be described as the doctrine of the hermaphroditism of nuclei and the theory of restitution.

Many investigators consider that the cells possess hermaphrodite nuclei, that is to say, nuclei with both male and female properties. For instance, according to van Beneden's hypothesis, which has been the most clearly worked out, immature egg and sperm-cells are hermaphrodite; they only gain their sexual character after the egg-cell has lost its male, and the sperm-cell its female constituents of their normal hermaphrodite nuclear apparatus. The male nuclear constituents are expelled from the egg in the nuclear segments of the polar cells. The reverse process occurs in a similar manner with sperm-cells. Thus the egg and sperm-nuclei, being halved, become pronuclei, and possess opposite sexual characteristics

Regarded from this point of view, fertilisation consists essentially in the replacement of the male elements, which have been expelled from the egg, by an equal number of similar elements, which are introduced by the spermatozoon.

More careful investigation shows that these theories are not tenable. For the empirical foundation, upon which they were based, is destroyed by the fact which was proved on p. 237, namely that the polar cells are morphologically nothing but egg-cells, which have become rudimentary. This follows from a comparison of the development of egg and sperm-cells in Nematodes. Hence the nuclear segments, expelled from the egg in the polar cells, cannot be the discharged male constituents of the germinal vesicle, as is stated in the restitution theory.

Apart from this, we are unable, with the methods of investigation at our command, to discover the least difference between the nuclear substances of the male and female cells. Nuclein and centrosomic substance are identical, both as regards quantity and composition. There is no specific male or female fertilising material. The nuclear substances, which come into contact with one another during the process of fertilisation, differ only in this, that they are derived from two different individuals.

Now, if, in consequence of this, it can no longer be allowed that the egg and sperm-nuclei are sexually opposed in the way understood by the supporters of the restitution theory, what meaning must be attached to the terms male and female sexual cells or male and female nuclei?

These terms do not really touch the essential part of fertilisation, and do not express an opposition based upon fundamental processes of reproduction; they refer rather to secondary differences of minor importance which have developed between the conjugating individuals, between the sexual cells and their nuclei, and which must be classed as secondary characteristics. Hence we will state at once that the formation of two separate sexes is not the cause of sexual generation, as might be concluded from a superficial investigation, but that the reverse is really true. All sexual differences, if we trace them back to their sources, have arisen because the union of two individuals of one species, which originally were similar, and hence sexless, is advantageous to the maintenance of the vital processes; without exception, these differences only serve one purpose, namely to facilitate the combination of two cells. On this account solely have the cells developed the differences which are termed male and female.

The theory built up by Weismann, Strasburger, Maupas, Richard Hertwig, and myself may be worked out more in detail in the following manner. During fertilisation two circumstances must be considered, which work together and yet are opposed to one another. In the first place, it is necessary for the nuclear substances of the two cells to become mixed; hence the cells must be able to find one another and to unite. Secondly, fertilisation affords the starting point for a new process of development and a new cycle of cell divisions; hence it is equally important that there should be present, quite from the beginning, a sufficient quantity of developmental substance, in order to avoid wasting time in procuring it by means of the ordinary processes of nutrition.

In order to satisfy the first of these conditions, the cells must be motile, and hence active; in order to satisfy the second, they must collect these substances, and hence increase in size, and this of necessity interferes with their motility. Hence one of these causes tends to render the cells motile and active, and the other to make them non-motile and passive. Nature has solved the difficulty by dividing these properties—which cannot of necessity be united in one body, since they are opposed to one another—between the two cells which are to join in the act of fertilisation, according to the principle of division of labour. She has made one cell active and fertilising, that is to say male, and the other passive and fertilisable, or female. The female cell or egg is told off to supply the substances which are necessary for the nourishment and increase of the cell protoplasm during the rapid course of the processes of development. Hence, whilst developing in the ovary, it has stored up yolk material, and in consequence has become large and non-motile. Upon the male cell, on the other hand, the second task has devolved, namely of effecting a union with the resting egg-cell. Hence it has transformed itself into a contractile spermatozoon, in order to be able to move freely, and, to as large an extent as possible, has got rid of all substances, such as yolk material or even protoplasm itself, which would tend to interfere with this main purpose. In addition it has assumed a shape which is most suitable for penetrating through the membrane which protects the egg, and for boring its way through the yolk.

We may transfer the terms male and female from the cell elements, which are thus differentiated sexually, to the nuclei which they contain, even when these are equal both as regards mass and composition. Only we must understand by the expression male or female nucleus nothing more than a nucleus derived from a male or female cell. In the same way, in Infusoria, the migratory nucleus may be termed male and the stationary nucleus female, in the sense of the above definition, since the former seeks the latter.

This difference, which has developed in sexual cells for the purpose of division of labour, and to fit them for their special work, is repeated in the whole organic kingdom, whenever the individuals in which the male and female sexual cells develop differ from one another in sexual characteristics. In all the arrangements referring to sex, one and the same object is aimed at: measures are taken on the one hand to facilitate the meeting of the sexual cells, and on the other to arrange for the nourishing and protection of the egg. The one organisation we call male, and the other female. All these relationships are secondary, and have nothing to do with the process of fertilisation itself, which is a true cell phenomenon.

Fertilisation is an union of two cells, and, above all, a fusing of two equivalent similar nuclear substances, which are derived from two cells, but it is not a combination of sexual opposites, for the differences depend solely upon structures of subsidiary importance.

The truth of the above law may be still more clearly demonstrated, if we compare the generative processes throughout the whole organic kingdom, and thereby endeavour to determine how the differences have gradually developed between the cells which unite for the purpose of fertilisation. Amongst unicellular organisms and plants, we find innumerable instructive examples of the elementary and primitive forms of sexual generation and of the origin of sexual differences in the plant and animal kingdoms.

5. **Primitive and fundamental modes of sexual generation and the first appearance of sexual differences.** The study of the lowest organisms, such as *Noctilucæ, Diatomaceæ, Gregarinæ, Conjugatæ*, and other low *Algæ*, shows that in many of them the conjugation of two individuals occurs in regular cycles, and this we must regard as a process of fertilisation.

In *Noctiluca* conjugation commences by two individuals, which are of the same size, and do not differ from one another in any respect, placing themselves side by side, with their mouth apertures opposite one another, and beginning to fuse, whilst their cell membranes become dissolved. A connecting bridge, which continually grows broader, develops; after which the protoplasmic masses stream together from all sides, until the two individuals become transformed into a single large vesicle. The two nuclei, each accompanied by a centrosome, travel towards each other, and place themselves in contact, but, according to Ishikawa, do not fuse (VII. 25). After a time, the conjugating pair of *Noctilucæ* again divide into two cells, a partition membrane having developed between them. At the commencement of this division, the pair of nuclei, which have united together, become extended; they then become constricted in the middle, and divide into two, after which they separate again, the result being that each *Noctiluca* contains half of each nucleus. Thus the result of conjugation is the production again of two individuals, each of which possesses a nucleus of twofold origin. Fertilisation is followed sooner or later by active multiplication by means of budding and spore formation.

The *Conjugatæ* (VII. 11) are of especial importance in the study of primitive modes of fertilisation. This order is subdivided into three families: the *Desmidiaceæ*, the *Mesocarpeæ*, and the *Zygnemaceæ*.

Klebahn (VII. 27) has discovered the minute details of the process of fertilisation in two species of Desmidiaceæ: the *Closterium* and *Cosmarium*.

Two *Closterium* cells, which are shaped somewhat like bent sickles, lie lengthwise against each other, being kept in contact by a gelatinous secretion; each then develops a protuberance near its centre. The two protuberances come closely into contact and fuse, whilst the wall separating them dissolves, to form a conjugation canal common to both. Here all the protoplasm from both the conjugating *Closterium* cells gradually collects, and, detaching itself from the old cell membrane, fuses to form a single globular body, which finally becomes surrounded by a membrane of its own.

This *zygospore* or *zygote*, which has been produced by the fusion of two similar individuals, now passes through a resting stage, which lasts for several months (Fig. 150). It contains two nuclei, which were derived from the two cells, and which remain apart

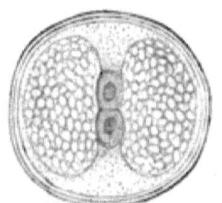

Fig. 150.—Zygote of *Closterium*, just before germination. (After Klebahn, Pl. XIII., Fig. 3.)

during the whole of the resting period. It is not until the spring, when a new vegetative period recommences, that the nuclei come close together, and fuse to form a germinal nucleus.

At this period the zygote, which is surrounded by a delicate membrane, makes its way through the old cellulose wall, whilst its germinal nucleus transforms itself into a large spindle, of somewhat unusual appearance (Fig. 151 *I*). This divides into two half-spindles (Fig. 151 *II*), which, however, do not enter into the resting condition, but immediately prepare to divide again

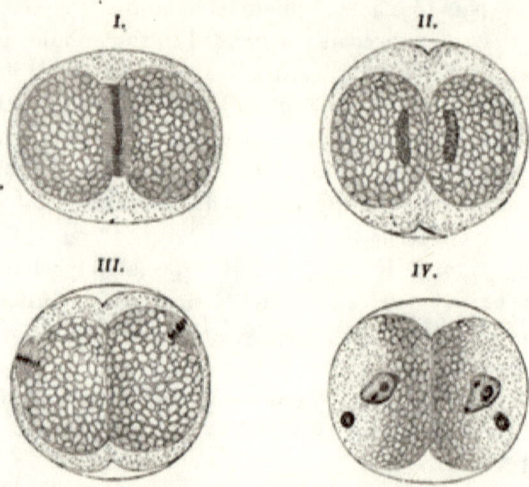

Fig. 151.—Various germinal stages of *Closterium*. (After Klebahn, Pl. XIII., Figs. 6b, 8, 9, 11, 13.)

(Fig. 151 *III*). Thus the germinal nucleus divides into four nuclei, by means of two divisions, the second of which succeeds the first without a pause (Fig. 151 *IV*).

Meanwhile the protoplasm of the zygote has divided into two hemispheres (Fig. 151 *IV*), each of which contains two nuclei, which have been produced by the division of one spindle. The two nuclei soon develop differences in appearance, the one (according to Klebahn, the large nucleus) becoming large and vesicular, whilst the other (the small nucleus) remains small, and finally quite disappears. The small nucleus becomes much more intensely stained than the large one. It seems to me that the former disintegrates and dissolves, just like the fragments of the chief nucleus and the subsidiary spindles in Infusoria. Before the process of dissolving is quite completed, the two halves of the zygote gradually assume the shape of a *Closterium* cell (Fig. 152).

What is the significance of this second division, which occurs immediately after the first, without any intermediate resting stage? It appears to me that by its means the same result is obtained, although in a different manner, as is produced by the division, with reduction, which occurs during the maturing of egg and sperm-cells. In both cases by means of the double division the nuclear substance is reduced to one half of that contained by a normal nucleus, and thus an increase of

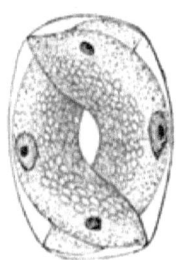

FIG. 152. — Two *Closteria*, which have developed from a zygospore, before they have escaped from their enclosing membrane.

nuclear substance is avoided when, in consequence of fertilisation, two nuclei coalesce. Similarly in *Desmidiaceæ* a reduction of nuclear substance occurs after fertilisation, and thus the double amount of nuclear substance, produced by the conjugation of two complete, fully developed nuclei, is reduced to a normal quantity. The germinal nucleus, instead of dividing into two daughter-nuclei, splits up in consequence of the two divisions, which follow immediately upon one another, into four granddaughter-nuclei. The protoplasmic body, however, is halved, each portion containing only one functional nucleus; the other two, being useless, disappear.

This supposition might be proved to be correct, if the nuclear segments were accurately counted at the various stages. One circumstance, which may be mentioned in its support, has frequently been observed by Klebahn, namely that in *Cosmarium* the four granddaughter-nuclei, which are derived from the germinal nucleus, are distributed unequally between the halves of the zygote, the one half containing one single active nucleus, and the other containing three, two of which degenerate. It does not matter whether the two degenerating nuclei fall to the share of one or both cells during division, since they behave like yolk contents.

In *Desmidiaceæ* we have observed conjugation as it occurs in isolated living cells; the *Zygnemaceæ* teach us its method of procedure in a colony of cells, where several individuals have joined together in rows to form long threads.

When, in the thick felt-like masses with which the Algæ cover the top of the water, two threads lie in contact with one another for any considerable portion of their length, conjugation occurs between neighbouring cells. As a rule all the cells prepare for reproduction at the same time by sending out lateral processes towards each other. These fuse at the point of contact, whilst the separating wall dissolves, and thus transverse canals are formed, which connect the conjugating threads at regular distances, and

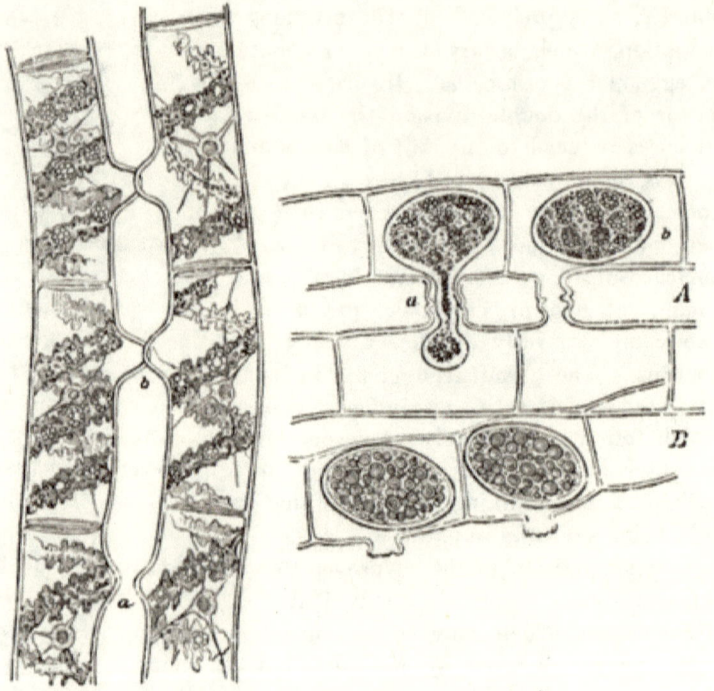

FIG. 153.—*Spirogyra longata* (after Sachs, Fig. 410). To the left, several cells of two filaments, which are about to conjugate: they show the spiral chlorophyll bands, in which crown-like arrangements of starch grains are lying, as well as small drops of oil. The nucleus of each cell is surrounded by protoplasm, from which threads stretch to the cell-wall, b, preparatory to conjugation. To the right, A, cells engaged in conjugation: the protoplasm of the one cell is just passing over into the other at a; in b the two protoplasmic masses have already united. In B, the young zygotes are surrounded by a wall.

resemble the rungs of a ladder (Fig. 153). The protoplasmic bodies of the cells then contract away from their cellulose wall, and after a time fuse together.

Differences which in themselves are trifling, but which on that very account are interesting, are seen in various species of *Zygnemaceæ*; they are worth noticing, for they teach us the way in which sexual differences may at first develop.

For instance, in *Monjeotia*, as in the *Desmidiaceæ*, the two protoplasmic bodies enter the conjugation canal and there fuse together to form a zygote, which becomes globular, expresses fluid, and surrounds itself with a membrane. In this case both cells behave exactly alike; neither can be termed male or female.

In other species, such as *Spirogyra* (Fig. 153), one cell remains passively in its membrane, and is sought out by the other, which in consequence may be called the male. It wanders into the conjugation canal, and, passing through it, reaches the female cell, as though attracted by it; they then fuse to form a zygote (Fig. 153 *A*, *a*). When the zygote is treated with reagents and staining solutions, it can be further established, that soon after the union of the cells their nuclei approach each other, and unite to form the germinal nucleus. Since in a thread all the cells act either as males or females, one of the two conjugating threads generally has all its cells empty, whilst the other contains a zygote in each cavity (Fig. 153 *B*). The zygote surrounds itself with a separate cell-wall, after which it generally rests until the next spring, when it commences to germinate, and finally, by means of transverse divisions, develops into a long *Spirogyra* thread.

The above-mentioned distinction between male and female *Spirogyra* threads by no means invariably occurs. For instance, it may happen that a thread bends back on itself, so that one end comes into the neighbourhood of the other. Under such conditions, cells situated at the opposite ends of the same thread conjugate together, so that those which under other circumstances would have functioned as male cells now play the part of female cells.

In the above-mentioned families of *Noctilucæ* and *Conjugatæ* and in others, such as *Diatomaceæ*, *Gregarinæ*, etc., the large protoplasmic bodies are enclosed in membranes; these pair, after having passed through periods of vegetative multiplication by simple division. A second series of primitive modes of sexual reproduction is afforded us by lower plant organisms, such as some of the *Algæ*. For purposes of reproduction they develop special cells, the swarm-spores, which are distinguished from the vegetative cells by their small size, by the absence of a cell membrane, and by the presence of two flagella or numerous cilia, by means of which they move about independently in the water. They are of especial interest, for they show us how, by means of gradual differentiation and division of labour in opposite directions, they have developed more highly differentiated forms, namely, typical eggs and typical antherozoids.

Swarm-spores are small, motile, naked cells, generally pear-shaped (Figs. 154, 155, 157, 158). The pointed end is anterior and goes in front, whilst the spore moves through the water; it consists of hyaline protoplasm, and frequently contains a red or brown pigment spot (the eye-spot); the remainder of the body is hyaline, or coloured green, red, or brown with colouring matter, according to the species; it contains one or two contractile vacuoles (Fig. 154). The swarm-spore moves along by means of flagella, which spring from the hyaline anterior portion; there are generally two flagella (Fig. 154), but sometimes there is only one; occasionally there are four or more (Fig. 14).

Fig. 154.—Swarm-spore of *Microgromia socialis*. (After R. Hertwig.)

The swarm-spores are derived at certain times from the contents of a mother-cell, either by means of repeated bipartitions, or by the splitting up of the mother-cell into several portions (pp. 232-234). When division into two occurs, the number of swarm-spores is small, being 2, 4, 8, or 16; when, however, many cells are produced, the number is very great, for in that case the mother-cell is of considerable size, and may produce as many as from 7,000 to 20,000 daughter-cells. When the wall of the mother-cell ruptures at one place, the broad end of the swarm-spore escapes first to the exterior.

There are two kinds of swarm-spores, which are developed at different times. The one kind multiply asexually, giving rise to young *Algæ*, whilst the others require fertilisation. The mother-cell, from which the former are derived, is termed by botanists the *sporangium*, that giving rise to the latter *gametangium*.

We will only consider sexual spores or gametes here. In many of the lower Algæ conjugating swarm-spores (Fig. 155 *a, b, c, d*) cannot be distinguished from one another in any respect, either as regards their sizes, mode of movement, or behaviour (*Ulothrix, Bryopsis, Botrydium, Acetabularia*, etc.). On the other hand, in other species sexual differences develop, which enable us to distinguish between male and female gametes. In the first case we speak of isogamous, and in the second oogamous fertilisation.

We may take either *Botrydium* or *Ulothrix* (Fig. 155) as an example of isogamous fertilisation. If minute swarm-spores from different sources are placed in a drop of water and examined with a high power of the microscope, some of them are seen to approach each other immediately, their hyaline anterior ends (*b*) coming into contact; and after a short time they commence to fuse together. At first they touch each other laterally (*c*), after which they grow together, the fusion commencing at their anterior ends and gradually extending backwards.

The couple (*d*) hurry about for some time in the water with an intermittent and staggering movement. After a short time the fusion is so far advanced that the two gametes form a single thick oval body, which, however, betrays its derivation from two individuals by containing two pigment spots and four flagella (*e*,*f*). The zygote now gradually slackens its movements, until finally it comes to rest; it then loses its four flagella, which are either drawn in or thrown off, becomes globular in shape, and surrounds itself with a cell-wall.

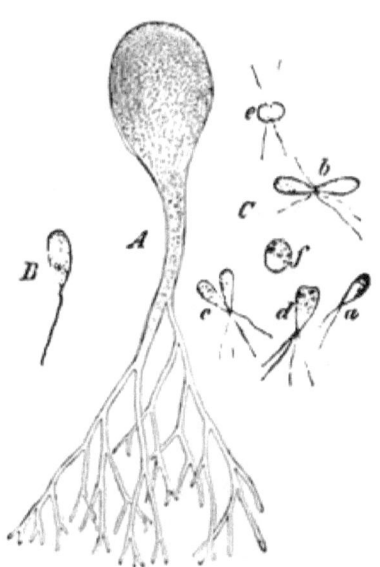

FIG. 155.—*Botrydium granulatum* (after Strasburger, Fig. 139): *A* free plant of medium size (×26); *B* swarm-spore, fixed with iodine solution (×540); *C* isogametes. *a* a single individual; *b* two isogametes which have just come into contact; *c*, *d*, and *e* the same lying side by side; *f* zygote, produced by the complete fusion of the gametes (×540).

Frequently the resting stage begins only a few minutes after the commencement of pairing; in other cases, however, the zygote may swim round in the water with its four flagella for three hours, in a naked condition, without a membrane, until finally it draws in its flagella, and sinks to the ground.

The gradual appearance of sexual differentiation can be followed still better in the very numerous species of lower Algæ, in which the fertilisation of gametes occurs.

As in *Spirogyra* (Fig. 153), one of the two individuals, which in other respects are absolutely similar, may be called female, since it remains at rest, and must be sought for by the other for the purposes of conjugation. Thus a relationship, similar to that seen in *Phæosporeæ* and *Cutleriaceæ*, is produced.

In some species of *Phæosporeæ*, the male and female swarm-spores cannot be distinguished from one another when they are evacuated from the mother-cell; they are of the same size, and are each provided with a pigment spot and two flagella; they do not pair whilst they are swimming about. However, a difference between the gametes soon becomes apparent. Some come to rest earlier than others; each of these attaches itself by the point of one of its flagella to some solid object, to which it draws up its protoplasmic body by shortening and contracting the connecting flagellum; it then retracts its second flagellum. These resting swarm-cells may be termed female; their capacity for becoming fertilised is only retained for a few minutes; they appear to exert, as Berthold expresses it, " a powerful attraction " upon the male gametes, which are swimming about in the water, so that in a few seconds one egg may be surrounded by hundreds of swarm-spores, one of which fuses with it (VII. 51).

Sexual differentiation is still more marked in *Cutleriaceæ*. Here the sexual swarm-cells become different in size before they are separated from the parent, the female one developing singly, and the male in groups of eight. In this genus the difference in size of the sexual cells is fairly striking. Both kinds of gametes swim about in the water for a time; fertilisation, however, can only occur after the female swarm-spore has come to rest, has drawn in its flagella, and has become spherical. Upon the egg, which is now capable of becoming fertilised, a hyaline spot appears, which was produced by the drawing in of the anterior beak-like end. This is the so-called reception spot. It is the only point at which one of the small male swarm-spores, which soon come to rest around the female cell, can fertilise it. When fertilisation is complete, the zygote surrounds itself with a cellulose cell-wall.

In *Fucaceæ*, *Characeæ*, and other *Algæ* the difference is still more marked than in *Cutleriaceæ*. Here the female cells, which attain a considerable size, do not even pass through the swarm-spore stage. They are either expelled to the exterior in a mature condition as globular immotile egg cells (*Fucaceæ*, Fig. 156 *G*), or they are fertilised at the place where they originated, that is, in the oogonium. The male cells, on the contrary (Fig. 156 F), are even smaller and more motile than those already described, and have assumed the characteristic properties of antherozoids;

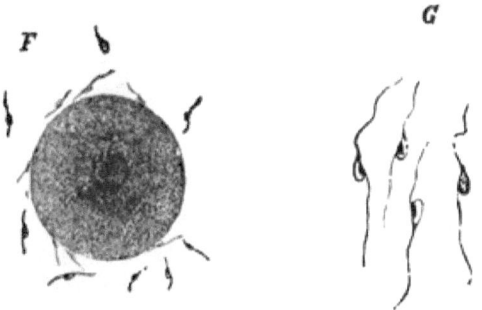

Fig. 156.—Spermatozoids of Fucus (× 540). Egg, with adhering spermatozoids. (After Strasburger, Fig. 87 G and F.)

they are composed almost entirely of nuclear substance, and are provided with two flagella, which function as organs of locomotion.

The view that eggs and spermatozoids of the higher *Algæ* are derived genetically from swarm-cells, which differentiate themselves sexually in opposite directions, and gradually assume a specific male and female form, is still more strongly supported by the phenomena observed in the little family of *Volvocineæ* than by comparing various species of *Algæ*.

This family is especially interesting and important in the consideration of the problem in question, since some of the various species, which in their whole appearance are extremely similar (*Pandorina morum*, *Eudorina elegans*, *Volvox globator*), exhibit marked differences in their sexual cells, whilst others show no difference at all, and in yet others an intermediate stage can be observed. The whole relationship is so clearly demonstrated that it is worth while to consider it more in detail.

Pandorina morum, which is especially well known—for as early as 1869 Pringsheim (VII. 35) discovered the pairing of its swarm-spores—forms small colonies of about sixteen cells, which are enclosed in a common gelatinous sheath (Fig. 157 *II*). Each cell bears two flagella on its anterior end; these stretch out beyond the surface of the gelatinous sheath, and are used for locomotion.

During sexual reproduction each of the sixteen cells splits up generally into eight portions, which after a time are set free, and

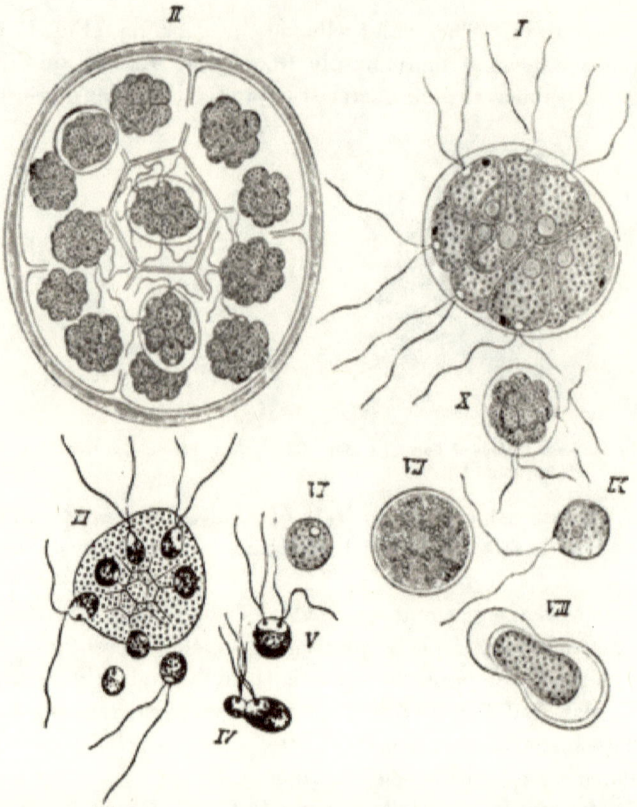

FIG. 157.—Development of *Pandorina morum* (after Pringsheim; from Sachs, Fig. 411): *I* a swarming family; *II* a similar family, divided into sixteen daughter-families; *III* a sexual family, the individual cells of which are escaping the gelatinous investment; *IV*, *V* conjugation of pairs of swarmers; *VI* a zygote, which has just been completed; *VII* a fully grown zygote; *VIII* transformation of the contents of a zygote into a large swarm-cell; *IX* the same after being set free; *X* a young family developed from the latter.

swim about independently (Fig. 157 *III, IV*). These swarm-cells, which are oval, and (with the exception of the anterior, somewhat pointed, hyaline end) are green in colour, possess a red pigment spot and two flagella; they are somewhat unequal in size. However, in this respect a marked sexual differentiation is not apparent in *Pandorina*. For when swarm-cells from two different colonies approach each other, it is seen amongst the crowd that sometimes two small ones, sometimes two large ones, and sometimes one large and one small unite together (Fig. 157 *IV, V*).

When two swarm-spores meet, they first touch each other with their points (*IV*), and then fuse together to form a biscuit-shaped body, which gradually draws itself up into a ball (*VI, VII, X*). This surrounds itself, a few minutes after fertilisation, with a cellulose cell-wall, and then, as a zygote, enters into a resting condition, during which its original green colour becomes brick-red.

A sexual difference is seen in *Eudorina elegans*, a species which is very similar in other respects to *Pandorina*, being also a gelatinous sphere containing from sixteen to thirty-two cells (Fig. 158). At the time of fertilisation the colonies become differentiated into male and female.

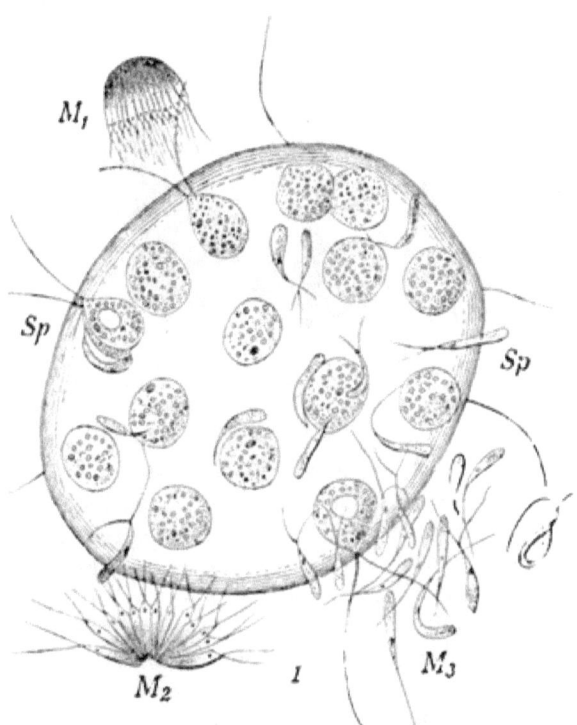

FIG. 158.—*Eudorina elegans*, female colony (*Coenobium*), around which antherozoids, *Sp*, are swarming (after Goebel; from Sachs, Fig. 412): M_1—M_3, bundle of antherozoids.

In the female colonies the individual cells transform themselves without further division into globular eggs; in the male colonies, on the contrary, each cell splits up by means of repeated divisions into a bundle of from sixteen to thirty-two spermatozoids (Fig. 158 M^1). They are "extended bodies, bearing anteriorly two cilia, the original green colour of which has been transformed into yellow." The individual bundles separate from the mother-colony, and swim about in the water. "If they meet a female colony, the cilia on both sides become entangled; by this means the male colony is fixed; it however soon falls to pieces, after which the individual spermatozoids, which become considerably longer, bore their way into the gelatinous vesicle of the female colony. They then make their way to the egg-cells, to which, after they have crept round them, they attach themselves, often in great numbers. We may assume that, as has been observed in many other cases, one of these spermatozoids makes its way into each egg-cell" (Sachs).

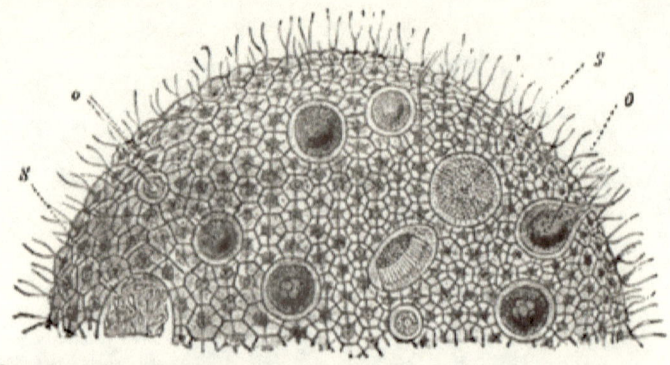

FIG. 159.—*Volvox globator*, sexual, hermaphrodite colony, somewhat diagrammatic representation constructed from figures by Cienkovsky and Bütschli (after Lang, Fig. 21): *s* male gamete (spermatozoid); *O* female gametes (eggs).

Finally, in *Volvox globator* (Fig. 159) the differentiation is greater than ever, for amongst the very numerous cells which constitute the globular colony some remain vegetative, whilst others become transformed into sexual cells. Further the eggs (O) are still larger than in *Eudorina*, and are fertilised by very small male elements (s), which swim about with two flagella.

If we take all these numerous facts into account, we may surely consider the following law as established, i.e. that egg and sperm-cells are derived from reproductive cells, which, to start with, are similar and not to be distinguished from one another, but which become differentiated by developing in opposite directions.

II. **The Physiology of the Process of Fertilisation.** Having discussed the morphological phenomena which have been observed in the organic kingdom during the process of fertilisation, we must now turn our attention to a still wider and more difficult subject—the examination of the properties which the cells must possess in order to unite themselves in the reproductive act, and thus to constitute a starting point for a new cycle of development.

In the first instance it is evident, that not all the cells of a multicellular organism are capable of fertilising or of becoming fertilised, and that even the sexual cells are only suitable for reproductive purposes for, in many cases, quite a limited time. Hence definite characteristics must be developed in the cells; these we will provisionally group under the common name of " need for reproduction."

This need for reproduction alone is in itself far from sufficient to ensure the occurrence of fertilisation. This is proved by the fact that, if mature eggs and spermatozoa from different organisms are brought together, they do not pair. Hence a second factor is necessary : the cells which are to unite sexually must suit one another in their organisation, and in consequence must have the inclination to combine with one another. This we will designate as *sexual affinity*.

Thus the physiology of the process of fertilisation may be separated into two parts :—

1. Investigation of the need for reproduction.
2. Examination of the sexual affinity of the cells.

In a third section various hypotheses, which have been started by various investigators, concerning the nature and aim of fertilisation, will be investigated.

1. **The "Need for Reproduction" of Cells.** By the expression " need for reproduction," we understand a condition of the cell, when it has lost the capacity of carrying on the vital processes by itself, although it regains the power to a still greater degree after it has fused with a second cell in the act of fertilisation. At present we entirely lack a deeper insight into the nature of this condition ; for it is one of the inherent properties of living matter, and as such is outside of the domain of our perceptive powers, since these properties can only be recognised by their results. Similarly the physiological side of the subject is completely unknown, since it as yet has not been subjected to systematic investigation. Hence we can only here mention certain observations, which must be extended and widened in future by means of physiological investi-

gation. We expect by this means to increase our knowledge by the study of the lowest organisms chiefly, since in them the individual cells possess an absolute, or at any rate a large, degree of independence, and are not, as in the higher organisms, related to and dependent upon the other cells of the body. Hence in them the fundamental vital phenomena are more clearly to be recognised.

The facts which we know at present may be summed up under the following heads:—

(1) The need for fertilisation occurs periodically during the life of the cell; (2) it invariably lasts only a short time; (3) it depends to a certain extent upon external conditions; and hence (4) in many cases it may be suspended and transformed into parthenogenesis and apogamy.

That the need for fertilisation is a phenomenon occurring periodically in the life of the cell may be demonstrated experimentally through the study of Ciliata. Maupas (VII. 30) has carried out a large number of very instructive experiments upon this subject.

During the life of one of the Ciliata, two periods can be distinguished—an asexual one and one of sexual maturity or need of fertilisation. The first commences after two animals have fertilised one another and moved apart; multiplication then occurs by the rapid and repeated division of the cells. During this period, individuals from different cultures may be brought together, and the most favourable conditions for conjugation be provided, and yet pairing never occurs. However, after a considerable time, they again experience a need for fertilisation. If at this time individuals from two cultures are brought together under suitable conditions, pairing occurs to a considerable extent for a few days.

Thus Maupas has established the fact, that in *Leucophrys patula* only individuals of the 300th to 450th generation after the act of fertilisation has taken place can reproduce themselves sexually. In *Onychodromus* this sexual period occurs between the 140th and 230th generations, and in *Stylonichia pustulata* between the 130th and 180th.

The second law runs: This condition of "need for fertilisation" is invariably of short duration. If cells capable of fertilisation are not fertilised at the right time, they soon perish. This may be demonstrated with Ciliata, swarm-spores of Algæ, and animal egg-cells.

If single individuals of *Onychodromus*, of a generation between the 140th and the 230th, or specimens of *Stylonichia pustulata* of a generation between the 130th and the 180th, do not have the opportunity of pairing, they become old sexually, or over-mature. It is true that they continue to multiply by means of division, and indeed are able to pair, but no result is produced. For, in spite of their pairing, they degenerate and succumb to a gradual decay of their organisations, as Maupas expresses it, "in consequence of senile degeneration." The commencement of this stage may be recognised by characteristic changes in the nuclear apparatus.

Swarm-spores or gametes of Algæ often die off, after swimming about in the water for a few hours, without having succeeded in pairing with suitable individuals. The receptive capacity of the large female gamete of the species *Cutleria*, after it has come to rest, and has become capable of functioning as an egg, only lasts for a comparatively short time. Falkenberg (VII. 10) has performed a large number of experiments which show "that, whilst on the third day after they have come to rest almost all the eggs are capable of becoming fertilised, on the fourth day only half are in that condition. Further, after this period all the eggs lose their receptive capacity, and although spermatozoids are placed in their neighbourhood, commence to die off, exhibiting the same changes as those eggs which were completely shut off from the fertilising cells."

Finally, mature animal egg-cells, even when under normal conditions in the ovary or in the oviducts, live only for a short time; they soon become over-mature (Hertwig VI. 32). Their normal functions become weakened, as is seen by the fact that, although they can still undergo fertilisation for a time, yet this occurs in an abnormal fashion; several spermatozoa make their way into the egg, the result being an abnormal process of development. Without doubt, this phenomenon is analogous to the senile degeneration of Ciliata which have been prevented from pairing at a suitable period.

The third law, that the commencement of the need for fertilisation may be hastened or postponed by external circumstances, may be clearly proved in some cases.

Thus, if nourishment be continually and abundantly supplied to cultures of Ciliata, pairing can be prevented (Maupas VII. 30). They continue to divide until the whole culture dies off in consequence of senile degeneration. On the other hand, cultures of

Infusoria, which are approaching sexual maturity, may be induced to pair by withholding nourishment. "Une riche alimentation," as Maupas observes, "endort l'appétit conjugant; le jeûne, au contraire, l'éveille et l'excite."

Similarly Klebs (VII. 28) has observed in *Hydrodictyon*, that changes in the environment influence the development of sexual cells, by either inducing or hindering the process.

Klebs has induced the formation of gametes in "nets," which were growing naturally, by cultivating them in a 7 to 10 per cent. solution of cane sugar. After from five to ten days, the net fell completely to pieces, gametes having developed in nearly all the cells. Further, the inclination for the formation of gametes was increased in the cells by cultivating fresh nets in shallow glass dishes, which contained a relatively small quantity of water, and which were placed in a sunshiny window. According to Klebs, the influence of chamber culture is "to arrest growth, but not to interfere with the production of organic compounds by means of assimilation; at the same time, however, a certain poorness in nutrient salts is produced."

On the other hand, sexual reproduction may be suppressed, as in Ciliata. For this purpose it is only necessary to place a net, the cells of which have just commenced to form gametes, in a 5 to 1 per cent. nutrient solution composed of 1 part sulphate of magnesia, 1 part phosphate of potassium, 1 part sulphate of potassium, and 4 parts sulphate of calcium. After a short time, asexual swarm-spores develop, especially if the net is put back into fresh water.

Eidam has observed that a small fungus, *Basidiobolus ranarum*, when cultivated from conidia in a nutrient medium, develops a firm mycelium, which produces simultaneously both asexual reproductive cells (conidia) and sexual cells. In an exhausted nutrient medium, on the contrary, the conidia produce only a loose mycelium, which immediately and exclusively gives rise to sexual cells, which unite together to form zygospores.

Abundant nourishment in plants is conducive to vegetative increase, as the experience of gardeners teaches us, but hinders the formation of seed, whereas the development both of bloom and seed is increased by restricting vegetative growth (cutting off roots and shoots), and thus diminishing the absorption of nourishment.

The same phenomenon has also been observed in animals, which multiply parthenogenetically. When nutriment is withheld from the *Phylloxera vastatrix*, the winged sexual forms, as Keller (VII. 26) has shown experimentally, soon make their appearance, and fertilised eggs are laid.

In many cases, especially amongst the lower organisms, the need for fertilisation is only relative.

When the female gamete of the Alga *Ectocarpus* (VII. 51) comes to rest, for a few minutes it becomes receptive. "If the egg is not fertilised at this time, it draws in its flagella completely, becomes spherical, and excretes a cellulose membrane. After from twenty-four to forty-eight hours parthenogenetic germination first begins to make its appearance." Even the male gametes are capable of spontaneous development, although in a less degree than the female. After they have swum round for several hours, they finally, as Berthold states, come to rest, " but only a portion of them develop slowly into very weak and tender embryonic plants, whilst the remainder become immediately, or after the course of one or two days, disintegrated."

A very peculiar facultative relation is seen in Bees, whose eggs, whether fertilised or not, develop into adults. But the unfertilised eggs produce drones, and the fertilised, female Bees (working and queen-Bees). Sometimes, as is stated by Leuckart, hermaphrodites are derived from eggs which were fertilised too late for the development in the male direction to be entirely set aside. The possibility of accelerating, or, on the contrary, of delaying the need of fertilisation in sexual cells by interference from without, throws light upon the phenomena of parthenogenesis and apogamy, which we are now about to discuss in detail.

a. **Parthenogenesis.** In most cases sexual cells, both in the animal and vegetable kingdoms, perish quickly, unless they are fertilised at the right time. Although they consist of a substance which is eminently capable of development, yet if this one condition fails they cannot develop.

Till a short time ago the majority of scientists were so convinced of the impossibility of the spontaneous development of the egg-cell, that they received the theory of parthenogenesis with incredulity, because they perceived in it an offence against a law of nature. And, indeed, it may be accepted as a law of nature for mammals, and for the majority of other organisms, that their male and female sexual cells are absolutely incapable of development by themselves. Any single species of mammal would unquestionably die out, if its male and female individuals did not unite in the act of generation. Nevertheless, it cannot be stated as a general law of nature, that ova are always incapable of development unless they are fertilised.

Both in the vegetable and the animal kingdoms, numerous instances occur of cells being formed in special sexual organs, which were, as far as we can judge by their design, originally destined to develop by means of fertilisation as eggs; but which have subsequently lost their need for fertilisation, and in consequence behave exactly like vegetative reproductive cells, that is to say, like spores.

Only female specimens of *Chara crinita*, one of the higher Algæ, are to be found in Northern Europe. In spite of this, ova, which develop without fertilisation into normal fruits, are formed in the oogonia.

Still more instructive are the cases of parthenogenesis which occur in the animal kingdom. They have been observed chiefly in small animals belonging to the *Arthropoda*, in *Rotatoria*, *Aphides*, *Daphnidæ*, *Lepidoptera*, etc. At one time females produce in their ovaries only ova which develop without fertilisation, and at another the same individuals form those which require fertilisation. Ova, with such different physiological attributes, generally differ in appearance. The parthenogenetic ones are exceptionally small and poor in yolk, and in consequence develop in a shorter time and in greater numbers; whilst, on the other hand, those which require fertilisation are much larger and contain much more yolk, and consequently require a longer time for their development. Since the former are only produced in summer and the latter at the commencement of the cold season, they have been distinguished as *summer* and *winter eggs*. The latter are also called retarded eggs (*Dauereier*), because they have to pass through a somewhat lengthy period of rest after fertilisation, whilst the summer eggs (*Subitaneier*) immediately enter upon the process of development.

The development of the parthenogenetic summer eggs, and of the winter eggs, which require fertilisation, may be affected by external conditions. In *Aphides*, abundant nourishment favours the formation of summer eggs, whilst a diminished supply of nourishment causes the production of ova requiring fertilisation. *Daphnidæ* are also evidently affected by the environment, although the individual factors can be less easily established experimentally.

This may be concluded from the fact, that, in certain species, the generation-cycle assumes a different appearance, according to the conditions of life under which the animals are living.

The inhabitants of small shallow pools, which readily dry up, produce only one, or at most a few generations of females, which multiply asexually; after this ova requiring fertilisation are produced, so that in the course of a year several generation-cycles (consisting of unimpregnated females and sexual animals) succeed each other. The inhabitants of lakes and seas, on the other hand, produce a long series of unimpregnated females before depositing ova, which require fertilisation; this occurs towards the end of the warm season. A generation-cycle, therefore, in this case occupies a whole year (polycyclical and monocyclical species of Weismann).

Weismann (VII. 39), who investigated the whole subject most thoroughly, remarks: "That asexual and bi-sexual generations alternate with one another in various ways in *Daphnidæ*, and that the mode of their alternation stands in a remarkable relation to their environment. According to whether the causes of destruction (cold, desiccation, etc.) visit a colony several times during the year, or once, or not at all, we find *Daphnoids* which exhibit several cycles within a year, others which have only one cycle, and finally there are species which do not exhibit any generation-cycle at all; hence we can distinguish between polycyclical, monocyclical and acyclical forms."

In many species, which are exposed to frequently changing conditions, we notice, that some of the ova, which are formed in the ovary, develop into summer eggs, whilst others have a tendency to become winter eggs. According to Weismann, "a war, as it were, goes on to a certain extent in the body of a female between the tendencies to form these two kinds of eggs."

In *Daphnia pulex*, the germ of a winter egg may often be recognised amongst several summer eggs in the ovary; this grows for a few days, even beginning to accumulate the finely granular, characteristic yolk; but then it is arrested in its development, becomes gradually dissolved, and finally completely disappears. If winter eggs have been developed, but owing to the absence of the males, have not become fertilised, they disintegrate after a time, and summer eggs are again formed.

How can it be explained, then, that, amongst eggs which have been developed one after another in the same ovary, some should require fertilisation and others not? Weismann (VII. 40), Blochmann (VII. 44), Platner (VII. 47), and others, have made the interesting discovery, that parthenogenetic ova, and those requiring fertilisation, exhibit an important and fairly essential difference in the matter of the formation of the polar cells (*vide* p. 236); whilst in the case of the latter two polar cells are divided off in the usual manner, in that of the former the development of the second polar cell, and consequently also the reduction of the nuclear substance, which is otherwise connected with this process, do not occur. Hence the egg-nucleus of the summer egg, of a *Daphnia*, for instance, possesses without fertilisation the whole nuclein mass of a normal nucleus.

However, this interesting behaviour by no means explains the nature of parthenogenesis. For the summer egg has the tendency to develop without fertilisation, before it begins to form the polar cells, as is seen from the small amount of yolk it contains, the different nature of its membranes, etc. Hence the ovum does not become parthenogenetic because it does not form the polar cell; but, on the contrary, it does not form the polar cell because it is already destined for parthenogenetic development; it does not develop it because, under these conditions, the reduction of the nuclear mass, which presupposes subsequent fertilisation, is unnecessary.

Many peculiar phenomena connected with parthenogenesis have been observed, the closer study of which will probably contribute much to the explanation of this question. Such a phenomenon, the importance of which cannot at present be estimated, is the fact, that the preparatory process for fertilisation can be retraced, even after the polar cell has been formed.

In many animals, the ova, if they are not fertilised, commence to develop parthenogenetically, at the normal time. Attempts are made by the ova of many worms, of certain Arthropods and Echinoderms, and even of some Vertebrates (birds) to begin to segment in the absence of male elements, and eventually to form germinal discs; but at that point they come to a standstill in their development and die off. Abnormal external circumstances seem to favour the occurrence of such parthenogenetic phenomena in individual instances, as, for example, in *Asteracanthion*. The following remarkable occurrence has been observed by Boveri in *Nematodes* and *Pterotrachea*, and by myself in *Asteracanthion*, during the formation of the polar cells.

After the separation of the first polar cell, that half of the spindle, which was left behind in the ovum, develops into a complete spindle again, just as if the second polar cell were going to be divided off. However, this does not occur; for the second spindle only divides into two nuclei, which remain in the ovum itself. After some time they fuse together in this place, and drifting towards the middle of the yolk, again produce a single nucleus as it were by self-fertilisation; by means of this nucleus the parthenogenetic processes, which quickly follow, are introduced. Thus, in this case, the second division, the purpose of which is to reduce the nuclear mass and to prepare for subsequent fertilisation, is abortive. That by this means no sufficient compensation is made for the absence of fertilisation is evident from the subsequent course of the parthenogenetic process of fertilisation, *i.e.* from the more or less premature death of the ovum.

From the circumstance, that in parthenogenetic development the formation of the second polar cell does not occur or is abortive, we might conclude, that development invariably becomes impossible after the nuclear mass has been reduced to one half of its normal amount, unless a fresh stimulus is given to the organism by means of fertilisation. However, at present, this conclusion, which perhaps contains some truth, cannot be said to be generally applicable. For Platner (VII. 47), Blochmann (VII. 46), and Henking (VII. 17) have observed, that the ova of certain Arthropods (*Liparis dispar, Bees*) develop in a parthenogenetic manner into normal animals, although, like ova which require fertilisation, they have produced two polar cells. In these cases a more careful investigation of the circumstances with reference to the number of the nuclear segments is certainly desirable.

Hence, at any rate, it must be admitted, that it is possible for ova, which contain reduced nuclei as a result of the formation of the two polar cells, to develop further in a parthenogenetic manner; for nuclei, which contain a reduced amount of nuclein, have in no way lost their capacity for division, as may be easily supposed. An experiment, conducted by Richard Hertwig and myself (VI. 38, 32), upon the ova of the sea-urchin, proves this in a striking manner.

By shaking the ova of sea-urchins violently, they can be split up into small portions, which do not contain nuclei; these then become globular, and exhibit signs of life for a fairly long time; further they may be fertilised by spermatozoa. By this means we can definitely prove that the sperm-nucleus, or, as is more frequently the case, the sperm-nuclei, which have penetrated into one of the fragments of the ovum, become metamorphosed into small typical nuclear spindles with a radiation at each pole. The sperm-nucleus now splits up into daughter-nuclei, which for their part again multiply by indirect division, so that the fragment of the ovum breaks up into a number of small, embryonal cells. Boveri (VIII. 2) has pursued this observation further, and has discovered the important fact, that out of a rather large non-nucleated fragment of an ovum, which has been fertilised by a single spermatozoon, a normal, although proportionately small, larva can be developed.

b. **Apogamy.** The phenomena, which de Bary (VII. 2) has included under the name of apogamy, have a close relationship to parthenogenesis, and may be conveniently treated now.

Apogamy has been observed in certain Ferns; it is well known that in the course of their development there is an alternation of generations. Minute plants, the prothallia, are derived from the vegetative reproductive cells, or spores; the function of these prothallia is to develop male and female sexual organs, the latter of which produce egg-cells. These, when fertilised, produce an asexual Fern-plant, which develops spores in a vegetative manner.

In *Pteris cretica* and *Asplenium filix-femina cristatum* and *falcatum*, the law of alternation of generations, which is generally so constant in Ferns, is broken through. The prothallia of these three species either produce no sexual organs at all, or only such as are no longer functional, *i.e.* have become rudimentary; on the other hand, a new Fern arises from the prothallium by means of vegetative budding.

Since these three species of Ferns have been affected by cultivation, it is possible that the development of cells requiring fertilisation has been suppressed by excessive nourishment, whilst the vegetative mode of reproduction has been favoured.

2. **Sexual Affinity.** By sexual affinity we understand the reciprocal influences which are exercised by cells of related species requiring fertilisation upon each other. This takes place in such a manner, that, when the cells are brought within a definite distance of one another, they exert a mutual attraction upon each other, and combine, fusing into one, like two chemical bodies, between which unsatisfied chemical affinities existed. If both

sexual cells are able to move, they precipitate themselves upon each other; if however one cell, as ovum, has become fixed, the reciprocal attraction is evinced by the movements of the spermatozoon. But sexual affinity continues to operate even after the two cells have fused, being seen in the attraction which the egg and sperm-nuclei, with their centrosomes, exercise upon each other, the result of which is, that they come into contact and coalesce as described above.

Thus two points remain to be proved in this section: firstly, that reciprocal influences between cells requiring fertilisation really do exist; these we will designate by the name of sexual affinity; and secondly, that this affinity is only evinced between cells of a definite kind: and this suggests the question as to what are the special attributes which these cells requiring fertilisation must possess.

a. **Sexual Affinity in General.** That sexual cells at a certain distance from one another exert upon one another a definite influence may be concluded from numerous observations, made by reliable observers. I will confine myself to a few especially instructive examples, which have been described by Falkenberg, de Bary, Engelmann, Juranyi and Fol.

Falkenberg (VII. 10) investigated the process of fertilisation in a low species of Alga, *Cutleria*. To the receptive ova of *Cutleria adspersa* which have come to rest, he added actively motile spermatozoids of the nearly allied species *Cutleria multifida*; these two species can only be distinguished from one another by small external differences. "In this case the spermatozoids, as seen under the microscope, wandered aimlessly about, and finally died, without having fertilised the ova of the allied species of Alga. It is true, that individual spermatozoids, which by chance came into contact with the quiescent ova, remained attached to them for a few moments, but they soon detached themselves again. However, a very different result was obtained as soon as a single fertilisable ovum of the same species was introduced into the vessel containing the spermatozoids. After a few moments, all the spermatozoids from all sides had gathered around this ovum, even when the latter was several centimetres distant from the place at which they were chiefly collected." In doing this they even overcame the attractive force exerted by the rays of light falling upon them, and moved in a direction opposed to the one which they would otherwise have chosen.

Falkenberg concludes from his observations, that the attraction between the ova and spermatozoids of *Cutleria* makes itself felt at a relatively great distance, that this attractive force must have its seat in the cells themselves, and further that the attraction is only exerted between sexual cells of the same species.

De Bary (VII. 2 b), investigating the sexual reproduction of *Peronosporeæ*, observed that, in the interlacing hyphæ, the oogonia at first lie alongside of each other. Somewhat later, the antheridia develop, but this invariably occurs in the immediate neighbourhood of the egg-cell only; they are frequently derived from hyphæ, which have no connection with the one from which the oogonium is formed. De Bary concludes from this, that the oogonium must exert an influence over a limited area, and that this influence induces the hyphæ to form an antheridium. A peculiarly striking instance of this influence exerted at a distance is seen in the circumstance, that the branch on which the antheridium is developed is diverted from its natural direction of growth; for, in order to approach the oogonium, it bends over with its end towards it, and then lies close to it. De Bary estimates that the distance at which the oogonium is able to exert this attraction is almost as great as its own diameter, and remarks further, that "the above-described divergence of the lateral branches can be ascribed to no other cause than the special attributes of the oogonium."

Not less interesting, and worthy of notice, are the statements which Engelmann (VII. 9) has made about the conjugation of *Vorticella microstoma*. In this case small male zoospores are formed by budding (p. 228); these, just like spermatozoa, fertilise the large female individuals (p. 271). Engelmann succeeded four times in tracing the bud after its separation from the mother-cell, until it had united with another individual.

Engelmann describes his observations as follows: "At first the bud, always rotating on its longitudinal axis, wandered with fairly constant rapidity (cir. ·6–·1 mm. per sec.), and, as a rule, in a fairly straight line through the drop. This lasted for from five to ten minutes, or even longer, without anything especial happening; then the scene was suddenly changed. Coming by chance into the neighbourhood of an attached *Vorticella*, the bud changed its direction, occasionally even with a jerk, and dancing, like a butterfly which plays round a flower, approached the fixed form; it then, as if it were feeling it, glided round about it, meanwhile always rotating on its own longitudinal axis. After this had been going on for several minutes, and had been repeated with several fixed individuals one after the other, the bud at last attached itself to one of them, generally at the aboral end, near the stalk. After a few minutes the fusion might be definitely observed to be taking place."

In connection with the above-mentioned description, Engelmann remarks: "At another time I observed a still more striking physiological or even psychophysiological exhibition. A free bud crossed the path of a large *Vorticella*, which was travelling with great rapidity through the drop, and which had abandoned its stalk in the usual manner. At the moment of meeting, although there was absolutely no contact, the bud suddenly changed its course, and followed the *Vorticella* with the greatest rapidity; then a regular chase ensued, which lasted for about five seconds. During this time the bud kept at a distance of about $\frac{1}{15}$ mm. behind the *Vorticella*; however, it did not succeed in overtaking it, but lost it in consequence of its making a sudden side movement. Hereupon the bud continued its path at its original slower pace."

This phenomenon of influence exerted at a distance has also been observed by Fol (VI. 19 a) in animal cells, such as the ova of the Star-fish. Each ovum is surrounded by a thin gelatinous envelope. When fresh spermatozoa of the same species approach the surface of the envelope, the one which is most in advance exercises a distinctly perceptible influence upon the yolk (Fig. 160).

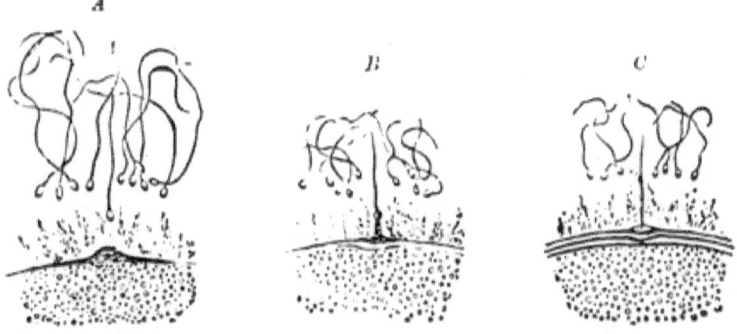

FIG. 160.—*A, B, C* Sections of ova of *Asterias glacialis*, after Fol. The spermatozoa have already penetrated the gelatinous envelopes of the ovum. In *A*, a prominence is just beginning to rise up to meet the most advanced spermatozoon. In *B*, the prominence and spermatozoon have met. In *C*, the spermatozoon has penetrated the egg which has formed a yolk-membrane with a crater-like aperture.

Its hyaline superficial layer raises itself up as a small protuberance, thus projecting a receptive prominence (*cone d'attraction*) towards the spermatozoon. Sometimes this protuberance is soft, and drawn out in the form of a needle or tongue, and sometimes it is broad and short. After contact with the spermatozoon has taken place, it is withdrawn.

Fol considers that it is impossible to doubt the accuracy of this observation, and remarks further: "Since we cannot deny the fact that the spermatozoon exercises an influence upon the yolk, from which it is separated by a relatively great distance, we must accept the theory that influence at a distance (*action à distance*) is a possibility."

I will confine myself to the above-mentioned observations, the number of which could be easily greatly multiplied, and will only quote the following words of the botanist Sachs (II. 33):—

"Influence at a distance, or the mutual attraction of sexual cells for one another, is one of the most startling facts connected with the processes of fertilisation. I have chosen this term for the facts about to be more minutely described, as it is not too long, and, at any rate, realistic. We must not, however, take the words, influence at a distance and mutual attraction, exactly in the same sense as in physics.

"In the numerous descriptions which various observers have given of the behaviour of antherozoids in the neighbourhood of the oosphere, and of wandering gametes and antherozoids in the neighbourhood of oogonia, we meet with the most definite assertions, that the sexual cells always exert a certain influence upon one another, which makes itself felt over a certain distance, and which always tends to induce the union of the two. This occurrence is the more remarkable, in that this mutual attraction immediately disappears after fertilisation has taken place."

The question naturally arises as to what are the forces to which the phenomena can be attributed. Pfeffer has expressed the view, based upon the above-mentioned experiments (p. 117), that in the objects examined by him the antherozoids are attracted to the egg-cells by chemical substances, which the latter secrete. Too great an importance, however, must not be attached to these opinions, as would be the case if we considered that the conjugation of two sexual cells was explained by them. In my opinion, the chemical substances, which are secreted by the egg-cells, only exert a secondary influence upon fertilisation; they play a part similar to that performed by the mucoid and gelatinous envelopes of many ova which retain the antherozoids.

On the other hand, they in no wise explain conjugation itself, i.e. the processes peculiar to fertilisation. This may be proved in a very simple manner. According to the researches of Pfeffer, malic acid is secreted in the archegonia of the most different Ferns. Nevertheless, only the antherozoids of the same species will fuse with the oosphere, those of a different species being as a rule unable to fertilise them. Thus we see, that there are relations existing between the sexual products which cannot be explained by the action of irritating chemical secretions. The same is true of the conjugation of gametes, of the formation of the receptive prominence in animal ova, and of the mutual attraction of egg- and sperm-nuclei.

Nägeli (IX. 20) suggests that electrical forces may be the cause of sexual attraction, and this seems to me to be an explanation of far-reaching importance. But, until this conjecture has been definitely proved, it is better to attribute the sexual phenomena in general to the reciprocal action of two somewhat differently organised protoplasmic bodies, and to call this reciprocal action sexual affinity. We must be content with such a general expression, since we cannot accurately analyse the forces which come into activity. Presumably it is not a question here of a simple phenomenon, but of a very complicated one.

This may be rendered still clearer by an investigation of the second point, namely, what is the nature of the cells requiring fertilisation, and between which there is sexual affinity?

b. **More minute discussion of sexual affinity, and its different gradations.** The possibility of the occurrence of fertilisation, and the results produced by it, are to a great extent determined by the degree of relationship which exists between the sexual cells. But since a near relationship implies a greater or less similarity in their organisation, these differences in their organisation must be the determining factor.

The degree of relationship between the two cells may vary considerably. It is nearest when both the cells to be fertilised are descended directly from the same mother cell; it is more distant where many cell-generations have developed asexually from the mother-cell, the final products at last producing sexual cells. Here, too, cases of nearer or more distant relationship are possible. If we take as an example one of the higher flowering plants, we see that the male and female sexual cells may be derived from the same sexual apparatus, *i.e.* from one blossom, or they may spring from different blossoms of the same shoot, or, finally, from different shoots; in this way, three different degrees of relationship are obtained. In hermaphrodite animals they may belong to the same individuals, or to different individuals of the same species.

The degree of relationship is still more distant when the sexual products are derived from two different individuals of the same species. In such cases also, many degrees of relationship are possible, according to whether the producing individuals are descendants of common parents, or are more distantly related. Finally, we may have the union of sexual products derived from parents which differ so much in their organisation, that they have been classified as varieties of a species, or as belonging to different species, or even to different genera.

The innumerable possibilities, which the above-mentioned series affords, are generally treated under three heads: (1) self-fertilisation and in-breeding, (2) normal fertilisation, and (3) hybridisation. There are, however, great differences of opinion concerning the classification of individual cases under one or other of the three heads. Further, there is no rule by means of which we can estimate the various degrees of relationship of the sexual cells, and which is equally applicable to all members of the organic kingdom.

A review of the facts connected with the subject teaches us, that when the relationship of the reproductive cells—I use the expression, relationship, in its widest sense—is either too near or too distant, sexual affinity is either lessened or entirely done away with; therefore we may state, as a general rule, that a moderate degree of relationship, which is more or less distant according to the species, is the one most likely to render fertilisation possible.

Further, we may also notice here, that sexual affinity is affected by the environment. We will first discuss the question of self-fertilisation, then that of hybridisation, and finally we will investigate the influence exerted by the environment upon these two.

a. **Self-fertilisation.** Self-fertilisation occurs under the most various conditions. In many cases there is no sexual affinity between cells needing fertilisation, which are nearly related to one another, being derived more or less directly from a common mother-cell or from the same highly differentiated multicellular mother-organism. Lower *Algæ*, *Infusoria*, *Phanerogamia* and all hermaphrodite animals supply us with a large number of examples of this.

In *Acetabularia*, sexual reproduction takes place in such a manner, that swarm-spores are derived in very great numbers from the contents of resting-spores. According to Strasburger and de Bary, conjugation only takes place between two swarm-spores if they are descended from two different resting-spores, whilst those that are derived from the same parent avoid each other.

Strasburger (VII. 38) says: "About mid-day I saw two neighbouring spores, which were absolutely indistinguishable from one another, rupture under my eyes, and the swarm-spores of both hurry straight to that edge of the drop which was nearest the window. Soon an extraordinary sight presented itself. I observed that the swarm-spores, which were derived from the same resting-spore, kept at equal distances from one another and evidently avoided each other, whilst at the same time conjugation groups,—if I may use the expression,—that is to say, heaped-up collections of conjugating-spores, were formed, into which the individual swarm-spores, as it were, precipitated themselves. From these conjugation centres, pairs of united swarm-spores were continually hurrying away."

In his investigations upon Infusoria, Maupas (VII. 30), by means of several hundred experiments on four different species (*Leucophrys*, *Onychodromus*, *Stylonichia*, *Loxophyllum*), has established the fact, that even when fertilisation is necessary conjugation only takes place when individuals of different generation cycles are brought together.

Maupas remarks: "In many pure cultures of nearly related individuals, the fast, to which I subjected them, resulted either in their becoming encysted, or in their dying of hunger.

"It was not until after senile degeneration had already begun to make inroads in the culture that I noticed that the conjugation of nearly related individuals occurred in the experimental cultivations. However, all such conjugations ended with the death of the Infusoria, which had paired, but which were unable to develop further, or to reorganise themselves after they had fused. Such pairings are, therefore, pathological phenomena due to senile degeneration."

Hence Maupas is of opinion that cross fertilisation between individuals of different origin is necessary for Infusoria also.

The ineffectuality of self-fertilisation has also been proved in certain cases amongst *Phanerogamia*. Hildebrandt (VII. 24, p. 66) says of *Corydalis cava*: "If a flower of this plant, in which the opened anthers lie close to the stigma, be protected from fertilisation by insects, no fruit is ever formed in it; that this is not due to the circumstances of the pollen not coming in contact with the susceptible part of the stigma may be seen from the fact that even those flowers, whose stigmata were powdered with the pollen of the surrounding anthers, were non-fertile."

"A perfect fruit can only develop when the pollen of the flowers of one plant is placed on the stigma of another; it is true that fruit is formed when the flowers of the same vine are crossed; but the resulting plants produce a much smaller number of seeds than is normal, and further they do not always come to perfect maturity."

A similar absence of result after self-fertilisation has been observed in a few other plants, *i.e.* certain species of *Orchids*, *Malvaceæ*, *Reseda*, *Lobelia* and *Verbascum*.

Unfortunately, no very thorough investigation concerning the behaviour of hermaphrodite animals has been made; the difficulties of such research would be very great. No doubt cases would be also found here in which no fertilisation occurs between the eggs and spermatozoa of the same individual when they are artificially brought into contact; with snails, for instance, this must be the case.

However, in opposition to the above-mentioned examples, we find others, which prove both that complete sexual affinity does exist, and also that normal development by self-fertilisation does take place between sexual cells, which are very nearly related to one another.

Thus in the case of certain *Conjugatæ* (*Rhynchonema*) sister-cells unite with one another, or, as in *Spirogyra*, cells which belong to the same filament conjugate together (*vide* p. 283).

Further, in many Phanerogams not only can the egg-cells be fertilised with the pollen of the same flower, but the resulting plants are strong and healthy; and, moreover, this in-breeding can be continued through many generations with equally happy results.

Between the two extremes—the absence of any sexual affinity

and the presence of strong mutual attraction in nearly related sexual cells—there are many gradations.

Amongst the numerous egg-cells which are contained in an ovary, only a few develop and become ripe seeds, where self-fertilisation with the pollen of the same flower is induced artificially. From this we may conclude that the individual egg-cells possess somewhat different sexual affinities; that is to say, that whilst some may be fertilised with the pollen of their own flower, others cannot; thus they exhibit differences similar to those which we shall come across in hybridisation.

Finally, it may be possible for egg-cells to be fertilised, to begin to develop, and then to die off prematurely. In support of this view, the phenomenon may be quoted, that many flowers, which have been induced artificially to fertilise themselves, fade more quickly than those which have been fertilised in a natural manner. Indeed, the flowers of certain Orchids become black and necrotic when treated in this fashion. This is probably due to the premature death and disintegration of the embryos which were about to be developed (Darwin VII. 8).

The seeds, which develop as a result of self-fertilisation, frequently produce only weakly plants, showing some defect or other in their constitution; further, the pollen grains are often imperfectly developed.

From these three facts, namely, that in many organisms nearly related sexual cells do not combine; that in others, even if fertilisation does take place, the embryo is arrested in its development, and soon dies; and that finally, even if development proceeds uninterruptedly, the evolved organisms are weakly; we are able to draw the general conclusion, that self-fertilisation on the whole acts disadvantageously. It is true, that in individual cases this disadvantage cannot be perceived, yet these exceptions do not disprove the accuracy of this statement any more than the occurrence of parthenogenesis can be taken as an argument against the theory, that great advantage is to be derived from fertilisation.

That there must be something detrimental in self-fertilisation may be inferred from a cursory glance over the organic kingdom. As Darwin (VII. 8) says, nature evidently abhors frequent self-fertilisation, for we see constantly on every side, that most complicated arrangements have been made in order to prevent its occurrence.

These arrangements are: (1) the distribution of the sexual organs over two different individuals, so that one produces only female sexual cells, and the other only male; (2) the reciprocal fertilisation of hermaphrodite individuals; (3) the different times at which the maturation of the ova and spermatozoa occurs, as in *Pyrosoma*, many molluscs, etc.; and (4) the peculiarities in the organisation of hermaphrodite flowers of phanerogams (both dichogamy and heterostylism), and the part played by insects, which, in carrying the pollen from one flower to the other, induce cross fertilisation, as has been observed and described by Koelreuter, Sprengel, Darwin (VII. 8), Hildebrandt (VII. 24), H. Müller (VII. 49), and others. These arrangements for the prevention of self-fertilisation are so many-sided and striking, especially in flowering plants, that Sprengel was able, in his book, to speak of "the discovered secret of nature, the fertilisation of flowers by insects," and to say: "Nature does not seem to have wished that a single hermaphrodite plant should be fertilised with its own pollen."

β. **Bastard Formation, or Hybridisation.** The opposite of self-fertilisation and in-breeding is hybridisation. By this is meant the union of several products of individuals, which are so different in their organisation, that they are classified into different varieties, species, or genera.

As a rule, the principle, that the sexual products of individuals, which are very different from one another, do not unite with one another, is correct. Everybody considers it impossible for the ovum of a mammal to be fertilised by the spermatozoon of a fish, or for that of a cherry-tree by the pollen of a conifer. But as the individuals become more closely related, whether they belong to different families or species, or even only to different varieties of the same species, the more difficult does it become to prophesy *à priori* as to the result of cross-fertilisation. This can only be discovered by means of experiment, which has shown that the various species in the animal and vegetable kingdoms do not always behave in a similar manner towards hybridisation, in that individuals which resemble one another in their form, down to the minutest details, often cannot be crossed, whilst *between others which are much more dissimilar bastard fertilisation is possible.*

Briefly, sexual affinity does not always march *pari passu* with the external similarity which can be perceived between the individuals in question.

Although the only difference between *Anagallis arvensis* and *A. cærulea* is in the colour of their blossoms, they cannot be induced to fertilise each other. No hybrids have been obtained from apple and pear-trees, or from *Primula officinalis* and *P. elatior*; whilst, on the other hand, hybrids have been successfully obtained between species which belong to different orders, such as *Lychnis* and *Silene*, *Rhododendron* and *Azalea*, etc.

Sachs says: "The absence of correspondence between sexual affinity and systematic relationship is shown in a more striking manner, in that occasionally varieties of the same species are either quite unable to fertilise each other, or can only do so to a partial extent; thus *Silene inflata* var. *alpina* cannot conjugate with var. *angustifolia*, nor var. *latifolia* with var. *litoralis*, and so on."

In both the animal and the vegetable kingdoms we find certain orders the species of which can be easily crossed, whilst there are others whose species offer the most obstinate resistance to all attempts. In the vegetable kingdom, Liliaceæ, Rosaceæ, Salicaceæ; and in the animal kingdom, Trout, Carp, Finches, etc., readily produce hybrids. Many dogs, too, which differ considerably in bodily structure, such as the dachshund and the pointer, the retriever and the St. Bernard, produce mongrels.

Further we see how unaccountable are the factors which are dealt with in hybridisation when we consider the following phenomenon: very frequently the ova of species A may be fertilised with the spermatozoa of species B; whilst, on the other hand, the ova of B cannot be fertilised with the spermatozoa of A. Thus sexual affinity between the sexual cells of two species is present in the one case and absent in the other. It seems to me that the determining factor should be sought for in the organisation of the ovum, as may be concluded from the experiments cited below.

A few examples of one-sided crossing may be quoted. The ova of *Fucus vesiculosus* may be fertilised with the antherozoids of *Fucus serratus*, but the reverse cannot occur. *Mirabilis Jalapa* produces seed when fertilised with the pollen of *Mirabilis longiflora*, whilst the latter remains unfruitful, if the attempt be made to fertilise it with pollen from the former.

Similar cases often occur in the animal kingdom, and amongst these the most interesting are met with in those species in which fertilisation can be induced artificially by mixing the sexual products.

My brother and I (VII. 20) attempted to cross different species of Echinoderms, and found that when the ova of *Echinus microtuberculatus* were mixed with the spermatozoa of *Strongylocentrotus lividus*, fertilisation took place in every case after a few minutes, the egg-membrane raising itself up from the yolk. After an hour and a half all the ova were regularly divided into two. On the following day glistening germ vesicles had appeared; on the third, gastrulæ; and on the fourth, the calcareous skeleton had developed. Cross-fertilisation in the opposite direction yielded varying results. When the spermatozoa of *Echinus micro-tuberculatus* were mixed in a watch-glass with the ova of *Strongylocentrotus*, the greater number of the ova remained unchanged, the egg-membrane raising itself from the yolk in only a few cases. After two hours only a few isolated ova were divided into two. Amongst these the egg-membrane lay fairly close to the yolk in some, and in others was raised a little. The next day a few glistening germ vesicles were apparent in the watch-glass, but the majority of the ova were quite unchanged.

Pflüger (VII. 50) observed a similar relationship between *Rana fusca* and *Rana esculenta*. Ova of the former species, when suspended in a watery extract of the testis of *Rana esculenta*, always remained unfertilised. When, however, the ova of *Rana esculenta* were mixed with the spermatozoa from the testis of *Rana fusca*, the greater number of the former developed in a regular manner, only a few dividing abnormally; however, after the blastula-stage was reached, they all, without exception, died.

In many respects the results of hybridisation, seen later in the development of the product of crossing, resemble those of self-fertilisation. For instance, when fertilisation does take place, the embryos in many cases die prematurely, or are of a weakly constitution.

The embryos, which develop when certain Echinoderms are crossed, do not live beyond the gastrula-stage. In the same way, Pflüger saw the ova of *Rana fusca*, which had been fertilised with the spermatozoa of *Rana esculenta*, die as germ vesicles. The reproductive organs of animal hybrids generally atrophy before the age of sexual maturity is reached, and hence the animals are sterile.

A still larger number of examples is to be found in the vegetable world. It is true, that seeds may develop, as a result of hybridisation, but they are defective in their development, and sometimes even incapable of germination. If, however, germination does take place, the seedlings may be either weakly or vigorous. Hybrids of widely different species are often very delicate, especially in youth, so that it is difficult to rear them. On the other hand, hybrids of nearly related species are usually uncommonly luxuriant and vigorous; they are distinguished by their size, rapidity of growth, early blooming, long life, wealth of blossoms, strong powers of multiplying, the unusual size of individual organs, and similar properties.

Hybrids of different species develop a smaller quantity of normal pollen grains in their anthers than plants of pure descent; frequently they produce neither pollen nor ovules. In hybrids of nearly related species, this weakening of the sexual reproductive powers is not usually to be observed.

As a general rule, the closer the relationship of the parents, and the greater their sexual affinity, the better does their hybrid product thrive. In individual cases it may get on even better than that of a normally fertilised ovum. For example, when egg-cells of *Nicotiana rustica* are crossed with pollen of *N. Californica*, a plant is produced which, as regards height, stands to its parents in the ratio of 228 : 100 (Hensen VII. 18).

γ. **The Influence of the Environment upon Sexual Affinity.** We have seen in our experiments upon self-fertilisation and hybridisation, that the sexual affinity of the egg and sperm-cells is a factor which cannot be reckoned upon with certainty, and with which a series of the most different resulting phenomena is connected; such as fertilisation or non-fertilisation, development which has been prematurely hindered and weakened, or which has been rendered more vigorous, etc. We shall find, however, that the phenomenon of sexual affinity is still more complicated by the fact that in many cases it may be influenced by external circumstances.

Most peculiar facts concerning hybridisation have been discovered by means of experimental researches upon certain Echinoderms (VII. 20). The unfertilised ova are naked, but in spite of this, fertilisation does not usually take place when spermatozoa, which are of nearly related species, and are exactly similar in appearance, are placed in their neighbourhood, although these latter settle upon the surface of the ova, and make boring movements. In this case the non-fertilisation can only be explained by imagining, that the ovum, if I may use the expression, refuses to admit the unsuitable spermatozoon.

This, however, does not invariably occur. In cross-fertilisations, which were made between *Strongylocentrotus lividus* and *Sphærechinus granularis*, it was found, that out of the hundreds of ova, which were experimented upon at various times, a varying number of eggs was produced, which had been fertilised by the strange spermatozoa, whilst the large majority of ova were unaffected. Thus we see, that the ova of the same animal differ from one another, just as swarm-spores of the same species may react differently to light, some seeking the positive edge of the drop, others the negative, and others, again, oscillating between the two (*vide* p. 101). As swarm-spores exhibit different light reactions, the ova of the same animal present different sex reactions, and what is still more extraordinary, these sex reactions can be largely influenced and altered by external circumstances.

The experiment is a very simple one. The mature ova of Echinoderms, after their evacuation from the ovaries, can be preserved in sea water in an unfertilised condition for 24-48 hours without losing their capacity for development. But, during this time, changes take place in them, which manifest themselves in their behaviour towards foreign spermatozoa.

Two different methods were adopted in the experiments, one of which may be described as the method of successive after-fertilisations. It consisted in this, that the experimenters crossed the same egg-mass several times with foreign spermatozoa. In doing this the following important result was obtained: all the ova, which were crossed immediately after their evacuation from the distended and full ovary, with extremely few exceptions, refused the foreign spermatozoa; but after 10, 20, or 30 hours, that is to say, after the second, third, or fourth crossing, an increasingly large proportion of the ova behaved differently, becoming cross-fertilised, and subsequently developing normally. The same result was always produced, whether the ova of *Strongylocentrotus lividus* were covered with the spermatozoa of *Sphærechinus granularis*, or of *Echinus micro-tuberculatus*, or whether the ova of *Sphærechinus granularis* were crossed with the spermatozoa of *Strongylocentrotus lividus*.

The success or failure of hybridisation cannot in these cases be attributed to a difference in the spermatozoa, since they were each time taken afresh from a distended and full testis, and may, therefore, be considered to be a relatively constant factor in the experiments. In this case, without doubt, it was the egg-cell alone that altered its behaviour towards the foreign spermatozoa.

Hence, if changes take place, or can be induced artificially to take place, in the egg-cell, by means of which hybridisation is rendered practicable, we must conclude, from a theoretical point of view, that it is also possible to induce so complete a hybridisation between the sexual products of two species, which have a certain degree of sexual affinity for one another, that scarcely any ovum should remain unfertilised. Thus, according to the conditions under which the sexual products are brought together, a maximum or a minimum of hybridisation may be obtained.

In order to establish these relations, it is best, in making the experiments, to divide the egg-material of a female into several portions, which are fertilised at different times. The smallest percentage of hybrids is always obtained when the foreign spermatozoa are added to the ova immediately after these latter have been evacuated from the ovaries. The later fertilisation takes place, whether after 5, 10, 20 or 30 hours, the greater is the percentage of the hybridised ova, until the maximum of hybridisation is reached. This is called the stage at which the addition of foreign spermatozoa produces normally the greatest possible number of eggs. This period is of short duration, since imperceptible changes in the ova are uninterruptedly taking place. After that, the percentage of the ova which, in consequence of the bastard fertilisation, develop normally, begins to decrease; and this is due to the fact, that a steadily increasing number of ova are caused to segment in an abnormal fashion and to become malformed, in consequence of several spermatozoa having penetrated into each of them.

The results obtained by fertilising eggs at different times may be represented by a curved line, the summit of which corresponds to the maximum of hybridisation. The results obtained by crossing the ova of *Sphærechinus granularis* with the spermatozoa of *Strongylocentrotus* serve as an illustration. When fertilisation takes place a quarter of an hour after the eggs have been evacuated from the ovary (minimum hybridisation), only a very few individual ova are developed. After two and a quarter hours 10 per cent. can be fertilised, after six and a quarter hours about 60 per cent., whilst after ten and a quarter hours almost all the ova, with the exception of about 5 per cent., are affected; in the latter case they generally develop normally (maximum hybridisation). If the ova are fertilised after twenty-five hours, some develop normally, and a not inconsiderable number irregularly, in consequence of multiple fertilisation, whilst a small number remain unaffected.

The results obtained with *Echinoderm* ova seem to me to offer an explanation of the fact, that domesticated animal and vegetable species are generally more easily crossed than nearly related species in the state of nature. The entire constitution seems to be altered and rendered less stable by domestication. The changes are most evident in the sexual products, since the generative apparatus is sympathetically affected by any variations which take place in the body.

In self-fertilisation, as in hybridisation, sexual affinity is influenced by the environment. Darwin (VII. 8) has pointed out, that *Eschscholtzia californica* cannot be induced to fertilise itself in Brazil, whilst it can in England; moreover, if seeds from England are taken back to Brazil, they quickly become useless for self-fertilisation. Further, various individuals behave in different manners. Just as in *Echinoderms*, in which some of the ova of an ovary may be fertilised with foreign spermatozoa, and others not, so we find experimentally, that some individuals of *Reseda odorata* can fertilise themselves whilst others cannot. In a similar manner we must attribute to individual differences of the egg-cells of an ovule the circumstance that in many plants far fewer seeds are produced by self-fertilisation and hybridisation than by normal fertilisation. A certain number of egg-cells either are not receptive to the foreign pollen, or if they do become fertilised, die prematurely.

Recapitulation and attempted Explanations. If we now review the facts described in the last chapter, there can be no doubt but that the necessity of fertilisation of the sexual cells and sexual affinity, which is closely connected with it, are extremely complicated, vital phenomena. The factors which are influential here are beyond our knowledge. Many circumstances seem to point to the fact, that the conditions, under which the egg-cells are able to develop either parthenogenetically or in connection with a sperm-cell, must be sought for in small differences of molecular organisation. Similarly, we can only explain the facts, that sometimes self-fertilisation and cross-fertilisation are possible, and at others not, that the egg-cells of the same individual often behave differently during self-fertilisation and cross-fertilisation, that the need for fertilisation and parthenogenesis, or the success of self-fertilisation and cross-fertilisation, may often be influenced by external circumstances, and that the well-being of the products of generation is dependent upon the mode of fertilisation, by the presence of these same differences of molecular organisation.

What now must be the molecular organisation of the sexual cells which renders them suitable for the purposes of fertilisation? Some help towards solving this problem may be obtained by comparing the phenomena of self-fertilisation and bastard fertilisation with normal fertilisation.

As is evident from numerous observations, the result of fertilisation is essentially determined by the degree of relationship which the male and female sexual cells bear to one another. The process of fertilisation is prejudiced by a relationship which is either too near or too distant; or, as we may express it, by a too great similarity, or a too great difference. Either the sexual cells do not unite at all, since they exhibit no sexual affinity towards each other, or the mixed product of both, *i.e.* the embryo produced by fertilisation, is unable to develop in a normal manner. In the latter case the embryo may either die during the first stages of development, or it may live as a weakly product; or further, this weakly product, owing to the destruction of its capacity for reproduction, may be useless for the preservation of the species. In all cases the product of reproduction thrives best when the generative individuals, and consequently their sexual cells, differ only slightly in their constitution and organisation.

Darwin (VII. 8) rendered science a great service when, by means of his extensive experiments and investigations, he laid the foundations of this knowledge, and first clearly formulated these theories. I will quote three of his sentences: "The crossing of forms only slightly differentiated favours the vigour and fertility of their offspring . . . and slight changes in the conditions of life add to the vigour and fertility of all organic beings, whilst greater changes are often injurious." The act of crossing in itself has no beneficial effect, but "the advantages of cross-fertilisation depend on the sexual elements of the parents having become in some degree differentiated by the exposure of their progenitors to different conditions, or from their having intercrossed with individuals thus exposed, or lastly from what we call in our ignorance 'spontaneous variation.'" The need of fertilisation consists in "mixing slightly different physiological units of slightly different individuals."[1]

Herbert Spencer (IX. 26) availed himself of these experiments of Darwin's, in order to build up a molecular theory of the nature of fertilisation, which deserves notice as a preliminary attempt.

Spencer, to a certain extent, states as an axiom, that the need of fertilisation of the sexual cell "recurs only when the organic units (micellæ) are approximating to equilibrium—only when their mutual restraints prevent them from readily changing their arrangements in obedience to incident forces."[2]

If this hypothesis, which appears to me to be at present but a possibility, could be proved, we could certainly accept without further consideration Spencer's explanation: "Gamogenesis (sexual reproduction) has for its main end, the initiation of a new development by the overthrow of that approximate equilibrium arrived at amongst the molecules of the parent organism."[3] For "by uniting a group of units from the one organism with a group of slightly different units from the other the tendency towards equilibrium will be diminished, and the mixed units will be rendered more modifiable in their arrangements by the forces acting on them; they will be so far freed as to become again capable of that redistribution which constitutes evolution."[4]

In this sense, fertilisation may be considered to be a process of rejuvenation, to employ the expression used by Bütschli (VII. 6), Maupas (VII. 30), and others.

Spencer's statement at present lacks an exact and scientific foundation, but it seems to deserve notice as a preliminary attempt to solve this extremely difficult question.

An important conclusion may be deduced from the above-mentioned principle, that the process of fertilisation consists in the "mixing of slightly different physiological units of slightly different individuals." If sexual reproduction is a mingling of the properties of two cells, it must result in the development of intermediate forms.

Thus reproduction, so to speak, strikes a balance between

---

[1] The first of these quotations is taken from Darwin's *Origin of Species*, p. 432, and the second and third from Darwin's *Cross- and Self-fertilisation of Plants*, pp. 462, 463.

[2] *Principles of Biology*, by Herbert Spencer, vol. i. p. 275.

[3] *Ibid.*, p. 284.

[4] *Ibid.*, p. 277.

differences by producing a new individual, which occupies a mean position between its parents. By this means numberless new varieties are developed, which only differ slightly from one another. Hence Weismann (IX. 34) is of opinion that fertilisation is an arrangement by means of which an enormous number of varying individual combinations arise; these supply the material for the operation of natural selection, the result being that new varieties are produced.

Whilst agreeing with the first part of this principle, I cannot support the second. The individual differences which are called into being by fertilisation, and which furnish the basis for natural selection, are as a rule only of an insignificant nature, and are always liable to become suppressed, weakened, or forced into another direction, by some subsequent union. A new variety can only be formed, if numerous members of a species vary in a definite direction, so that a summation or strengthening of their peculiarities is arrived at, whilst other individuals of the same species, which preserve their original characters, or vary in another direction, must be prevented from uniting sexually with them. Such a process presupposes the presence of an environment which always acts in a constant manner, and the existence of a certain intervening space between the two sets of individuals belonging to the species, which is destined to divide into two new species.

Sexual reproduction, therefore, seems to me to influence the formation of a species in a manner opposed to that suggested by Weismann. By creating intermediate forms, it continually reconciles the differences which are produced by external circumstances in the individuals of a species; thus it tends to make the species homogeneous and to enable it to retain its own peculiar features. Here, too, sexual affinity, that mysterious property of organic substance, by preventing a combination, or at any rate a successful one, between substances which are either too similar or too dissimilar, acts as an important factor. For, if the sexual products, on account of their different organisation and their slight sexual affinity, cannot mingle successfully, the species and orders in question are kept apart.

Darwin and Spencer express the same opinion. According to the former, "intercrossing plays a very important part in nature, in keeping the individuals of the same species or of the variety true and uniform in character." And Spencer remarks : " In a species there is, through gamogenesis, a perpetual neutralization of those contrary deviations from the mean state, which are caused in its different parts by different sets of incidental forces; and it is similarly by the rhythmical production and compensation of these contrary deviations that the species continues to live."[1]

Literature VII.

1. AUERBACH. *Ueber einen sexuellen Gegensatz in der Chromatophilie der Keimsubstanz, etc. Sitzungsber. d. kgl. Preuss. Akad. d. Wissensch.* Nr. 35.
2A. A. DE BARY. *Ueber apogame Farne u. die Erscheinungen der Apogamie im Allgemeinen. Botanische Zeitung. Bd. XXXVI. 1878.*
2B. A. DE BARY. *Beiträge zur Morphologie u. Physiologie der Pilze. Abhandl. d. Senkenberg. naturf. Gesellschaft. 1881.*
3. VAN BENEDEN. *Siehe Capitel VI.*
4. BÖHM. *Ueber Reifung u. Befruchtung des Eies von Petromyzon. Arch. f. mikrosk. Anatomie. Bd. XXXII.*
5. BÖHM. *Die Befruchtung des Forelleneies. Sitzungsber. d. Gesellsch. f. Morph. u. Physiol. zu München. 1891.*
6. BÜTSCHLI. *Ueber die ersten Entwicklungsvorgänge der Eizelle, Zelltheilung u. Conjugation der Infusorien. Abhandl. der Senkenberg. naturf. Gesellsch. Bd. X. 1876.*
7. CALBERLA. *Befruchtungsvorgang beim Ei von Petromyzon Planeri. Zeitschrift für wissenschaftl. Zoologie. Bd. XXX.*
8. DARWIN. *Results of Cross and Self-Fertilisation in the Vegetable Kingdom.* London. 1879.
9. ENGELMANN. *Ueber Entwicklung und Fortpflanzung von Infusorien. Morpholog. Jahrbuch. Bd. I.*
10. P. FALKENBERG. *Die Befruchtung und der Generationswechsel von Cutleria. Mittheilungen aus der zoologischen Station zu Neapel. 1879.*
11. P. FALKENBERG. *Die Algen im weitesten Sinn. Schenk's Handb. der Botanik. Bd. II. 1882.*
12. FOCKE. *Die Pflanzen-Mischlinge. Botanische Zeitung. 1881.*
13. H. FOL. *Archives des sciences physiques et naturelles. Génève, 15. Oct., 1883.*
14. H. FOL. *Le quadrille des centres, un épisode nouveau dans l'histoire de la fécondation. Archives des scienc. phys. et nat. Génère. Troisième pér. Tom. XXV. 1891.*
15. L. GUIGNARD. *Nouvelles études sur la fécondation: Comparaison des phénomènes morpholog. observés chez les plantes et chez les animaux. Annales des sciences natur. Tom. XIV. Botanique. 1891.*
16. M. HARTOG. *Some Problems of Reproduction: a Comparative Study of Gametogeny and Protoplasmic Senescence and Rejuvenescence. Quar. Journ. Mic. Soc. 1891.*

---

[1] *Principles of Biology*, by Herbert Spencer, vol. i. p. 286.

17. Henking. *Untersuchungen über die ersten Entwicklungsvorgänge in den Eiern der Insekten.* Zeitschr. f. wissenschaftl. Zoologie. Bd. 49, 51, 54.
18. Hensen. *Die Physiologie der Zeugung.* Handb. der Physiologie. Bd. VI.
19. Oscar Hertwig. See Cap. VI., Nr. 30a, 32, 33, 34.
20. Oscar Hertwig u. Richard Hertwig. *Experimentelle Untersuchungen über die Bedingungen der Bastardbefruchtung.* Jena. 1885.
21. Richard Hertwig. *Ueber die Conjugation der Infusorien.* Abhandl. der bayer. Akad. der Wissensch. Cl. II. Bd. XVII. 1889.
22. R. Hertwig. *Ueber die Gleichwerthigkeit d. Geschlechtskerne bei den Seeigeln.* Sitzungsber. d. Gesellsch. f. Morphol. u. Physiol. in München. Bd. IV. 1888.
23. R. Hertwig. *Ueber Kernstructur u. ihre Bedeutung f. Zelltheilung u. Befruchtung.* Ebenda.
24. Hildebrand. *Die Geschlechter-Vertheilung bei den Pflanzen, etc.* Leipzig. 1867.
25. Ishikawa. *Vorläufige Mittheilungen über die Conjugationserscheinungen bei den Noctiluken.* Zoolog. Anzeiger. Nr. 353. 1891.
26. Keller. *Die Wirkung des Nahrungsentzuges auf Phylloxera vastatrix.* Zoolog. Anzeiger. Bd. X. p. 583. 1887.
27. Klebahn. *Studien über Zygoten: Die Keimung von Closterium und Cosmarium.* Pringsheim's Jahrbücher f. wissenschaftl. Botanik. Bd. XXII.
28. Klebs. *Zur Physiologie der Fortpflanzung.* Biolog. Centralblatt. Bd. IX. 1889.
29. E. L. Mark. *Maturation, Fecundation, and Segmentation of Limax campestris.* Bullet. of the Museum of Comp. Zool. at Harvard College. Vol. VI. 1881.
30. E. Maupas. *Le rajeunissement karyogamique chez les ciliés.* Arch. de Zool. expér. et génér. 2e série. Vol. VII.
31. C. Nägeli. *Die Bastardbildung im Pflanzenreiche.* Sitzungsber. der kgl. bayer. Akad. d. Wissensch. zu München. 1865. Bd. II. p. 395.
32. C. Nägeli. *Die Theorie der Bastardbildung.* Sitzungsber. der kgl. bayer. Akad. der Wissensch. zu München. 1866. Bd. I.
33. Nussbaum. *Zur Differenzirung des Geschlechts im Thierreich.* Arch. f. mikroskop. Anatomie. Bd. XVIII.
34. Oppel. *Die Befruchtung des Reptilieneies.* Arch. f. mikroskop. Anat. Bd. XXXIX. 1892.
35a. Pringsheim. *Ueber die Befruchtung der Algen.* Monatsber. d. Berliner Akad. 1855.
35b. Pringsheim. *Ueber Paarung von Schwärmsporen, die morphologische Grundform der Zeugung im Pflanzenreich.* Ebenda. 1869.
36. Rückert. *Ueber physiologische Polyspermie bei meroblastischen Wirbelthiereiern.* Anat. Anzeiger. Jahrg. VII. Nr. 11. 1892.
37. Selenka. *Befruchtung der Eier von Toxopneustes variegatus.* Leipzig. 1878.
38. Strasburger. *Neue Untersuchungen über den Befruchtungsvorgang bei den Phanerogamen als Grundlage für eine Theorie der Zeugung.* Jena. 1884.

39. WEISMANN. *Beiträge zur Naturgeschichte der Daphnoiden. Zeitschr. f. wissenschaftl. Zoologie. Bd. XXXIII.*
40. WEISMANN. *On the Number of Polar Bodies and their Significance in Heredity, trans. by Schönland; Essays upon Heredity, trans. by Poulton, Schönland, and Shipley. Oxford.* 1889.
41. WEISMANN u. ISHIKAWA. *Ueber die Bildung der Richtungskörper bei thierischen Eiern. Berichte der naturforsch. Gesellsch. zu Freiburg. Bd. III.* 1887.
42. WEISMANN and ISHIKAWA. *Weitere Untersuchungen zum Zahlengesetz der Richtungskörper. Zoolog. Jahrbücher. Bd. III., Abth. f. Morph.*
43. OTTO ZACHARIAS. *Neue Untersuchungen über die Copulation der Geschlechtsproducte und den Befruchtungsvorgang bei Ascaris megalocephala. Archiv f. mikroskop. Anat. Bd. XXX.* 1887.
44. BLOCHMANN. *Ueber die Richtungskörper bei Insecteneiern. Morphol. Jahrb. Bd. XII.*
45. BLOCHMANN. *Ueber die Reifung der Eier bei Ameisen u. Wespen. Festschr. zur Feier des 30 0jähr. Bestehens der Univers. Heidelberg.* 1886. *Med. Theil.*
46. BLOCHMANN. *Ueber die Zahl der Richtungskörper bei befruchteten und unbefruchteten Bieneneiern. Morphol. Jahrb. Bd. XV.*
47. PLATNER. *Ueber die Bildung der Richtungskörperchen. Biolog. Centralblatt. Bd. VIII.* 1888–89.
48. WEISMANN. *On Heredity, trans. by Shipley; The Continuity of the Germ-Plasm as the Foundation of a Theory of Heredity, trans. by Schönland; Essays on Heredity. Oxford.* 1889.
49. HERM. MÜLLER. *Die Befruchtung der Blumen durch Insecten.* Leipzig. 1873.
50. PFLÜGER. *Die Bastardzeugung bei den Batracheiern. Archiv f. die ges. Physiologie. Bd. XXIX.*
51. BERTHOLD. *Die geschlechtliche Fortpflanzung der eigentl. Phaeosporeen. Mittheil. aus der zool. Station zu Neapel. Bd. II.* 1881.
52. DARWIN. *The Origin of Species.* London. 1869.
53. DARWIN. *Variation of Animals and Plants under Domestication.* London. 1875.
54. HERBERT SPENCER. *Principles of Biology.* London. 1864.
55. RAY LANKESTER. *Art. Protozoa, Encyclopædia Britannica.* London. 1891.
56. HERBERT SPENCER. *First Principles.* London. 1870.

CHAPTER VIII.

METABOLIC CHANGES BETWEEN PROTOPLASM, NUCLEUS, AND CELL PRODUCTS.

ALL the morphologically different parts of a living organism necessarily stand to one another in a definite relation, as regards metabolic changes. In most cases it is extremely difficult to understand these relations, on account of the complexity of the vital processes. However, some knowledge has already been gained upon the subject, by means of observation and experiment, and the fact that protoplasm takes part in all formative processes, such as the formation of the cell-wall, of intercellular substance, etc., is indicated by various circumstances, which can scarcely be explained in any other manner.

In plants the main portion of the protoplasm is always massed together at those parts, where growth is chiefly taking place: *e.g.* at the ends of growing root-hairs, in the growing hyphæ, with Fungi, etc., and at the growing points of multicellular and unicellular plants, such as Caulerpa. Further, the protoplasm, in individual cells, always accumulates in the regions of greatest activity.

Sometime before the cellulose membrane of a plant-cell forms thickenings or sculpturings, the protoplasm undergoes preparatory changes, by collecting in the places where the most rapid growth is taking place. Further, whilst these thickenings are being formed, continuous streams of granular protoplasm are seen to pass along them.

If a small portion of *Vaucheria* is cut off, the protoplasm immediately tries to repair the injury. "Granular plasma can be seen to collect in dense masses about the wound, and to close up to form a layer, which is sharply defined externally. A cell-membrane immediately commences to develop upon this layer." (Klebs.)

If the protoplasm of a plant-cell has by means of plasmolysis been separated from its membrane, without damage having been done to its vital functions, it soon develops upon its surface a new cellulose layer, which becomes stained red when congo-red is added to the water.

As long as cells are young and growing vigorously, they contain a large quantity of protoplasm, whilst older cells, especially those in which formative activity has been arrested, only contain a small quantity of it. For instance, the protoplasmic layer, on the inner surface of the cellulose membrane of large and fully developed plant cells, may be so extremely thin that its presence, as a distinct stratum, can only be demonstrated by means of plasmolysis. Similarly, only minute traces of protoplasm are present in the notochordal cells of animals, etc.

The relations that the nucleus bears to the remaining component parts of the cell are at present attracting great attention. It has already been shown (p. 214) that very remarkable metabolic interactions take place between the nucleus and the protoplasm during the processes of division. But it is evident, that the nucleus plays an important physiological part at other times, as well, in the life of the cell; all the formative and nutritive processes seem to be dependent upon it, and to bear a close relationship to it. The true nature of this relationship, however, cannot at present be more exactly defined, as may be deduced from the observations of Haberlandt and Korschelt, which will be described later, as well as from the experiments of Gruber, Nussbaum, Balbiani, Klebs and Hofer.

I. **Observations on the position of the nucleus, as an indication of its participation in formative and nutritive processes.** According to the extensive and important observations of Haberlandt (VIII. 4) the nucleus of young and developing plant-cells is " situated in that portion of the cell where growth is most active, or lasts longest. This is true both for the growth of the cell as a whole and for the increase in volume and superficial area of the cell-membrane in especial. If the cell is growing in more than one place, the nucleus takes up a central position, so that it is about equidistant from the regions of most active growth (Fig. 161, *II*). Occasionally the nuclei are connected with the places of most active growth by means of protoplasmic strands, which are as short as possible. The nucleus only rarely retains its original position in fully developed cells. As a rule it has left the place which it occupied in the growing cell, and generally has no definite position. In other cases, however, its position is fixed."

I will cite a few especially instructive examples from the numerous observations, on which Haberlandt has based his laws.

The epidermal cells of many plants often exhibit thickenings on the surface of their walls; this may occur either on those pointing outwards or on those pointing inwards. The nucleus here lies near to the one in which the thickening occurs, being always close to the middle of the latter. The examples given in Fig. 161 show this very distinctly: No. I., a row of cells from the epidermis of a foliage-leaf of *Cypripedium insigne*; No. III. an epidermal cell of the fruit-scale of *Carex panicea*, and No. IV. a young epidermal cell of a foliage-leaf of *Aloë verrucosa*.

A second series of investigations have been made upon the development of plant-hairs, growing both above and below ground.

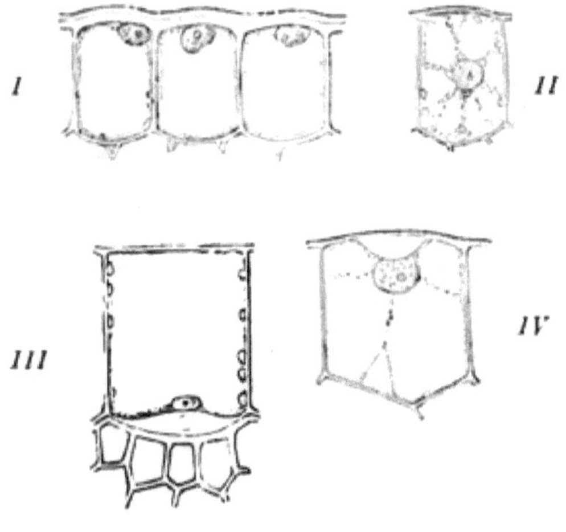

FIG. 161.

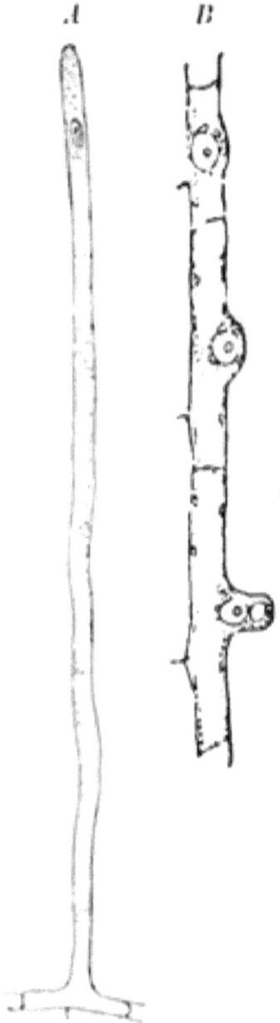

FIG. 162.

FIG. 161.—*I* Epidermal cells of a foliage leaf of *Cypripedium insigne* (after Haberlandt, Pl. I., Fig. 1). *II* Epidermal cells of *Luzula maxima* (after Haberlandt, Pl. I., Fig. 3). *III* Epidermal cells of the fruit-scale of *Carex panicea* (after Haberlandt, Pl. I., Fig. 14). *IV* Young epidermal cells of a foliage leaf of *Aloë verrucosa* (after Haberlandt, Pl. I., Fig. 7).

FIG. 162. *A* Root-hair of *Cannabis sativa* (after Haberlandt, Pl. II., Fig. 20). *B* Formation of root-hairs of *Pisum sativum* (after Haberlandt, Pl. II., Fig. 22).

The tender root-hairs of plants exhibit a characteristic structure at their growing points. Hence the nucleus, as long as growth continues (Fig. 162 *A*), is situated at the free end, whilst when the hairs are old and fully developed, it is higher up. When a root-hair is developing out of an epidermal cell, a protuberance is always formed upon that part of the external wall, which is situated over the cell-nucleus (Fig. 162 *B*). In many plants (*Brassica oleracea*) the root-hair cell may form branches, into one of which the single nucleus enters. This one becomes at once the richest in protoplasm and also the longest, whilst the other branches leave off growing.

The hairs that grow above ground, differ from the root-hairs, in that they exhibit a basipetal, or intercalary growth, as Haberlandt has established by measurements. In consequence of this, the nucleus is not situated at the apex, but near to the place, where the secondary, basal growing-point is situated, and where longitudinal growth persists longest.

Stellate hairs (Fig. 163) are peculiar, unicellular structures, which split up at their peripheral end into several radially divergent branches. Under these circumstances the nucleus, as long as the formative processes continue, is situated in the middle of the radiation, but after growth is finished it returns to its former position near to the base.

Confirmatory evidence of this participation of the nucleus in the formative processes is furnished us by the examination of Fungi and Algæ. In the multi-nucleated hyphæ of *Saprolegnia* lateral branches develop; these are always found immediately over a nucleus, which is situated close to the cell-wall. In *Vaucheria* and other multi-nucleated Algæ, as in the higher plants, special growing points are present, at which growth chiefly occurs; at each of these, immediately underneath the cellulose membrane, there is an accumulation of small nuclei, after which comes a layer of chromatophores; in the remaining portions of the cell the positions of these bodies are reversed.

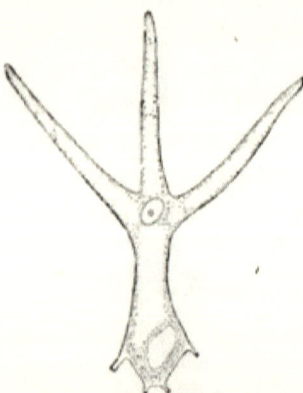

Fig. 163.—Young stellate hair of *Aubrietia deltoidea* (after Haberlandt, Pl. II., Fig. 28).

Phenomena, which are still more remarkable, and which indicate the part played by the nuclei in the formation of the cell-wall, are to be observed during the healing of wounds in *Vaucheria*. Numerous small nuclei appear in the protoplasm, which collects round about the wound, thus approaching the upper surface, whilst the grains of chlorophyll are forced back in exactly the opposite direction. By this means the nuclei and chlorophyll grains exchange places. This observation immediately refutes the objection, which might otherwise easily be raised, namely, that the nucleus or nuclei are present in those places to which the protoplasm flows in greater quantities, because they are carried along by the protoplasmic stream. For, if this were the case, we should expect to find the chlorophyll grains also in the same places, since these are much smaller than the nuclei, and may even be induced to change their positions by variations in illumination, which have no effect upon the nuclei.

"Thus we see," as Haberlandt remarks, "that the nuclei and chlorophyll grains exhibit quite independent changes of position, which, if we assume that they are passive, cannot in any way be influenced by the movements of the granular plasma as a whole. These phenomena—that the streaming protoplasm to a certain extent selects the bodies, which it carries along with it, in the one case taking the larger cell-nucleus, and leaving the smaller chromatophores and neglecting the cell nuclei, which are as small or even much smaller—can only be explained by supposing, that their rôle is to effect definite accumulations, which depend upon the functions of the nuclei and the chromatophores."

Korschelt (VIII. 8) has demonstrated, that relations, similar to those described by Haberlandt, as existing between the position and the function of the nuclei in plant cells, are also present in animal cells.

Ova increase considerably in size, by absorbing large quantities of reserve materials. In these, the germinal vesicle is frequently found in that place, where the absorption of material must of necessity take place. Thus, for instance, in one species of *Coelenterates*, the ova are derived from the endoderm and are nourished by the gastrovascular system by means of endodermal cells. In conformity with the above-stated law the germinal vesicles of young ova are situated superficially near to the surface of that wall, which is turned towards the gastric cavity (Fig. 164). In many *Actiniæ* (Hertwig, VIII. 5b) the ova, for a considerable period, protrude a stalk-like (peduncular) process right up to the surface of the intestinal epithelium (Fig. 165). This process has a regular fibrillary (rodded) structure, as is always seen, when an active exchange of material takes place in definite directions; it may, therefore, be considered to be a special nutrient apparatus of the ovum. In this case, too, the germinal vesicle is always situated in immediate contact with the base of the nutrient apparatus.

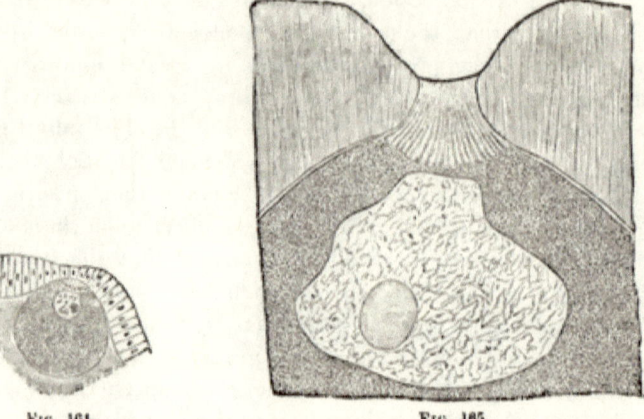

Fig. 164. Fig. 165.

Fig. 164.—Immature ovum of *Actinia parasitica*. (× 145; after Korschelt, p. 47, Fig. 8.)

Fig. 165.—Transverse section through the peripheral end and through the stalk of egg-cells of *Sagartia parasitica* (after O. and R. Hertwig); from Korschelt, Fig. 10. The striated stalk of the egg-cell has penetrated into the epithelium at the top of the figure.

A similar condition is found in the tubular ovaries of Insects, which are divided into germ compartments and yolk compartments. In this case the germinal vesicle is either again placed close to the yolk compartment, or, which is more interesting, it extends towards this compartment numerous pseudopodic processes, by which means it considerably increases its superficial area in that region, where the absorption of material is taking place. Here, too, the yolk in the neighbourhood of the germinal vesicle begins to separate off numerous dark granules, which have been derived from the nutritive cells.

In most animals the ova are nourished by means of the follicular cells. Thus Korschelt has found that, as long as the formation of the yolk and chorion is proceeding, the nuclei of the follicular cells in Insects are situated in immediate contact with that surface which is directed towards the ovum, whilst after the chorion has been completed, they retreat into the middle of the cell.

Still more striking is the behaviour of the nuclei in the so-called double cells, which occur in the eggs of water-bugs (*Ranatra* and *Nepa*, Fig. 167 *A*, *B*). These develop radiating chitinous processes on the chorion. The protoplasmic bodies of the two cells, between which a radiation figure develops, coalesce. During this process both of the very large nuclei extend numerous fine processes towards that side, which is turned towards the radiated figure.

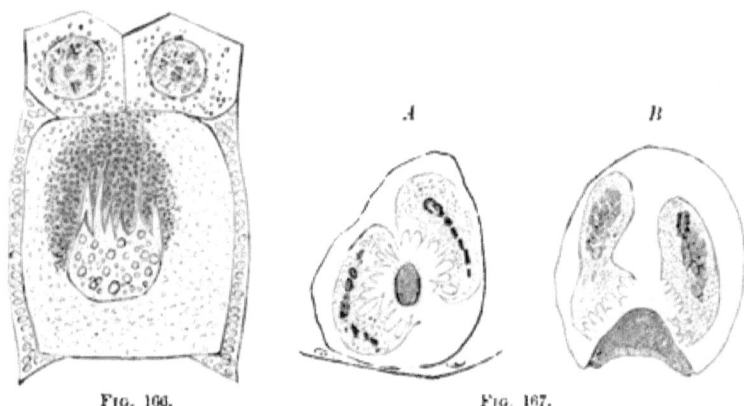

Fig. 166.

Fig. 167.

Fig. 166.—Egg-follicle of *Dytiscus marginalis* with neighbouring yolk compartment, in which a large number of granules are being separated off. The germinal vesicle of the ovum is extending processes towards the accumulations of granules. (After Korschelt, Pl. I., Fig. 20.)

Fig. 167.—*A* Transverse section of a secreting double cell from the egg-follicle of *Nepa cinerea* L. The formation of the radiation figure is still taking place (× 270; after Korschelt, Pl. V., Fig. 120). *B* Longitudinal section of a double cell from the egg-follicle of *Nepa*. Commencement of the development of the radiation figure (× 195; after Korschelt, Pl. V., Fig. 121).

From these and similar observations, Haberlandt and Korschelt draw the following conclusions, respecting the function of the cell-nucleus:—

1. "The fact that the nucleus is generally found in a definite position in the immature and developing cells, indicates that its function is connected chiefly with the developmental processes of the cell." (Haberlandt.)

2. "From its position it may be concluded that the nucleus plays a definite part during the growth of the cell, especially during the thickening and increase in superficial growth of the

cell-wall. This does not prevent it from eventually fulfilling other functions in the fully developed cell." (Haberlandt.)

3. The nucleus takes part both in the excretion and absorption of material. This is shown by its position, and also by the fact that the nucleus increases its superficial area by extending numerous processes towards the place where excretion and absorption are occurring.

II. **Experiments proving the reciprocal action of the nucleus and protoplasm.** The experimental researches of Gruber, Nussbaum, Hofer, Verworn, Balbiani, and Klebs have led to the same results. Their method was to divide by some means or other, a unicellular organism or a single cell into two portions, one nucleated and the other non-nucleated, and then to follow and compare their future behaviour.

By means of plasmolysis in 16 per cent. sugar solution, Klebs was enabled (IV. 14; VIII. 7) to divide the cells of *Spirogyra* threads into one nucleated part and several non-nucleated portions. Although these latter sometimes live for six weeks before they disintegrate, the vital processes occurring in them differ considerably from those taking place in the nucleated ones, the latter continuing to grow and to surround themselves with a new cell-wall, which stains easily with congo red, and can thus be rendered visible. The former on the other hand remain globular in form, do not increase in size, and develop no cell-wall. That the latter process is considerably influenced by the presence of the nucleus, is clearly shown by the fact that, when the fragments obtained by means of plasmolysis, are connected by a thin bridge of protoplasm, the non-nucleated part is able to form cellulose.

However, certain metabolic processes take place in protoplasm without the presence of the nucleus; for instance, the non-nucleated parts are still able to assimilate, to dissolve, and to form starch, provided that they contain a portion of the chlorophyll-band. If they are kept for a considerable time in the dark, they become free from starch, because they have used up the stock of stored-up granules; when they are brought back again into the light, the chlorophyll bands recharge themselves with newly-assimilated starch; indeed, in this case the accumulation of starch is even greater than in the nucleated part, probably because its consumption, whilst all the other vital functions are in abeyance, is reduced to a minimum.

Non-nucleated portions of *Funaria hygrometrica* behave somewhat differently, in that they are able to dissolve starch, but cannot develop it, even if they remain alive for six weeks.

When a *Vaucheria* thread is divided into various sized masses of protoplasm, some of which contain nuclei, we find that the vital activity of these, as well as the separation of a new cellulose membrane, depends upon the presence in each, of at least one cell-nucleus. (Haberlandt, VIII. 4.)

Results, which are no less important than those obtained with plants, are observed when *Amoeba*, *Reticularia* and *Ciliata* are cut up. Nussbaum (VIII. 9), Gruber (VIII. 3), Hofer (VIII. 6), and Verworn (VIII. 10) all agree that only nucleated parts are able to replace organs which they had lost, and thus to reconstruct themselves into normal individuals, that grow and multiply. Non-nucleated portions, even when they are larger than the nucleated ones, are unable either to replace the lost organs or to grow, but for some time, often for more than fourteen days, appear to lead a kind of pseudo-existence; eventually, however, they disintegrate. Thus the formative activity of protoplasm appears to be primarily influenced by the nucleus. This is less certainly established in the case of the other functions of the cell, viz. power of movement, irritability and processes of digestion. As regards these the opinions of different observers vary.

Hofer observed that a non-nucleated portion of an *Amoeba*, after the first stage of irritability occasioned by the operation had passed off, exhibited for from fifteen to twenty minutes, movements which were nearly normal. He ascribes this to an after-effect of the nucleus, which, he considers, exerts a regulating influence upon the movements of the protoplasm. For whilst, further, the nucleated part extends pseudopodia like a normal individual, and propels itself forwards, the non-nucleated part contracts up into a round body, and only occasionally, after pauses of many hours' duration, makes abnormal, jerky movements; it does not attach itself to the bottom of the glass, as crawling *Amoebae* do, and in consequence vibrates upon the slightest movement of the water.

Verworn discovered that the protoplasm in *Difflugia* was still more independent of the nucleus. Even small non-nucleated portions extended long finger-like pseudopodia in a manner characteristic of an uninjured Rhizopod, and continued their movements even for five hours. Further, they were unimpaired as regards irritability, reacting to mechanical, galvanic, and chemical stimuli by contracting their bodies.

According to Verworn, *Ciliata*, too, which have developed special locomotive organs, such as cilia, flagella, cirrhi, etc., assume, when cut up, a complete autonomy and independence of the nucleus.

In *Lacrymaria*, each part, when deprived of its nucleus, exhibits, after its separation from the body, the same movements as it was performing before. Small portions of *Stylonichia*, which are furnished with a number of ventral cilia, continue to make with them the movements peculiar to their species. Even the minutest portion of protoplasm, which is furnished with only one bristle-like cilium, continues to make with it characteristic movements. If it was directed backwards, it is suddenly from time to time jerked forwards, by which movement the portion receives a short jerk backwards; thereupon the cilium returns again to a state of rest, and so on.

The contractile vacuoles of the *Protista* are, like cilia and cirrhi, remarkable for complete autonomy. Even in non-nucleated portions they can be observed to contract rhythmically for days together (Verworn).

Finally, an important difference is noticeable between non-nucleated and nucleated portions, as regards digestion. Whilst small *Infusoria*, *Rotifera*, etc., are normally digested by nucleated portions, in non-nucleated parts digestion is considerably diminished, both as regards time and intensity. It may, therefore, be concluded that protoplasm can only produce digestive secretions with the assistance of the nucleus (Hofer, Verworn).

It is not surprising that diversities of opinion, as mentioned in Chapter VII., should exist upon this subject, when the difficulty of the problems to be solved be taken into account.

Literature VIII.

1. BALBIANI. *Recherches expérimentales sur la mérotomie des Infusoires ciliés.* Prém. part. Recueil. Zool. Suisse. 1889.
2. BOVERI. *Ein geschlechtlich erzeugter Organismus ohne mütterliche Eigenschaften.* Gesellsch. f. Morphol. u. Pysiol. zu München. 1889.
3. GRUBER. *Ueber die Einflusslosigkeit des Kerns auf die Bewegung, die Ernährung u. das Wachsthum einzelliger Thiere.* Biolog. Centralblatt. Bd. III.

    GRUBER. *Ueber künstliche Theilung bei Infusorien.* Biolog. Centralbl. Bd. IV. u. V.

4. HABERLANDT. *Ueber die Beziehungen zwischen Function und Lage des Zellkerns bei den Pflanzen.* Jena. 1887.
5A. OSCAR u. RICHARD HERTWIG. *Ueber den Befruchtungs- u. Theilungsvorgang des thierischen Eies unter dem Einfluss äusserer Agentien.* Jena. 1887.
5B. OSCAR u. RICHARD HERTWIG. *Die Actinien, anatomisch und histologisch mit besonderer Berücksichtigung des Nervenmuskelsystems untersucht.* Jena. 1879.
6. HOFER. *Experimentelle Untersuchungen über den Einfluss des Kerns auf das Protoplasma.* Jenaische Zeitschrift f. Naturwissenschaft. Bd. XXIV.
7. KLEBS. *Ueber den Einfluss des Kerns in der Zelle.* Biolog. Centralbl. Bd. VII. 1887.
8. KORSCHELT. *Beiträge zur Morphologie u. Physiologie des Zellkerns.* Zool. Jahrbücher. Abth. f. Anatomie. Bd. IV. 1889.
9. NUSSBAUM. *Ueber die Theilbarkeit der lebendigen Materie.* Archiv. f. mikroskop. Anatomie. Bd. XXVI. 1886.
10. VERWORN. *Die physiologische Bedeutung des Zellkerns.* Archiv. f. d. ges. Physiologie. Bd. LI. 1891.
11. VINES. *Students' Text-book of Botany.* London. 1895.
12. CLARK, J. *Protoplasmic Movements and their relation to Oxygen Pressure.* Proceedings of the Royal Society, XLVI. 1889.
13. WOODHEAD AND WOOD. *The Physiology of the Cell considered in relation to its Pathology.* Edinburgh Medical Journal. 1890.

CHAPTER IX.

THE CELL AS THE ELEMENTAL GERM OF AN ORGANISM (THEORIES OF HEREDITY).

We are forced to the conclusion, that the cell is a highly organised body, composed of numerous, minute, different parts, and that hence it is in itself to a certain extent a small elementary organism, when we consider, that it is capable of executing movements, and of reacting in a constant manner to the most various external stimuli, which may be chemical, mechanical, or caused by heat or light; and further that it can execute complicated chemical processes and can produce numerous substances of definite composition.

This idea is still more impressed upon us, when we take into account the fact, that egg- and sperm-cells form by their union the elemental germ which develops into an organism, the latter reproducing on the whole the attributes of the parents, even often to the most insignificant characteristics. Hence we must conclude, that the egg- and sperm-cells possess all the constituent properties which are necessary for the production of the final result of the developmental process. It is true that these properties elude our perception, but that they are anything but simple, is evident from the complex composition which is attained by the final product of development in the highest organisms. The sexual cells must therefore, of necessity, possess a large number of attributes and characteristics, which are concealed from us, but whose presence renders the formation of the final product possible. These hidden or latent properties, which only gradually become evident during the process of development, are called fundamental constituent attributes. These attributes, taken collectively, to a certain extent foreshadow or potentially determine the matured organism.

At a certain stage of their development, when they are simple cells, all organisms are extremely alike. The ova of man, of rodents, of ruminants, and even of many invertebrate animals, do not differ from one another in any essential points; they resemble one another more closely than do the egg- and sperm-cells of the same animal.

However, these similarities and differences in form appear to be of less importance when we go more deeply into the subject. For, as men, rodents, ruminants, and invertebrate animals present to us more or less important external differences, the sexual cells originating from them must differ in a corresponding manner as regards their fundamental attributes, in so far as they represent the embryonic stage of the subsequent complete organism. The only thing is that, at present, the essential differences lie beyond our perception. On the other hand the egg- and sperm-cells of the same organism, although they differ so much in external appearance, must resemble one another in their essential properties, since they must contain potentially all the characteristics of the fully-developed animal.

Nägeli pertinently remarks (IX. 26): "The egg cells must contain all the essential characteristics of the mature organism, and hence they must differ as much from one another, when they are in this early stage, as when they are more fully developed. The Hen's egg must possess the characteristics of its species as completely as the Hen, and hence must differ as much from the Frog's egg as the Hen does from the Frog."

What is true of the egg is equally true of individual cells and collections of cells, which, being detached from the mother organism, either as spores or buds, are able to reproduce the parent. They, too, must possess all the essential properties of the whole, in an embryonic condition, although they are imperceptible to us.

What idea can we form to ourselves of these invisible properties of the cells, which predetermine the complex organism? What is the connection between the developed and undeveloped stage?

These problems are amongst the most difficult which the theory of life presents. Scientists and philosophers have occupied themselves with these questions for centuries, and have formulated their conclusions in hypotheses, which have frequently influenced enquiry. We will mention shortly those theories which are most important historically, since they are both of general interest, and will serve as a suitable introduction to the consideration of the views, which are suggested by modern research.

I. History of the older Theories of Development.

Two important scientific theories which are directly opposed to one another, were advanced up to the beginning of this century; viz., the *theory of Preformation or Evolution and the theory of Epigenesis.*

The *theory of Preformation* was embraced by such well-known authorities of the 17th and 18th centuries, as Swammerdam, Malpighi, Leeuwenhoek, Haller, Bonnet (IX. 3), and Spallanzani (cf. His IX. 14). They held the opinion, that the germ, as regards structure, absolutely resembles the mature organism, and that hence it must, from the very first, possess similar organs, which, although extremely minute, must be in the same positions and similarly related to one another. Since, however, it was impossible by means of the microscopes at their command, actually to observe and demonstrate these organs, which they assumed to be present in the egg at the beginning of its development, they took refuge in the theory, that certain parts, such as the nervous system, glands, bones, etc., were present not only in a minute, but also in a transparent condition.

In order to render the process more comprehensible, the development of the butterfly from the chrysalis, and the flower from the bud, were quoted as examples. Just as a small bud of green, tightly closed sepals, contains all the parts of the flower, such as stamens and coloured petals, and as these parts grow in secret, and then suddenly, when the sepals unfold, become revealed, so the "Preformists" considered, that the minute parts, which are supposed to be present in a transparent condition, grow, gradually reveal themselves, and become perceptible to our eyes.

Hence the old name of the "*theory of Evolution or Unfolding,*" in the place of which the more pertinent, intelligible, designation of the "*theory of Preformation*" has been adopted. For the peculiarity of this doctrine, is that nothing is supposed to be newly formed at any period of development, each part being present or preformed from the beginning, and that, therefore, the true nature of development or growth is denied. "There is no new development," says Haller, in his *Elements of Physiology*; "no part in the animal body is formed before the other; all are created at the same time."

The theory of Epigenesis is directly opposed to the theory of Preformation. Its chief supporter was Caspar Friedrich Wolff (IX. 36), who lived in the middle of the 18th century. In his important paper, entitled "*Theoria Generationis,*" published in the year 1759 (Germ. ed. 1764), he enunciated the following axiom, which was in opposition to the generally accepted dogma of preformation, namely, "that what cannot be perceived by the senses, is not present in a preformed state in the germ; that the germ at the outset is nothing but unorganised matter, excreted from the sexual organs of the parents, which in consequence of fertilisation, gradually becomes organised during the process of development." He states further that the organs differentiate themselves one after another out of this unorganised germinal substance, and he tried to actually demonstrate this process in individual cases. Thus he showed how various plant organs gradually differentiate themselves out of the germinal substance, and in so doing undergo alterations in their shape, and he pointed out that the intestinal canal of a chick develops out of a leaf-shaped embryonic structure.

By thus basing his arguments upon accurate observation, instead of upon preconceived notions, Wolff laid the foundation-stone of the important hypothesis, which, based upon the theory of development, has been gradually built up during the course of this century.

If we carefully compare these two theories, we see that neither can be accepted in its entirety. Both have their weak points.

The theory of Preformation is open to attack from the standpoint of the evolutionists, since, in the higher organisms, each individual is produced by the co-operation of two members of separated sexes. When, later on, Leeuwenhoek discovered the existence of spermatozoa as well as ova, an animated discussion arose as to whether the egg or the spermatozoon constituted the preformed germ.

The hostile schools of the Ovists and Animalculists existed for a century. The Ovists, such as, for instance, Spallanzani, stated that the unfertilised ovum of a Frog was a diminutive Frog, being of opinion that the spermatozoon only acted as a stimulating agent, exciting vital activity and growth. The Animalculists, on the other hand, by means of the magnifying glasses at their disposal, discovered the presence of heads, arms, and legs in the spermatozoon. They therefore considered that the egg was only a suitable nutrient medium, which was necessary for the development of the spermatozoon.

Further, the theory of Preformation, more logically worked out, leads to very serious difficulties. One such obstacle, which even Haller and Spallanzani did not think could be overcome, was the consideration that the germs of all the subsequent animals would have to be stored up or contained in one germ. This principle would necessarily follow from the fact, that sexual animals develop in unbroken sequence from one another. Therefore, the natural outcome of the Preformation theory, is the pill-box theory, or, as Blumenbach (IX. 2) expresses it, the theory of the "imprisoned germs." The eagerness of its supporters actually carried them so far, that they reckoned out how many human germs were boxed up in the ovary of mother Eve, and put down the number as, at the very least, 200,000 millions (*Elemente der Physiologie*, by Haller).

On the other hand, the theory of Epigenesis in its older form, when worked out more fully, also presents difficulties. For the question suggests itself how nature, with the forces that we know of at her command, can produce in a few days or weeks, out of unorganised matter, an animal organism resembling its progenitors. On this point no theory, which regards the organism as a completely new creation, can supply us with an acceptable and satisfactory solution.

Blumenbach (XI. 2), therefore, took refuge in the conception of a peculiar "nisus formativus," or formative instinct, which was supposed to cause the unformed or unorganised male and female fluids to assume a "formation," *i.e.* a definite form, and later on to replace any parts that had been lost. But if we accept the existence of an especial formative instinct, we have obtained nothing more than an empty expression, in the place of an unknown thing.

The cell theory, which has been gradually worked out during the latter half of this century, has furnished us with new fundamental facts, upon which to base more accurate theories of generation and heredity. These facts are, first, that ova and spermatozoa are simple cells, which free themselves from the parent organism for the purposes of reproduction, and that the developed organisms are only organised combinations of a very large number of such cells, which are able to function in various ways, and which are produced by the repeated division of the fertilised egg-cell. A second, and still more advanced principle, is, that the cell in itself is an extremely complex body, that is to say, that it is an elementary organism. Thirdly, we have gained a fuller knowledge of the process of fertilisation, of nuclear structure and nuclear division (longitudinal division and arrangement of the nuclear segments), whilst the discovery of the fusion of the egg and sperm nuclei, of the equivalence of the male and female nuclear masses, and of their distribution amongst the daughter-cells, has given us a greater insight into the complicated processes of egg and sperm maturation, and the reduction of the nuclear substance thus produced.

II. More recent Theories of Reproduction and Development.

The new theories of generation have been worked out chiefly by Darwin (IX. 6), Spencer (IX. 26), Nägeli (IX. 2), Strasburger (IX. 27, 28), Weismann (IX. 31-34), de Vries (IX. 30), and myself (IX. 10-13). The sharp antagonism which existed between the theories of Preformation and Epigenesis has been diminished in these theories, in that in certain respects they resemble both; so that they could be designated from one point of view, as the continuation of preformatory, and from another, as a further extension of epigenetical views. The new theories, although they hardly deserve more than the name of hypotheses, differ from the old, in that they are based upon a large collection of well-substantiated facts, which are to a certain extent fundamental.

It would take too long to mention the different views of the above-mentioned scientists, who, though they agree in many essential points, differ considerably as to details. I will, therefore, limit myself to a short description of what seems to me to be the essential part of the modern theories of generation and development.

All the numerous attributes of the developed organism are present in an embryonic condition in the sexual products since they are passed on from the parent to the offspring. They may be considered to constitute an hereditary mass (idioplasm, Nägeli). Each act of generation or development, therefore, does not result in a new formation, or epigenesis, but produces a transformation or metamorphosis of an elemental germ, or of a substance which was provided with potential forces, converting it into a developed organism; this, again, in its turn produces elemental germs, similar to those from which it was derived.

If the matured organism be considered to be a macrocosm, the hereditary mass on the other hand represents a microcosm, composed of numerous regularly arranged particles of material of different kinds, which, each being provided with its own peculiar forces, are the bearers of the hereditary properties. Just as the plant or animal can be divided into milliards of elementary parts, viz. cells, so each cell is composed of numerous, small, hypothetical elementary particles.

Darwin, Spencer, Nägeli, and de Vries have called these hypothetical units by different names, although they mean the same thing by them. Darwin (IX. 6) in his provisional hypothesis of Pangenesis, calls them little germs or gemmulæ; Spencer (IX. 20), in his Principles of Biology, speaks of physiological units; Nägeli (IX. 20), of particles of idioplasm or groups of micellæ; and de Vries, in his essay upon Darwin's Pangenesis, calls them Pangenæ.

What then are these small elementary portions of the cell, which I will in future call idioblasts, in accordance with Nägeli's views, who, in my opinion, has most ably criticised the subject in question?

It must be borne in mind, in answering this question, that no precise definition of an idioblast can at present be given, like that given by chemists and physicists of the terms atoms and molecules. We are still on unknown ground, like the scientists of the eighteenth century, who tried to prove that animal bodies were constructed out of elementary units. Naturally, the danger of going astray increases, the more we try to work this hypothesis out in detail. I will, therefore, confine myself as far as possible to the most general considerations.

The hypothetical idioblasts are the smallest particles of material into which the hereditary mass or idioplasm can be divided, and of which great numbers and various kinds are present in this idioplasm.

They are, according to their different composition, the bearers of different properties, and produce, by direct action, or by various methods of co-operation, the countless morphological and physiological phenomena, which we perceive in the organic world. Metaphorically they can be compared to the letters of the alphabet, which, though small in number, when combined form words, which, in their turn, combine to form sentences; or to sounds, which produce endless harmonies by their periodic sequence and simultaneous combinations.

De Vries remarks that "just as physicists and chemists have been obliged to resort to atoms and molecules, the biologist has been forced to presuppose the existence of certain units, in order to explain by means of them the various vital phenomena."

In Nägeli's opinion, "the characteristics, organs, structures, and functions, all of which are only perceptible to us collectively, are resolved into their true elements in the idioplasm." Such elements, according to de Vries, are the particles which are able to form chlorophyll, the colouring matter of flowers, tannic acid or essential oils, and we may add muscular tissue, nerve tissue, etc.

> Similar ideas are expressed in a somewhat different form, and regarded from other points of view, by Sachs (IX. 25) in his essay "Stoff und Form der Pflanzenorgane." Here he says, "we are forced to assume the presence of as many specific formative materials as there are definite forms of organs to be distinguished in a plant." We must therefore imagine that "very small quantities of certain substances are able so to influence those masses of materials, with which they are mixed, that they induce them to set into different organic forms."

Although at present we cannot with any degree of certainty define the specific nature of a single idioblast, we are able to draw fairly definite conclusions regarding some of their common properties.

It is, of course, first necessary to consider, that the hypothetical idioblasts must possess the power of multiplying by means of division, like the higher elementary units, the cells. For the egg imparts to each of the two cells into which it divides, and these again to the daughter-cells, which are derived from them, certain particles, which are the bearers of specific properties. Hence a multiplication of these particles must take place during the different processes of development; they must further be able to go on dividing, and in consequence must possess also the power of growth, without which continuous divisibility is inconceivable. Darwin, Nägeli, and de Vries, therefore, logically assume that their gemmulæ, particles of idioplasm, and pangenæ, are both able to grow and to divide.

This assumption enables us to draw another conclusion about the nature of the idioblasts, viz. that by their very nature they cannot be identical with the atoms and molecules of the chemist and physicist; for the former are indivisible, and the latter, although divisible, split up into portions, which no longer possess the properties of the whole. A definite molecule of albumen cannot grow without changing its nature, for when it takes up new groups of atoms, it enters into new combinations, by which means its properties are altered. Neither can it break itself up into two similar molecules of albumen, since the portions obtained by dividing a molecule, consist of groups of atoms of unequal value. On this account idioblasts are not identical with the plastidules, the existence of which is assumed by Elsberg and Hæckel (IX. 8 b). For, according to Hæckel, the latter possess all the physical properties, which physicists ascribe to molecules, or to collections of atoms, in addition to especial attributes, which belong exclusively to themselves, viz. "the vital properties which distinguish the living from the dead, and the organic from the inorganic."

Our units, therefore, the gemmulæ of Darwin, the pangenæ of de Vries, and the physiological units of Spencer, must be complex units, or, at any rate, groups of molecules. In this fundamental view, all the above-mentioned scientists agree. Thus, according to Spencer, there is nothing left but to assume, that chemical units combine together to form units of an infinitely more complex nature than their own, complex though this be, and that in every organism the physiological units, produced by such combinations of highly complex molecules, possess various characters."

If Nägeli's hypothesis of the molecular structure of organised bodies be accepted, it is easy to imagine that the nature of the idioblasts is as follows: "They can as little be single micellæ (crystalline molecule-groups), as molecules; for even if, as a mixture of different modifications of albuminates, they possess different properties, they would still lack the capacity of multiplying and forming new similar micellæ. Insoluble and stable groups of albuminous micellæ alone afford all the necessary conditions for the construction of the gemmulæ; they alone, in consequence of their varying composition, can acquire all the necessary properties, growing indefinitely by storing up micellæ, or multiplying by means of disintegration. Hence, the pangenæ or gemmulæ must consist of small masses of idioplasm."

Now comes the question: What is the size and number of the idioblasts contained in a complete germ?

As regards size, the idioblasts must certainly be exceedingly small, since all the hereditary elemental germs of a highly-developed organism must be present in the minute spermatozoon. Nägeli has attempted to make an approximate calculation on this important point. He starts with the assumption, that the hypothetical albumen formula of chemists, with seventy-two atoms of carbon ($C_{72}H_{106}N_{18}SO_{22}$), does not represent a molecule of albumen, but a micella of crystalline construction composed of several molecules. Its absolute weight is the trillionth part of 3·53 mg. The specific weight of dry albumen is 1·344. Hence, 1 cubic micro-millimetre contains about 400 million micellæ. Nägeli, basing his calculations on some further hypotheses, considers that the volume of such a micella is ·0000000021 cub. mic. mil. Further, upon the supposition that micellæ are prismatic, and are only separated from one another by two layers of molecules of water, 25,000 micellæ would occupy a superficial area of ·1 sq mic. mil. Hence, in a body of the size of a spermatozoon there would be room for a considerable number of micellæ, united together in groups. Thus, no difficulties present themselves on this point.

Logically thought out ideas are especially valuable, when they harmonize with perceptible facts. The following observations are in support of the above-mentioned hypothesis, *i.e.* that idioblasts multiply by growth and sub-division ; the capacity of self-division does not only apply to the individual cell as an elementary organism, but also to the above-mentioned masses of special material, which are enclosed in the cell. Chlorophyll, starch, and pigment formers multiply by direct division : the centrosomes, which are only just perceptible with the microscope, also divide, when nuclear segmentation occurs ; the nuclear segments split up longitudinally into daughter-segments, and this is attributed by many to the presence in the mother-thread of qualitatively different units (mother-granules), which are arranged in a row one behind the other; each of these is supposed to divide directly into two, after which the daughter-granules thus obtained, distribute themselves evenly amongst the daughter-segments.

Even if the idioblasts, which we have supposed to be of a much smaller size, do not themselves take part in these divisions, we may assume that groups of idioblasts are so concerned; the importance of these observations, as concerns our theory, consists in this, that they teach us how small masses of material grow in the cell by themselves, and are able to multiply by division.

Finally, another aspect of this theory may be mentioned here. If the elemental germs, taken in the aggregate, give rise to a definite organism, the individual constituents must evolve in regular sequence, during the process of development. As sentences, with logical meanings, are formed of words, and these of letters ; and similarly, as harmonies, and whole musical compositions, consist of individual notes, suitably arranged, so we must also assume that the idioblasts are arranged in a constant regular manner. This portion of the theory is the most difficult to understand.

In the above, certain logical principles for the formation of a physiological molecular theory of generation and heredity have been deduced, in accordance with Nägeli's views. We must leave the proof of the correctness of these assumptions to future observers and experimenters, who will thereby establish the relation between the theory, and the facts which are perceptible to our senses. The physiological idea of the creation of the organic world from elementary units, and of the essential agreement in the structure of plants and animals, have been of real service in building up the cell and protoplasm theories; in a similar manner we must hope to obtain a corresponding position for the theory of heredity. Several attempts have already been made in this direction, connected with the observations made upon the fertilisation in animals, plants, and Infusoria.

III. The Nucleus as the transmitter of Hereditary Elemental Germs.

The hypothesis that the nuclei are the transmitters of the hereditary properties, was suggested to both Strasburger and myself by the study of the process of fertilisation and of the theoretical considerations connected with it; thus we have assigned to the nuclear substance a function, which is different from that of protoplasm. A short time before, Nägeli had been compelled, solely on logical grounds, to assume, that two different kinds of protoplasm were present in the sexual cells, the one sort which occurs in exactly equal proportions in the egg and sperm cell, conveying the hereditary properties, and the other, which is stored up in great quantities in the ovum, functioning chiefly as a nutritive medium. He calls the first idioplasm, and the second somatoplasm, and assumes that the former is more solid in consistency, the micellæ being regularly arranged, whilst the latter contains more water, and hence its micellæ are less closely united. He imagines that the idioplasm is extended like a fine network throughout the whole cell body.

If it be admitted, that the assumption of a separate idioplasm is logically justifiable, it cannot be denied that the nuclear substance probably constitutes the hereditary mass.

Further, by means of this theory, a practical interpretation has been given to Nägeli's deduction, which was based simply upon reasoning, and which in consequence could neither be verified by observation nor developed further.

In order to establish the hypothesis, that the nucleus is the transmitter of the hereditary elemental germs, four points have to be considered:—

1. The equivalence of the male and female hereditary masses.
2. The equal distribution of the multiplying hereditary mass upon the cells, which are derived from the fertilised ovum.
3. The prevention of the summation of the hereditary masses.
4. The isotropism of protoplasm.

1. The Equivalence of the Male and Female Hereditary Masses. It is evidently true, and hence must be accepted as an axiom, that the egg and sperm cells are two similar units, each of which, being provided with all the hereditary properties of its kind, transmits an equal quantity of hereditary material to the offspring. The offspring is in general a mixed product of both its parents; it receives from both father and mother an equal number of idioblasts, or active particles, which are the bearers of hereditary attributes.

However, it is only in the lowest organisms that the sexual cells resemble each other in size and composition; in the higher organisms, they present in both respects the greatest differences, so that in extreme cases an animal spermatozoon may be even smaller than the hundred-millionth part of an egg. It is, however, inconceivable, that the carriers of the elemental germs, which, *a priori*, must be assumed to be equal both as to number and attributes, can present such differences in their volume. On the contrary, the fact that two cells, which are quite different as regards mass, can possess equal hereditary potentialities, can be easily explained by the assumption, that they may contain at the same time substances of very different hereditary value, *i.e.* for idioblastic and non-idioblastic substances.

We must, therefore, endeavour to find this idioplasm in the egg and spermatozoon, and to isolate it from the other substances.

First of all, there is no doubt that the reserve materials—fat globules, yolk platelets, etc., must be included in the category of germ substances, which are useless as regards heredity. But even if we discard these, the egg and sperm cells still remain unequal, as regards the quantity of their other constituents. For the protoplasm which is present in a large egg-cell, even after all the contents of the yolk have been abstracted, is much greater in volume than the total substance of a spermatozoon; hence protoplasm cannot be the idioplasm. Only one substance fulfils all the necessary conditions, namely, the nuclear substance. The study of the phenomena of fertilisation in the animal and vegetable world proves this irrefutably.

As was described in chapter seven, the essence of the process of fertilisation consists in this, that the sperm and egg nuclei, *i.e.* one nucleus derived from the spermatozoon, and one derived from the egg-cell, each accompanied by its centrosome, place themselves in contact, and, fusing together, form a germ-nucleus, from which subsequently, one after another, all the nuclei of the developed organism are obtained by repeated divisions. In *Ciliata*, two individuals only lay themselves alongside of each other for a short time, so as to exchange migratory nuclei, each of which subsequently fuses with the stationary nucleus of the other organism.

As far as the most careful observation shows, the egg and sperm nuclei contribute exactly equal quantities of material towards the formation of the germ-nucleus, that is to say, equal quantities of nuclein, and of polar substance, which I include amongst the nuclear substances.

Fol (VII. 14) has proved the equivalence of the polar substance, which is contributed by the two conjugating individuals, whilst the observations of van Beneden (VI. 4 b) upon the process of fertilisation, as seen in *Ascaris megalocephala*, demonstrate irrefutably the equivalence of the nuclein so obtained.

We, therefore, draw the following important conclusion from the facts observed during the process of fertilisation: since in fertilisation the nuclear substances (nuclein and polar substance) are the only materials which are equivalent in quantity, and which unite to form a new fundamental structure, the germinal nucleus, they alone must constitute the hereditary mass which is transmitted from parent to child. We cannot at present decide what is the exact relation borne by the nuclein and the polar substance to the idioplasm.

2. **The equal Distribution of the multiplying Hereditary Mass, amongst the Cells, proceeding from the fertilised Egg.** We are obliged to assume that the multiplying hereditary mass is evenly distributed amongst the descendants of the egg-cell, when we consider the various phenomena of reproduction and regeneration; for instance, the circumstance that each new organism produces numerous egg or sperm cells, which contain the same hereditary

mass as the sexual cells, from which the organism was derived, renders this assumption absolutely necessary.

Secondly, we are forced to this conclusion, when we consider the fact, that in many plants and lower animals, even an extremely small group of cells is able to reproduce the complete organism. When a *Funaria hygrometrica*, is chopped up into very small pieces, and placed upon damp soil, a complete plant grows out of each minute fragment. Similarly, if the fresh water *Hydra* is cut up into small portions, each develops into a complete *Hydra*, possessing all the properties of its species. Buds may be formed from the most different parts of a tree by the growth of the vegetative cells; these buds develop into shoots, which, if separated from the parent, and planted in the earth, can take root and grow into complete trees. In *Cœlenterata*, in many worms and *Tunicates*, the asexual mode of multiplication is similar to the vegetative mode, since at each part of the body a bud can be formed, which is able to develop into a new individual. In *Bougainvillea ramosa*, for instance (Fig. 168), new animals are developed, not only as side branches of the hydroid stock, but also as stolons, which extend themselves like roots upon any surface, and serve to attach the colony.

Thirdly, many processes of regeneration, or replacement of lost parts, prove that in addition to the properties, which are evidently exercised, there must be others which are latent, but which are capable of development under abnormal conditions. For instance, if a willow twig is cut off and placed in

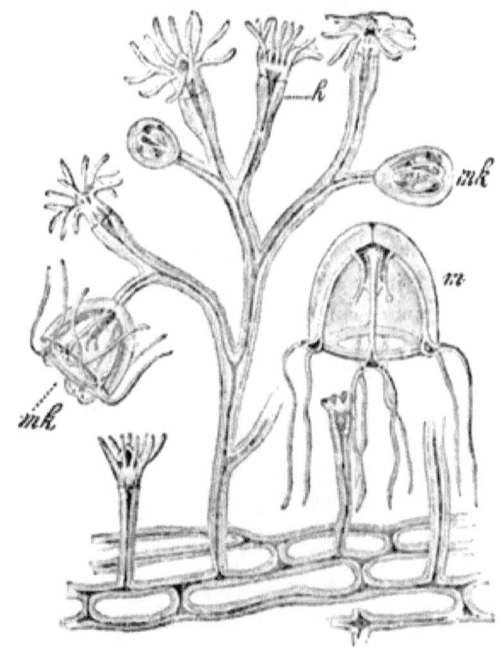

FIG. 168. — *Bougainvillea ramosa* (from Lang): *h* hydranths, which develop into medusa buds *m k*; *m* free medusa *Margelis ramosa*.

water, it develops root-forming cells at its lower extremity; thus the cells are here executing functions, very different from their original ones, which proves that they possessed this capacity potentially. Further, on the other hand, shoots can develop from severed roots, and even subsequently can produce male and female sexual products. In this case, therefore, sexual cells proceed directly from the component parts of a root-cell, and hence serve for the reproduction of the whole. Certain hydroid polyps, according to von Loeb (IX. 17), display similar powers.

Most botanists agree with the theory, recently advanced by de Vries (IX. 30), in opposition to Weismann, which states that all, or at any rate by far the greater number, of the cells of a vegetable body contain all the hereditary attributes of their species in a latent condition. The same is true of the lower animal organisms, although we are unable to prove it for the higher ones. However, on this account, it is not necessary to conclude that the cells of the higher and lower organisms differ so much from one another, that the latter possess all the attributes in a latent condition, and therefore the whole hereditary mass, whilst the former only contain a part of it. For it is quite as likely that the incapacity of most of the cells of the higher animals to develop latent properties, is due to their external conditions, which have produced a great differentation of the cell-body, in which the hereditary mass is enveloped, or to other similar conditions.

Johannes Müller (IX. 18), has raised the question: "How does it happen, that certain of the cells of the organised body, although they resemble both other cells and the original germ-cell, can produce nothing but their like, *i.e.* cells which are capable of developing into the complete organism? Thus epidermal cells can only, by absorbing material, develop new epidermal cells, and cartilage cells only other cartilage cells, but never embryos or buds." To which he has made answer: "This may be due to the fact, that these cells, even if they possess the power of forming the whole, have, by means of a peculiar metamorphosis of their substance, become so specialised, that they have entirely lost their germinal properties, as regards the whole organism, and when they become separated from the whole, are unable to lead an independent existence."

Whatever opinion is held as regards the conditions present in the higher animal, it is quite sufficient for our purpose to acknowledge, that in the plants and lower animals, all the cells which are derived from the ovum, contain equal quantities of the hereditary mass. Hence this must grow and multiply in the cell before division takes place. All idioblasts must divide and must be transmitted to the daughter-cells, in equal proportions both as regards quality and quantity.

Nägeli (IX. 20, p. 531) has enunciated the same view: "Idioplasm, by continuously and proportionately increasing, splits itself up during cell-division—by means of which the organism grows into as many parts as there are individual cells." Therefore, "each cell of the organism is capable, as far as the idioplasm is concerned, of becoming the germ of a new individual. Whether this potentiality ever becomes a reality, depends upon the nature of the nutrient plasm (somatoplasm)."

If we look upon the vital processes of the cells from this second point of view, there can be no doubt that the nuclear substance is the only one amongst all the constituents of the cell, which is able to fulfil all the conditions in every respect.

The nucleus is strikingly uniform in all plant and animal elementary tissues. If we disregard a few exceptions, which require a separate explanation, the nuclei of all the elementary tissues of the same organism resemble each other closely, as regards shape and size, whilst the protoplasm differs in quantity to a marked degree. In an endothelium cell, or in a portion of muscle or tendon, the nucleus has almost the same characters and contains the same substances as an epidermal, liver, or cartilage cell, whilst, in the former case, the protoplasm is barely distinguishable, and, in the latter, is present in large quantities.

The striking and complicated phenomena of the process of nuclear division, are both more important and more comprehensible, when regarded in the light of our theory. The arrangement of the substance into fibrillæ, which consist of small microsomes, arranged alongside of each other, the formation of loops and spindles, the longitudinal halving of the fibrils, and the mode of their distribution amongst the daughter-nuclei, can only serve one purpose, namely, to halve the nuclear substance and to apportion it equally amongst the daughter-cells.

Roux, from another stand-point, has already pertinently denominated "the nuclear division-figures as mechanisms, by means of which it is possible to divide the nucleus, not only according to its own volume, but according to the volume and nature of its special constituents. The essential part of the process of nuclear division is the division of the mother-granule; all the other processes only serve to convey one of the daughter-granules, which have been derived by division from the same mother-granule, into the centre of each daughter-cell." If we replace the term "mother-granule" by the expression "idioblast," we have established a connection between the process of nuclear segmentation and the theory of heredity.

This conception of the nuclear substance as an hereditary mass is important, since it offers some explanation of the facts that the nuclear substance takes less part in the coarser processes of metabolism, than the protoplasm does, and that, for its better protection, it is enclosed in a vesicle provided with a special membrane.

3. **The Prevention of the Summation of the Hereditary Mass.** I consider the third point, viz. the prevention of the summation of the hereditary mass, during sexual reproduction, to be a most important point in the argument. In consequence of the nature of the process of nuclear division, each cell receives the same quantity of nuclear substance as the fertilised egg-cell, A. Now when two of its descendants unite, as sexual cells, the product of generation, B, ought to contain twice as much nuclear substance as the cell A originally did. Then when members of the third generation conjugate, the product C ought to contain twice as much nuclear substance as B, or four times as much as A, and thus with each new act of fertilisation the nuclear mass would increase by geometrical progression. Such a summation, however, must be prevented by nature in some way or other.

This would also be true of the idioplasm, if the full quantity of it were transmitted to each cell, and if it were doubled each time by the act of fertilisation. By this means, its nature, *per se*, would not be changed. For instead of twice, each individual elemental germ would be represented four, eight, or even more times. Thus, although the quantity would be increased, the quality would always remain the same. But it is self-evident that the mass cannot thus increase to an unlimited extent. Nägeli, and especially Weismann, have laid stress upon this difficulty, and have tried to solve it.

Nägeli remarks: " If during each act of reproduction by means of fertilisation, the volume of the idioplasm of whatever constitution it may be, were to become doubled, after a few generations the idioplasmic bodies would have increased so much, that there would not be room for them in a spermatozoid. It is, therefore, unavoidable, that in bisexual reproduction, the union of the parental idioplasmic bodies must take place without causing a corresponding and permanent increase of their substance."

Nägeli has attempted to overcome this difficulty by assuming, that idioplasm consists of strands, which are fused together in such a peculiar way, that the transverse section of the product of fusion remains the same as that of the simple thread, whilst the length of the whole is increased (IX. 20, p. 224).

Weismann (IX. 32–34) has investigated this subject most carefully, and has attempted to demonstrate, that a summation of the hereditary mass is prevented by means of a process of reduction, it being halved before each act of fertilisation. He considers that theoretically it is so absolutely necessary for reduction to take place in each generation, "that the processes by which it is brought about must be discoverable, even if they are not to be deduced from the facts already mentioned."

Weismann has been led to these conclusions by considering the nature of idioplasm; however, his views do not agree with the ones I have mentioned above. He groups them under the common name of "ancestral plasma theory," to the essential points of which I will refer later.

The enquiry into the processes of fertilisation and of nuclear division proves logically, on the one hand, that the two hereditary masses must fuse, and must subsequently be re-distributed amongst the cells, and on the other that a summation of the nuclear substance of the hereditary mass must be avoided. The unanimity of opinion as regards the assumption, that the nuclear substance is the hereditary mass sought for, may certainly be taken as evidence in its favour, especially if, during the fusion of the nuclei, processes can be demonstrated, which correspond in every respect to the necessary conditions.

A priori, there are only two possible means of preventing the sum of the equal quantities from being greater than either of the added parts. Either the quantities, which are to be added together, must be halved beforehand, or their sum must be halved subsequently. Both methods appear to have been adopted during the process of fertilisation.

The one course occurs in phanerogamous plants and in animals. When the male and female sexual products are mature, the nuclear mass of both the egg and sperm mother cell, as was described at length on p. 235, under the title of division with reduction, is so distributed amongst the four grand-daughter cells, that each of them only contains half the nuclear mass of an ordinary cell, and hence only half the normal number of nuclear segments.

The second course occurs during the process of fertilisation in *Closterium*. Here, according to the observations of Klebahn (VII. 27), the germinal nucleus, formed by the fusion of two nuclei, divides consecutively twice without entering into a state of rest, just as when pole-cells are formed. Of the four vesicular nuclei, two disintegrate, so that each half of the original mother-cell contains only *one* nucleus, which possesses only a fourth part of the germ-nucleus, instead of one half, as in normal division (see the description and figures on pp. 280, 281).

If, according to our assumption, the nuclear mass is identical with the hereditary mass, we must conclude, arguing from the process of division with reduction, that *the hereditary mass may be divided up to a certain point, without losing its power of reproducing the whole out of itself.* The question then arises, as to how far this conception is admissible.

Weismann and I both lay emphasis upon the necessity of a reduction of mass, but we have arrived at different conclusions as regards particulars.

In his ancestral germ-plasm theory, Weismann starts with the supposition, that in the hereditary mass the paternal and maternal portions having kept themselves apart, form units, which he calls ancestral germ plasms. He assumes that these are very complicated in structure, being composed of extremely numerous biological units. At each new act of fertilisation still more numerous ancestral germ-plasms come together. Supposing that we revert to the beginning of the whole process of fertilisation, then in the tenth generation 1024 different ancestral plasms must have taken part in the formation of the hereditary mass. But since the total mass of the latter does not double itself with each act of fertilisation, Weismann makes the ancestral plasms divisible in the first stages of the process, and supposes that they are transmitted to the following generation, reduced each time by one half; "at last, however," he continues, "the limit of this constant diminution of the ancestral plasms must be reached, and this must occur when the mass of substance, which is necessary in order that all elemental germs of the individual may be contained therein, has reached its minimum."

After this period, which, by the way, would be reached in a few years in the case of low, quickly-multiplying organisms, formation of the hereditary mass would be obliged to take place with each fresh act of fertilisation, in consequence of the impossibility of diminishing the ancestral plasms any further, unless some other arrangement be made. Weismann considers, that this new arrangement consists in this, that, when the sexual products are mature, half of the ancestral plasms are ejected from the hereditary mass in the pole-cells, before fertilisation occurs. In place of the division of the individual ancestral plasms, therefore, the division of the total number of plasms takes place after they have become no longer divisible as units.

Thus, according to Weismann's assumption, the hereditary mass is an extremely complicated piece of mosaic, composed of innumerable units, the ancestral plasms, which, by their very nature are indivisible and incapable of mixing with other units, and each of which in its turn is composed of numerous elemental germs, which are necessary for the production of a complete individual.

Thus, every hereditary mass, in consequence of its composition, would have to produce countless individuals, if each ancestral plasm were to be active. The essential nature of the process of fertilisation lends itself to a combination and elimination of ancestral plasms. Further, if the ancestral plasm theory were true, elemental germs of equal value would accumulate in the hereditary mass. In fact the generative individuals belonging to the same species are essentially similar in their properties, if we disregard small individual differences of coloration. All the ancestral plasms must, therefore, contain essentially the same elemental germs. These various germs are represented in the hereditary mass as many times as there are ancestral plasms, the majority being similar to one another, and only presenting differences of shade. But all these similar, or slightly different, elemental germs would stand in no direct relation to each other, since they must remain integral component parts of the ancestral plasms, for which we have assumed indivisibility.

The question of heredity, instead of being simplified by Weismann's theory of ancestral plasms, is rendered more complicated by it, especially by the assumption that the paternal and maternal hereditary masses are incapable of mixing with one another.

I cannot see that this theory of Weismann's is of any great use, since it leads to so many difficulties, which appear to be entirely superfluous. Neither Nägeli nor de Vries consider that the ancestral plasms have this construction; they assume rather that the units contained in the two hereditary masses are capable of mixing with one another. Neither can I imagine that, during the process of hereditary transmission, the idioblasts of paternal and maternal origin continue as parts of two separated elemental germs, it seems more likely that they unite together in some way or other to form a compound elemental germ.

How then, on this supposition, is the summation of the hereditary mass, occasioned by the act of sexual generation, to be avoided? I do not think that there is the slightest difficulty if we assume the divisibility of the hereditary mass as a whole. Even Weismann has assumed that this is possible at the beginning of sexual generation, otherwise, a summation of the ancestral plasms, could not have taken place without causing an increase of the hereditary mass.

But the hereditary mass can only be divided, without its properties being altered, if several individual units of each different kind are present in it. Since the progeny are produced from two almost equal combinations of elemental germs, derived from the parents, there must be at least two individuals of every kind of idioblast in the embryo. Nothing prevents us, however, from conceiving that, instead of two individuals of each kind, there may be four, eight, or speaking generally, a number of equivalent idioblasts in the hereditary mass. Then it is self-evident, that a reduction of mass, without the essential nature of the idioplasm itself being altered, is possible in the same manner, as has been observed during the maturation of the sexual products, and therefore any further complicated hypotheses are superfluous.

In order to explain the so-called reversion to an ancestral type, we need not assume the existence of ancestral plasms, for, as will be seen later, the elemental germs may themselves remain latent.

4. **Isotropy of Protoplasm.** Various investigators have attempted to ascribe to the whole egg a very complex organisation, namely, that it is composed of very minute particles, the arrangement of which corresponds to that of the organs of the mature animal. The clearest conception of this subject is that formulated by His in his "*Princip der organbildenden Keimbezirke.*" According to this author, "on the one hand, every point in the embryonal area of the germinal disc must cor-

respond to an organ which develops later, or to part of such an organ, and on the other hand, every organ developed from the germinal area must have its preformed germ in a definite region of this area. The material for the germ is already present in the flat germinal disc, but it is not morphologically distinct, and hence is not to be recognised as such at this stage. By tracing the mature organs back to their elemental form, we shall be able to discover the situation of each during the period of incomplete morphological separation, and indeed, if we wish to be consistent, we must apply this method to the fertilised and even to the unfertilised ovum also."

It is hardly necessary to emphasise how sharply opposed this principle of the formation of organs in the germinal area is to the above-mentioned theory of heredity. One of the first points to be noticed is, that the influence of the paternal elemental germs, upon the formation of the embryo, is entirely left out of account. For this reason alone, the theory is evidently untenable. But, in addition, various experimental facts, which, as Pflüger has pointed out, indicate that the egg is isotropous, entirely disprove it.

By the term isotropy of the egg, Pflüger (VII. 50), wishes to imply, that the contents of the egg are not arranged in such a manner as that the individual organs can be traced back to this or that portion of it. He draws his conclusions from experiments made upon Frog's eggs. The Frog's egg is composed of two hemispherical portions, one of which, the animal half, is pigmented black, whilst the other, or vegetative portion, is clear or colourless, and is, at the same time, specifically heavier. In consequence of this difference in specific gravity, the eggs, immediately after fertilisation, assume a definite position in the water, the pigmented portion always being directed upwards, so that the egg-axis, which connects the animal with the vegetative pole, is vertical. It is possible, however, to experimentally force the eggs which have just been fertilised to take up an abnormal position, that is to say, to prevent them from rotating in the yolk-membrane by applying friction to it. The experimenter, for instance, can force the egg to assume such a position that the egg-axis shall lie horizontally, instead of vertically. Now when the process of division begins, the first division plane, in spite of the changed position of the egg, is in a vertical direction, for its position depends on that of the nuclear spindle, as shown on p. 219. As Born (IX. 37), has minutely described, however, although the nucleus and the specifically lighter portion of the egg have been forced to change their position, the first division plane takes anew a vertical direction. This plane cuts the horizontal egg-axis at various angles. For instance, Pflüger often saw that it separated the egg into a black and a white hemisphere. Under such circumstances, therefore, the hemispheres evidently do not contain the same particles of material, as when they are under normal conditions. Nevertheless, a normal embryo is developed out of the egg. Even after the formation of the notochord and spinal cord, one half of the body can be seen to be darker than the other. Thus, according to the position of the original cleavage plane, the individual organs must be composed of different parts of the egg contents. The experiments made by Richard Hertwig and myself (VI. 38), by Boveri (IX. 4), by Driesch (IX. 7), and by Chabry (IX. 5), all furnish additional proof of the isotropy of the egg.

Richard Hertwig and I found, that the ova of Echinoderms can be divided by violent shaking into small portions; these become spherical in form, and may be fertilised by spermatozoa. Boveri indeed has succeeded in raising a few dwarf larval forms from such small fertilised portions. Driesch, by shaking normally developed and dividing Echinoderm ova, was able to separate from one another the two first cleavage segments; these he then isolated, and was thus able to establish the fact that a normally shaped though somewhat small blastula, followed by a gastrula, and even in some cases by a pluteus, developed from each half.

Chabry has obtained a corresponding result. He destroyed, by pricking it, one of the two, or, when it had divided into four, one of the four cells of the ovum of an Ascidian. In many cases he succeeded in raising from such mutilated ova, absolutely normal larvæ, which only occasionally, were without subordinate organs, such as otoliths or attachment papillæ. From all these experiments the fundamental proposition is proved, that the cell-nucleus, which may be enclosed in any part of the yolk, is able to produce a complete organism. This isotropy of the egg negatives the hypothesis that there is a germinal region from which organs are developed. Moreover, at the same time, it supplies an additional proof that the idioplasm is not to be found in the protoplasm, but in the nucleus; and further, it allows us to draw some conclusions as to the construction of protoplasm and nuclear substance.

Protoplasm must consist of loosely-connected particles of micellæ, which are more similar to one another than those of the nucleus. For, firstly, fragments of a cell, which contain the nucleus, are capable of normal development (*vide* experiments, p. 330). Secondly, the first division plane can be induced, by means of external influences, to divide the contents of the egg in the most various directions, without causing any deviation from the normal, in the product of development. Thirdly, considerable changes of position of the egg substance may be induced, by means of gravity, in Frog's ova which have been forced into an abnormal position, without causing any difference in their subsequent development. Fourthly, we are able to infer, that the micellæ are loosely connected together from the streaming movements of protoplasm, in which, of necessity, the groups of micellæ are obliged to push past one another in the most different directions, and apparently without any method. On the other hand the complicated phenomena of the whole process of nuclear segmentation indicate a more stable arrangement of the nuclear substance.

Nägeli has assumed that there is a similar difference between his hypothetical trophoplasm and idioplasm. He states (pp. 27, 41): "If the arrangement of the micellæ determines the specific properties of the idioplasm, the latter must be composed of a fairly solid substance, in order that the micellæ may not be displaced in consequence of active forces in the living organism, and in order to secure to the new micellæ, which become deposited during multiplication, a definite arrangement. On the other hand, ordinary plasma consists of a mixture of two kinds, fluid and solid, the two modifications easily merging into one another, whilst the micellæ, or groups of micellæ of the insoluble form, are more easily able to push past one another, as must be assumed to be the case when the streaming movements occur." Nägeli, therefore, makes the assumption, which however cannot be proved off-hand, that the idioplasm is spread out like a connected net throughout the whole organism.

IV. Development of the Elemental Germs.

Having assumed that there is a special germ substance or idioplasm in the cell, we must next enquire how the individual idioblasts become active, and thus determine the specific properties or the character of the cell as a result of their development.

It has been suggested, that during the process of development of the ovum, the idioplasm is qualitatively divided unequally by means of the process of nuclear division, so that different parts of the cells acquire the different properties, which are subsequently developed in them. According to this view, the essential nature of development would consist in gradually separating all the elemental germs, taken collectively, which the idioplasm or the fertilised egg contains, into constituent parts, and of distributing them differently, both as regards time and place. Only those cells, which function in the reproduction of the organism, are supposed to be exceptions to this rule, and to receive again the whole collection of the elemental germs during the processes of development. Hence a twofold mode of distributing the idioplasm is assumed to occur, one by the growth and halving of similar germs, and one by the resolution into different component parts of dissimilar ones.

It is difficult to imagine how such a process can actually take place in any concrete case. Further, this assumption does not agree with the above-mentioned facts of reproduction and regeneration; for instance, in plants and in the lower animals, almost any collection of cells is able to reproduce the whole; and again, cells may alter their functions, as seen in the phenomena of regeneration.

Therefore, the views which I have frequently upheld (IX. 10-13), and which agree with those held by Nägeli and de Vries, etc., seem to be more probably true, that as a rule each cell of an organism receives all the different kinds of elemental germs from the egg-cell, and that its especial nature is solely determined by its conditions, only certain individual elemental germs or idioblasts becoming active, whilst the others remain latent.

But in what manner can individual idioblasts become active, and thus determine the nature of the cell? Two hypotheses have been suggested in answer to this question, a dynamic one by Nägeli (IX. 20), and a material one by de Vries (IX. 30). In order to explain the specific activity of idioplasm, Nägeli assumes that "occasionally a definite colony of micellæ, or a combination of such colonies, become active," that is, "are thrown into definite conditions of tension or motion," and he considers that "this local irritation, by means of dynamic influence, and the transmission of peculiar conditions of oscillation acting at a microscopical distance, governs the chemical and plastic processes." " It produces fluid trophoplasm in enormous quantities, and by its help effects the formation of non-albuminous constructive material, of gelatinous, elastic, chitinous, cellulose-like substances, etc., and it gives to this material the desired plastic form. Which micella group of the idioplasm becomes active during development depends upon its shape, upon the stimulation it has previously received, and finally, upon the position in the individual organism in which the idioplasm is placed."

In place of this dynamic hypothesis, de Vries (IX. 30) assumes that the character of the cell is affected in a more material fashion. He is of opinion that, whilst the majority of the idioblasts or "pangenæ" (de Vries) remain inactive, others become active, and grow and multiply. Some of these then migrate from the nucleus into the protoplasm, in order to continue here their growth and multiplication in a manner corresponding to their functions. This outwandering from the nucleus can, however, only take place in such a fashion as to allow of all the various kinds of idioblasts remaining represented in the nuclear substance.

This hypothesis of de Vries appears at present to be a simpler explanation and to be more in accordance with the many phenomena that have been observed. Thus, for instance, as described above, there are separate starch-forming corpuscles, chromatophores, and chlorophyll grains, which function in a specific manner and multiply independently of the rest of the cell, and are transferred at each cell-division from one cell to another. De Vries calls this "transmission outside the cell-nuclei." According to his hypothesis, some of the transmitted idioblasts are those which have become active, have reproduced themselves in the protoplasm, and have united together to form larger units, whilst in addition there are similar idioblasts present in the nucleus (in the germinal substance). The same would be true of the centrosomes, if it were not that the balance of proof is already in favour of their belonging to the nucleus.

By means of the hypothesis of "intracellular pangenesis," the intrinsic difference, which was apparently revealed by the theory of heredity, between nuclear substance and protoplasm, is more or less modified, without the fundamental character of the theory being interfered with ; further, it has been shown how a cell can contain the whole of the attributes of the complex organism, in a latent condition, whilst at the same time it can discharge its own special functions.

The transmission and development of characteristic potentialities are, as de Vries rightly remarks, very different. The transmission is the function of the nucleus, and the development, that of the protoplasm. In the nucleus all the various kinds of idioblasts of the individual in question are represented; therefore, the nucleus is the organ of heredity; the remaining protoplasm of the cell contains practically only those idioblasts which have become active in it and which can multiply rapidly in an adequate manner. We have, therefore, to distinguish between two modes of multiplication of the idioblasts; the one referring to all of them, which results in nuclear division and in their equal distribution amongst the two daughter cells; and the other, which to a certain extent, is a multiplication connected with function; and this latter only affects those idioblasts which have become active; moreover, it is connected with the material changes which occur in them and it takes place chiefly in the protoplasm, outside the nucleus.

This conception is another indication that the protoplasm is composed of small elementary units of substance, as has been assumed latterly by several investigators, who have started various theories; as for instance Altmann (II. 1), in his theory of bioblasts, and Wiesner (IX. 35), in his recent work "*Die Elementarstructur und das Wachsthum der lebenden Substanz.*" The protoplasm, like the nucleus, consists of a large number of small particles of material, which differ as to their chemical composition, and which have the power of assimilating material, of growing and of multiplying by division. (*Omne granulum e granulo*, as Altmann expresses it.) Material for growth is supplied by the fluid, which bathes the nucleus and protoplasm, and in which plastic materials of the most different kinds (albumen, fats, carbohydrates, salts) are dissolved.

In order to distinguish the idioblasts of the nucleus from those of the protoplasm, we will call the latter "plasomes," a name which has been used by Wiesner.

As the plasomes (or as it were the active idioblasts) are, according to the theory of "intracellular pangenesis," supposed to be derived from the idioblasts of the nucleus, so they may also form the starting-point of the organic products of the plasma, since according to their specific characters, they join to themselves various substances; for instance, certain kinds of plasomes, by combining with carbo-hydrates, might produce the cellulose membrane, or by combining with starch the starch granules; hence they might be designated, the cell-membrane formers or starch formers.

Thus the most different occurrences in cell life may be regarded, from a common point of view, as vital processes taking place in the most minute organised, dissimilar particles of matter, which multiply indefinitely and which are found in the nucleus, in protoplasm, and in the organised plasmic products, according to the different phases of their vital activity.

Wiesner has formulated his conception, which is in accordance with the above, in the following sentences: "The assumption, that protoplasm contains organised separate particles, which are capable of division, and that it, in fact, entirely consists of such living, dividing particles, is forced upon us as the result of recent enquiry." By means of the division of these particles "growth is brought about," and "all the vital processes occurring in the organism depend on them." "They must, therefore, be considered to be the true elementary organs of life."

Literature IX.

1. R. S. BERGH. *Kritik einer modernen Hypothese von der Uebertragung erblicher Eigenschaften.* Zoolog. Anzeiger. 1892.
2. BLUMENBACH. *Ueber den Bildungstrieb und das Zeugungsgeschäft.* 1781.
3. BONNET. *Considérations sur les corps organisés.* Amsterdam, 1762.
4. BOVERI. *Ein geschlechtlich erzeugter Organismus ohne mütterliche Eigenschaften.* Gesellschaft f. Morphol. u. Physiol. zu München. 1889.
5. CHABRY. *Contribution à l'embryologie normale et teratologique des Ascidies simples.* Journal de l'anat. et de la phys. 1887.
6. DARWIN. *Animals and Plants under Domestication.* Vol. II.
7. DRIESCH. *Entwicklungsmechanische Studien. Der Werth der beiden ersten Furchungszellen in der Echinodermenentwicklung. Experimentelle Erzeugung von Theil- und Doppelbildungen.* Zeitschr. f. wissenschaftl. Zoologie. Bd. LIII. Leipzig, 1891.
8. HAECKEL. *Generelle Morphologie. Der Perigenesis der Plastidule.*
9. V. HENSEN. *Die Grundlagen der Vererbung nach dem gegenwärtigen Wissenskreis.* Landwirthschaftl. Jahrbücher. Bd. XIV. 1885.
10. OSCAR HERTWIG. *Das Problem der Befruchtung und der Isotropie des Eies, eine Theorie der Vererbung.* Jena, 1884.
11. O. HERTWIG. *Vergleich der Ei- und Samenbildung bei Nematoden. Eine Grundlage für celluläre Streitfragen.* Archiv f. wissenschaftl. Anatomie. Bd. XXXVI. 1890.
12. O. HERTWIG. *Urmund und Spina bifida.* Archiv f. mikrosk. Anatomie. Bd. XXXIX. 1892.
13. O. HERTWIG. *Aeltere und neuere Entwicklung theorieen.* 1892.
14. W. HIS. *Die Theorieen der geschlechtlichen Zeugung.* Archiv f. Anthropologie. Bd. IV. u. V. 1871, 1872.
15. W. HIS. *Unsere Körperform u. das physiologische Problem ihrer Entstehung. Briefe an einen befreundeten Naturforscher.* 1874.
16. KÖLLIKER. *Bedeutung der Zellkerne für die Vorgänge der Vererbung.* Zeitschrift f. wissenschaftl. Zoologie. Bd. XLII.

KÖLLIKER. *Das Karyoplasma und die Vererbung. Eine Kritik der Weismann'schen Theorie von der Continuität des Keimplasmas. Zeitschr. f. wissenschaftl. Zoologie. Bd. XLIV.* 1866.
17. LOEB. *Untersuchungen zur physiologischen Morphologie der Thiere. Organbildung u. Wachsthum.* 1892.
18. JOHANNES MÜLLER. *Handbuch der Physiologie des Menschen.*
19. JOSEPH MÜLLER. *Ueber Gamophagie. Ein Versuch zum weiteren Ausbau der Theorie der Befruchtung u. Vererbung.* Stuggart, 1892.
20. NÄGELI. *Mechanisch-physiologische Theorie der Abstammungslehre.* München. 1884.
21. NUSSBAUM. *Zur Differenzirung des Geschlechts im Thierreich. Archiv f. mikrosk. Anatomie. Bd. XVIII.*
 NUSSBAUM. *Ueber die Veränderungen der Geschlechtsprodukte bis zur Bifurchung, ein Beitrag zur Lehre von der Vererbung. Archiv f. mikrosk. Anatomie. Bd. XXIII.*
22. PFLÜGER. *Loc. citat. Cap. VII.*
23. ROUX. *Beiträge zur Entwicklungsmechanik des Embryo im Froschei. Zeitschrift f. Biologie. Bd. 21.* 1885.
24. ROUX. *Ueber die künstliche Hervorbringung halber Embryonen durch die Zerstörung einer der beiden ersten Furchungskugeln. Virchow's Archiv. Bd. CXIV.* 1888.
25. SACHS. *Ueber Stoff und Form von Pflanzenorganen. Arbeiten des botan. Instituts. Würzburg. Bd. II u. III.*
26. SPENCER. *Principles of Biology.* 1864.
27. STRASBURGER. *Neue Untersuchungen über den Befruchtungsvorgang bei den Phanerogamen als Grundlage für eine Theorie der Zeugung.* Jena, 1884.
28. STRASBURGER. *Ueber Kern- und Zelltheilung im Pflanzenreich, nebst einem Anhang über Befruchtung.* Jena, 1888.
29. VÖCHTING. *Ueber Organbildung im Pflanzenreich.* Bonn, 1878.
30. HUGO DE VRIES. *Intracellulare Pangenesis.* Jena, 1889.
31. WEISMANN. *Ueber Vererbung.* 1883.
 WEISMANN. *The Continuity of the Germ-plasm as the Foundation of a Theory of Heredity,* 1885. *Translated by Schönland.* 1889.
32. WEISMANN. *The Significance of Sexual Reproduction in the Theory of Natural Selection,* 1886. *Translated by Schönland.* 1889.
33. WEISMANN. *On the Number of Polar Bodies and their Significance in Heredity,* 1887. *Translated by Schönland.* 1889.
34. WEISMANN. *Amphimixis, or the Essential Meaning of Conjugation and Reproduction,* 1891. *Translated by Poulton and Shipley, etc.* 1892.
35. WEISNER. *Die Elementarstructur und das Wachsthum der lebenden Substanz.* 1892.
36. CASPAR FRIEDR. WOLFF. *Theorie von der Generation.* 1764.
37. BORN. *Ueber den Einfluss der Schwere auf das Froschei. Arch. f. mikrosk. Anatomie. Bd. 24.*

INDEX

Abortive eggs, 238.
Acetabularia, 307.
Achromatin, 181.
Actinosphærium, 35.
Adventitious substances in the cell, 27.
Æthalium septicum, 17, 99, 111, 115, 117.
Affinity, sexual, 300.
   sexual influence of environment, 313.
Albumen, building up of, 150.
   circulation of, 31.
   crystals of, 150, 159.
   molecule, 17.
   peptonisation of, 151.
Algæ, 3, 6, 34.
Alternation of generations, 255.
Alveolar layer, 21.
Amitosis, 207.
Amœba, structure of, 27.
   movements of, 67.
   stimulation of, 107, 111.
Amphiaster, 193.
Amphipyrenin, 44.
Amyloplasts, 160, 164.
Anæsthetics, 112.
   action of, upon Mimosa, ova and spermatozoa, 113.
Analysis of pus corpuscles, 18.
   of ash of Fucus, 136.
Ancestral plasma theory, 351.
Aniline dyes, absorption of by living cell, 136.
Animalculists, 337.
Antheridia, deviation of, 302.
Anticlinal division walls, 220.
Antipolar area, 184.
Aphides, 296.
Apogamy, 295, 300.
Apposition, 164, 169.
Archoplasm, 190.
Aroideæ, formation of heat in germinating seeds of, 130.
Ascaris megalocephala, corps résiduel, 246.
   division with reduction in spermatozoa of, 235.
   division with reduction in ova of, 237.
   fertilisation of, 259.
   nuclear division of, 189.
Ascidians, multiplication of nuclei in immature eggs of, 213.
Asexual condition in Ciliata, 292.
Ash, analysis of, in Fucus, 136.
Asparagin, attractive effect of, upon Bacteria, 120.
Asplenium, apogamy of, 300.
Assimilation, 132.

Attraction centre, 245.
   sphere, 181, 190.
Aureole (Fol), 259.

Bacteria, anærobic, 129.
   as tests for oxygen, 116.
   traps, 121.
Basidiobolus ranarum, influence of nutriment upon formation of sexual cells, 294.
Bastard formation, 310.
Bees, 295.
Bibliography, 9, 61, 89, 123, 174, 246, 320, 332, 361.
Bioblasts of Altmann, 24.
Botrydium, 101, 285.

Cane sugar as a stimulant to antherozoids, 120.
Carbo-hydrates, 147.
Carbon dioxide, absorption of, 132.
Carica papya, 151.
Carnivorous plants, 151.
Cartilage cell, 31.
Cell-budding, 228.
   contents, 26, 27, 31, 35.
   definition of (Brücke), 8.
   definition of (Schleiden & Schwann), 5.
   definition of (Schulze), 8.
   division, equal, 224.
   division, influence of the environment upon, 239.
   division, partial, 230.
   division, unequal, 225.
   membrane, 5.
   nutritional substances of, 27.
   permanent substances of, 27.
   plate, 189, 198, 234.
   sap, 6, 31, 154.
   territories, 173.
   theory, history of, 2.
Cellular pathology, 1.
Cellulose, formation of, 152.
   reaction of, 166.
Cell-wall, 166.
   corky change of, 168.
   deposition upon, 168.
   formers, 199.
   growth of, 169.
   woody change of, 168.
Central corpuscles (see Centrosomes).
Central spindle, 202.
Centrolecithal eggs, 232.
Centrosomes, 55, 180.
   division of, 189, 199, 259.
   in Echinoderm fertilised ova, 258.
   in lymph corpuscles, 56.
   in over fertilised eggs, 244.

Centrosomes in ovum of Ascaris, 262.
   in Phanerogams, 264.
   in pigment cells, 56.
   in Radiolaria, 212.
   female, 258, 265.
   male, 258, 265.
   multiple division of, 212, 244.
   origin of, 263.
   quadrille of, 259.
Characeæ, nuclei of, 210.
   parthenogenesis in, 296.
   rotation in, 71.
Chemical stimuli, 111.
Chemistry of assimilation, 146.
Chemotaxis, 115.
Chemotropism, 92.
   in Æthalium, 115.
   in antherozoids, 119.
   in Bacteria and Infusoria, 116.
   in leucocytes, 121.
Chief nucleus in Infusoria, 267, 269.
Chief spindle in Infusoria, 269.
Chloral, effect upon nuclear division, 240.
   effect upon ova and spermatozoa, 113.
Chloroform, 113.
Chlorophyll, 161.
   corpuscles, 161.
   effects of chloroform upon, 113, 133.
   function of, 132, 146.
   movements of corpuscles under influence of light, 103.
Chorda dorsalis, 157.
Chromatic nuclear figures, 182.
Chromatin, 13, 181.
Chromatophores, 99.
Chromatoplasts, 160.
Chromosomes, 180, 200.
Cilia, 77.
   formation of, 77, 83.
   movements of, 77.
Ciliata, fertilisation of, 265.
   galvanotropism in, 108.
   need for fertilisation of, 292.
Circulation in protoplasm, 71, 72.
Cleavage line in segmentation, 225.
Cleavage nucleus, 259.
Closterium, 279, 352.
Cold rigor, 96.
Colloids, 59.
Colour granules in plants, 162.
Colouring matter, absorption of by living cell, 136.
Conjugation, 278.
   epidemics in Infusoria, 267.
Constant current, effect upon protoplasm, of 107.
Cork formation in cell-wall, 168.
Coronal furrow in Frog's egg, 196.
Corps résiduel in Ascaris, 246.
Corydalis cava, 308.
Cross-fertilisation in Acetabularia, 307.
   in Amphibia, 312.
   in Ciliata, 308.
   in Echinoderms, 312.
   in plants, 310.
   need for, 318.

Crystalloids, 59.
Cuticle, formation of, 172.
Cutleriaceæ, fertilisation of, 286, 293.
   sexual affinity in, 301.
Cytoblast, 177.
Cytoblastem, 6, 177.

Daphnoids, parthenogenesis in, 296.
Degeneration of animal egg-cells, 293.
   of Infusoria 267, 292, 307.
   of nuclei, 245.
   of swarm-spores of Algæ, 245.
Desmidiaceæ, 279.
Deutoplasm, 26.
Development, theories of, 339.
Diapedesis, 122.
Diastase, 150.
Directive corpuscles, 228 (see Pole-cells).
Division of centrosomes, 180, 199, 259.
   of chlorophyll granules, 161.
   of egg-cell, 223–232.
   of idioblasts, 341.
   of nuclei, direct, 207.
   of nuclei, indirect, 179.
   of plasomes, 360.
   of trophoplasts, 160.
   with reduction, 235.
   with reduction in Cosmarium, 279.
Division plane, position of, in division of egg-cell, 219.
Division plane, change of position through external influences, 355.
Drosera, 151.
Dumb-bell figure in egg-cell division, 19.

Echinoderms, division of egg-cells of, 192.
Ectocarpus, 295.
Ectoplasm, 15.
Egg-cell, division of nucleus in, 193.
   segmentation of, 223–232.
Electrical stimuli, 106.
Elementary organisms, 7, 24.
   particles, 1, 3, 24, 340, 361.
   units, 1, 3, 24, 340, 361.
Elemental germs, 334,
   of an organism, 334, 339, 344.
   development of, 357.
Embryo-sac of Phanerogams, 233, 263.
Endoplasm, 15.
Energy, kinetic, 126.
   potential, 126.
Epigenesis, 336.
Epistylis, fertilisation of, 271.
Equivalence of male and female hereditary masses, 345.
Equivalence of nuclear substances in fertilisation, 272.
Eudorina, 254.
   fertilisation of, 289.
Euglena viridis, reaction of light to, 100.

Fat, 151, 157.
Fertilisation, 252.
   isogamous, 284.
   methods of, 252.
   need for, 291.

Fertilisation of Algæ, 284.
  of Ascaris megalocephala, 259.
  of Botrydium, 285.
  of Ciliata, 265.
  of Cutleriaceæ, 286.
  of Desmidiaceæ, 279.
  of Echinoderm eggs, 256.
  of Fucaceæ, 286.
  of Infusoria, 265.
  of Monjeotiæ, 283.
  of Noctilucæ, 278.
  of non-nucleated portions of protoplasm, 299.
  of Phanerogams (Lilium martagon), 283.
  of Phæosporeæ, 286.
  of Spirogyra, 283.
  of Vorticella, 281.
  of Volvocineæ, 290.
  of Zygnemaceæ, 281.
  oogamous, 284.
  phenomena of, 289.
Filament theory (Flemming), 23.
Filamentous substance, 23.
Ferments, 128, 150.
  action of, 151.
Flagella, 77.
Foam theory of protoplasm (Bütschli), 20.
Foam, structure of, 21.
Formative instinct (Blumenbach), 338.
Formative activity of the cell, 145.
Framework theory of protoplasm, 19.
Fritillaria imperialis, nuclear division in the embryo-sac of, 196.
Fritillaria persica, nuclear division in pollen grain of, 198.
Fucaceæ, fertilisation of, 286.
Fucus, analysis of the ash of, 136.

Galvanotropism, 92, 108.
Gametangium, 284.
Gametes, 284, 293.
Gas chamber, 112.
Gemmulæ (Darwin), 340.
Generation cycle, 252, 297.
  theories of, 339.
Geotropism, 92.
Germinal nucleus, 259.
Germinal spot, 50.
  of Asteracanthion, 53.
  of Molluscs, 51, 52.
Germinal vesicle, 49.
Giant cells of bone marrow, 244.
Gliding movements of protoplasm, 70.
Goblet cells, 36.
Granula, 24, 25, 44.
  theory (Altmann), 24.
Granular plasm, 15, 68.
Granule and mass theory (Arnold and Purkinje), 8.
Granules, streaming movements of, 68.
Gravity, effect of upon egg-cell division, 214.
Gromia oviformis, 29.
  movements of, 69.

Growing point, arrangement of cells in, 221.
  heaping up of protoplasm at, 323.
Guanin crystals, 158.

Heat production, a vital process, 130.
Heat rigor, 94.
Heliotropism, 92.
Hereditary mass, 339.
  combination of, 353.
  distribution of, in the cell, 346.
  division of, 352.
  equivalence of male and female, 345.
  prevention of the summation of, 350.
Heredity, theories of, 334.
Hermaphroditism of the nucleus, 275.
History of the cell-theory, 2.
History of the protoplasmic theory, 6.
Honeycomb theory of protoplasm (Bütschli), 20.
Hyaloplasm, 15.
Hybrids, 313.
Hybridisation, 310.
Hydrocharis, 71.
Hydrodictyon, 294.
Hydrotropism, 117.

Idioblasts, 340.
  arrangement of, 344.
  division of, 341.
  size and number of, 342.
Idioplasm, 339, 342, 357.
Infusoria, fertilisation of, 265.
  galvanotropism of, 108.
  need for fertilisation of, 292.
Intercellular substance, 173.
Interfilamentous substance, 23.
Intergranula substance, 24.
Internal vesicle of Thalassicola, 212.
Intracellular digestion, 142.
  pangenesis, 359.
Intramolecular heat, 127.
  respiration, 131.
Intraplasmic products, 27.
Intussusception, 169.
Invertin, 150.
Irritability of the cell, 91.
  of protoplasm, 91.
Isogamous fertilisation, 285.
Isotropy of protoplasm, 354.

Karyokinesis, 179.
Karyolisis, 199.

Latent properties, 334.
Leucocytes, absorption and digestion of foreign bodies by, 143.
  chemotropism of, 121.
Leucophrys patula, 253, 292.
Leucoplasts, 160.
Life-cycle in animals and plants, 148.
Light, action of, upon Æthalium, Pelomyxa, chromatophores, and pigment cells of retina, 99.
  action of upon Euglena and swarm-spores, 100.

Light pictures produced upon leaves, 104.
    stimulation, 99.
    tone (phototonus), 101, 102.
Lilium martagon, 263.
Linin, 43.
Lymph corpuscles, centrosomes of, 203.
    division of, 209.
    movements of, 66.
    perforated nuclei of, 209.
    structure of, 28.

Macrocosm, 339.
Macro-gametes, 271.
Macro-nuclei of Ciliata, 266.
Malic acid as an attracting agent for Fern antherozoids, 119.
Mechanical stimuli, 110.
Membrane of the cell, 5.
Meroblastic segmentation, 230.
Merocytes, 233, 245.
Mesocarpus, action of light upon, 104.
Metabolic products of protoplasm, 18.
    of micro-organisms, 122.
    of the cell, 128.
Metabolism of the cell, 126-154.
    progressive, 126.
    retrogressive, 126.
Metastasis in plants, 150.
Micellæ, 58, 340, 343.
Micella theory, 19, 58.
    solution, 60.
Microcosm, 339.
Microgametes, 27.
Micro-nuclei of Ciliata, 266.
Micro-organisms, destruction of, by phagocytes, 144.
    metabolic products of, 122.
    nuclei of, 55.
Microsomes, 14, 19, 22.
Middle portion or neck of spermatozoon, 45, 56.
Migratory nuclei of Infusoria, 269.
Mimosa pudica, 113.
Mitome, 23.
Mitosis, 179.
Molecular structure, 58.
Monjeotia, 283.
Movements, changes in the cell during passive movements, 88.
    occurring in oil drops, 73.
    of contractile vacuole, 86.
    of flagella and cilia, 77.
    of protoplasm, 73-89.
    of protoplasm during heat stimulation, 94.
    of protoplasm due to light stimulation, 99.
Mucous cells, 36.
Multiple fertilisation in chloralised egg-cells, 114.
Multipolar giant cells, 244.
    mitoses, 243.
Muscle fibres, 173.
Mycoderma aceti, 147.
Myxomycetes, movements in, 67.
    structure of, 28.

Narcosis (of protoplasm, Mimosa, egg-cells, and spermatozoa), 112-115.
Neck or middle portion of spermatozoon, 45, 56.
Nematodes, nucleus of the fertilised egg-cell of, 218.
Nematocysts, 164.
Nerve fibres, 173.
Net-like structure of protoplasm, 23.
    of nucleus, 47.
Nisus formativus (Blumenbach), 338.
Noctiluca, 278.
Non-nucleated cells, 54.
Nuclear framework, 47.
    membrane, 44.
    sap, 43.
    spindle, 181.
Nuclein, 40, 41.
    bodies, 49.
    in division, 180.
    reaction of, 40.
Nucleoli, 42, 49, 52.
    fate of, 205.
Nucleus, connecting fibrils in, 187, 198.
    definition of, 37.
    degeneration of, 245.
    determination of position of in the cell, 214, 216, 217.
    discovery of, 3, 36.
    division of, direct, 207.
    division of, indirect, 179.
    division of, influence of the environment upon, 239.
    division of, in fertilised egg-cells, 263, 264, 273.
    division of, multipolar nuclear figures, 243.
    division of, pathological, 244.
    division of, with reduction, 235.
    fixed position of, in plant cells, 325.
    form, size, number of, 37.
    germinal, 259.
    history of, 37.
    importance of, in segmentation, 349.
    influence of, upon cell processes, 324, 330.
    in segmentation, 179.
    longitudinal splitting of segments, of 186, 191.
    migratory, of Infusoria, 269.
    multiplication of, 211.
    of animal cells, 327.
    of Bacteria, Oscillaria, etc., 54.
    of Chironomous larva, 49.
    of Ciliata, 47.
    of egg-cells, 50.
    of egg-cell of Dytiscus, 329.
    of Fritillaria, 48.
    of Salamander, 49.
    of secreting cells of Nepa, 329.
    of spermatozoa, 45.
    of sperm mother cells of Ascaris, 46.
    of Spirogyra, 49.
    segments during fertilisation, 263, 264, 273.

INDEX

Nucleus segments, number of, in division with reduction, 235.
   spindle, 180.
   spindle, derivation of, 200, 202.
   spindle, formation of, 185.
   staining of, 40.
   structure of, 44-54.
   transmitter of hereditary elemental germs, 344.
Nutrient plasm, 349.
   solutions, 147, 294.
   substances of the cell, 27.

Œdogonium, 34.
Onychodromis grandis, 253, 271, 293.
Oogamous fertilisation, 284.
Oogonium, 287, 302.
Osmosis, 138.
Over fertilisation, 243.
Over mature egg-cells, 293.
Ovists, 337.
Ovocentrum, 258, 274.
Oxygen, action upon Æthalium, 115, 128.
   action upon Bacteria and Ciliata, 116, 117.
   action upon cells, 112.

Pandorina, 254.
   fertilisation of, 287.
Pangenæ (de Vries), 340, 359.
Pangenesis, 340.
   intracellular, 359.
Paramœcia, need for fertilisation of, 267.
   need for oxygen of, 117.
Paramitome, 23.
Paranuclein, 42, 257.
Paranucleus of Ciliata, 267, 269.
Paranuclear spindle, 269.
Paraplasm, 26.
Parthenogenesis, 255, 295.
Pelomyxa, 99.
Pepsin, 151.
Perforated nuclei, 210.
Periclinal division walls, 220.
Peripheral layer of protoplasm of the cell, 15.
   of Frog's eggs, 15.
   rôle of in osmosis, 140.
Permanent material of the cell, 27.
Peronosporeæ, sexual affinity in, 302.
Phæosporeæ, fertilisation of, 285.
Phagocytes, 143.
Phagocytosis, 143.
Photophobic spores, 102.
Photophylic spores, 102.
Phototonus, 101, 102.
Phylloxera, 295.
Physiological units (Spencer), 340.
Phytogenesis, 3.
Pigment granules, 158.
Pill-box theory of development, 338.
Plane of division, position of, in egg-cell, 219.
Plant anatomy, 2.
Plasmic products, 27, 154.

Plasmolysis, 140.
Plasmodium, 68.
   light-stimulation of, 99.
Plasomes, 360.
Plastidule, 342.
Plastin, 17.
   reaction of, 17.
Podophrya gemmipara, 229.
Polar area, 184.
   differentiation, 214.
Pole cells, 228, 237, 269.
   of parthenogenetic ova, 298, 299.
Pollen grains, 263.
Pollen tube, 264.
Polyaster, 243.
Polyspermia, 114, 243.
Preformation theory, 336.
Primordial utricle, 32.
Pronuclei, 275.
Proteid substances, 17.
Protoplasm, adventitious substances in, 34.
   alkalinity of, 17
   chemico-physical and morphological properties of, 11.
   death from cold of, 95.
   double refraction of, 18.
   first use of the word, 6.
   formation of, 16.
   history of protoplasmic theory, 6.
   of Amœba, 28.
   of lymph corpuscles, 28.
   of Myxomycetes, 28.
   of Reticularia, 28.
   structure of, 18.
Protoplasmic movements, 68, 73-89.
   due to heat stimulation, 94.
   due to light stimulation, 99.
   metabolic products of, 18.
   of Amœbæ, 67.
   of flagella and cilia, 77.
   of Gromia oviformis, 69.
   of lymph corpuscles, 66.
   of Myxomycetes, 67.
   of plant cells, 71.
   simulated by drops of oil, 73-77.
   theories concerning, 73.
Protoplasmic threads, 23, 31.
Pseudopodia, 27, 28, 29, 66, 110.
Pteris cretica, 300.
Ptyalin, 151.
Pyrenin, reaction of, etc., 42.

Quadrille of the Centrosomes, 259.

Radiation figures in Echinoderm eggs, 192.
   figures in protoplasm, 55, 181.
Radiolaria, 212.
Receptive protuberance, 257, 304.
   spot in Algæ, 286.
Reduction of nuclear segments, 262, 264, 265.
   of nuclear segments in Ciliata, 270.
Regeneration, 346.

Reproduction of the cell, 177.
   theories of, 339.
Reserve materials, 26, 35, 150.
Respiration of the cell, 128.
   intramolecular, 131.
Restitution theory, 276.
Retarded eggs, 296.
Reticularia, 28.
   movements in, 69.
Rheotropism of Myxomycetes, 68.
Rotation in protoplasm, 71.
Rotatoria, 296.

Saccharomyces cerevisiæ, effect of chloroform upon, 114.
Salamandra maculata, nuclear division of, 183.
Sarcode, 7, 29.
Segmentation of the egg, 223–232.
   equal, 224.
   meroblastic or partial, 230.
   unequal, 225.
Selective powers of the cell, 135.
Self-fertilisation, 299.
Separation bodies (Flemming), 189.
Sexual affinity, 300.
   characters, 276.
   dimorphism in Vorticella, 272.
   generation, fundamental modes of, 278.
   maturity in Ciliata, 292, 293.
   nuclei, 266.
   swarm-spores, 284.
Skeleton of the cell, 159.
Somatoplasm, 349.
Specific energy, 92.
Sperm centrum, 258, 274.
   nucleus, 199, 243, 257.
   nucleus in non-nucleated fragments of egg-cells, 300.
   spindle, 244.
Spermatozoon, of Ascaris, 46.
   movements of, 82.
   narcosis of, 114, 147, 160.
   of Echinoderms, 257.
   structure of, 45.
Spindle aggregations, 245.
   fibrils, 181, 202.
Spirogyra, 283.
Sporangium, 234, 284.
Staphylococcus, 122.
Starch formation, 132.
   formation in plant cells, 160, 163.
   granules, 162.
Stationary nuclei of Infusoria, 269.
Stimulation, phenomena of, 91, 93.
   after-effects of, 94.
Stimuli, chemical, 111.
   electrical, 106.
   kinds of, 92.
   light, 99.

Stimuli, mechanical, 110.
   protoplasmic, 91.
   thermal, 94.
Streaming movements of oil drops, 73–77.
Stylonichia, 253, 292.
Suberin, 168.
Summer eggs, 296.
Swarm-spores, action of light upon, 100.
   formation of, 234.
   passing out from cell membrane of, 6, 34.
   sexual and asexual, 284.

Telolecithal eggs, 232.
Temperature, effect of, upon cell, 94.
   maximum, minimum, 94, 97.
Tension (potential energy), 126.
Tetraster, 243.
Thermal stimuli, 94, 239.
Tradescantia, 72, 94, 106.
Transverse division plane, 220.
Triænea bogotensis, 71.
Triaster, 243.
Trophoplasm, 357.
Trophoplasts, 159.
   division of, 160.
Tuberculin, mode of action of, 123.
Turgor (turgescence), 141, 155.

Ulothrix, 101.

Vacuoles, 31, 34, 154.
   contractile, 85.
Vallisneria, 71, 194.
Vaucheria, repair of, after injury, 323.
Vegetative reproduction, 255.
Vessels in plants, 2.
Vital elementary units, 1.
   force, 91.
   properties of the cell, 65, 126.
   processes, 128.
Vitalism, theory of, 91.
Vitelline membrane, 257.
Volvocineæ, 287.
Volvox globator, 290.
Vorticellæ, 271, 302.

Winter eggs, 296.
Woody change of cell-wall, 168.

Xanthophyll, 132.

Yolk, 158.
   granules, 158.
   nuclei, 233.

Zooglœa, 24.
Zygnemaceæ, fertilisation of, 281.
Zygote, 279, 281, 283.

www.ingramcontent.com/pod-product-compliance
Lightning Source LLC
Chambersburg PA
CBHW030358230426
43664CB00007BB/644